T0254476

Discrete Fourier and Wavelet Transforms

An Introduction through Linear Algebra with Applications to Signal Processing

Discrete Fourier and Wavelet Transforms

An Introduction through Linear Algebra with Applications to Signal Processing

Roe W. Goodman
Rutgers University, USA

World Scientific

NEW JERSEY · LONDON · SINGAPORE · BEIJING · SHANGHAI · HONG KONG · TAIPEI · CHENNAI · TOKYO

Published by

World Scientific Publishing Co. Pte. Ltd.

5 Toh Tuck Link, Singapore 596224

USA office: 27 Warren Street, Suite 401-402, Hackensack, NJ 07601

UK office: 57 Shelton Street, Covent Garden, London WC2H 9HE

Library of Congress Cataloging-in-Publication Data

Names: Goodman, Roe.

Title: Discrete Fourier and wavelet transforms : an introduction through linear algebra with
applications to signal processing / by Roe W. Goodman (Rutgers University, USA).

Other titles: Fourier and wavelet transforms

Description: New Jersey : World Scientific, 2016. |
Includes bibliographical references and index.

Identifiers: LCCN 2015043388| ISBN 9789814725767 (hardcover : alk. paper) |
ISBN 9789814725774 (pbk. : alk. paper)

Subjects: LCSH: Fourier transformations--Textbooks. | Algebras, Linear--Textbooks. |
Signal processing--Mathematics--Textbooks. | Wavelets (Mathematics)--Textbooks.

Classification: LCC QC20.7.F67 G66 2016 | DDC 515/.723--dc23

LC record available at http://lccn.loc.gov/2015043388

British Library Cataloguing-in-Publication Data

A catalogue record for this book is available from the British Library.

Cover designer: Chuan Ming Loo

Copyright © 2016 by World Scientific Publishing Co. Pte. Ltd.

All rights reserved. This book, or parts thereof, may not be reproduced in any form or by any means, electronic or mechanical, including photocopying, recording or any information storage and retrieval system now known or to be invented, without written permission from the publisher.

For photocopying of material in this volume, please pay a copying fee through the Copyright Clearance Center, Inc., 222 Rosewood Drive, Danvers, MA 01923, USA. In this case permission to photocopy is not required from the publisher.

Printed in Singapore

Preface

Signal processing has become an essential and ubiquitous part of contemporary scientific and technological activity, and the signals that need to be processed appear in most sectors of modern life. Signal processing is used in telecommunications (telephone and television), in the transmission and analysis of satellite images, and in medical imaging (echograph, tomography, and nuclear magnetic resonance), all of which involve the analysis, storage or transmission, and synthesis of complex time series. Signal processing occurs in most late-model automobiles, typically for some monitoring or control function. The record of a stock price is a signal, and so is a record of temperature readings that permit the analysis of climatic variations and the study of global warming.[1]

This textbook for undergraduate mathematics, science and engineering students introduces the theory of discrete Fourier and wavelet transforms and applies it to signal processing. The approach is through linear algebra (vector spaces, linear transform methods) as implemented in MATLAB to represent and manipulate signals (for example, sounds or two-dimensional images). Much of the mathematics is of recent origin. From a family of wavelet transforms created by Ingrid Daubechies and her collaborators, which are treated in detail in Chapters 3 and 4, two were chosen for the JPEG 2000 image-processing algorithms because of their superior performance compared to older Fourier transform block-encoding techniques. These new algorithms are used in digital cameras, medical imaging, and digital television, for example.

After thirty years of intensive development, wavelet transforms can now serve as the focus for an undergraduate course in applied mathematics with wide student appeal. The only prerequisite for the book is some experience with linear algebra and multivariable calculus, so it is also suitable for self-directed study by engineers and scientists interested in the mathematics behind wavelet algorithms and software. No prior knowledge of signal processing or advanced analysis is assumed.

In broad outline, the book shows how to use the Fourier matrix to extract frequency information from a digital signal and how to use circulant matrices as filters to pick out selected frequency ranges. It introduces discrete wavelet transforms for

---

[1]From [Jaffard, Meyer, and Ryan (2001), p. 1].

digital signals through the *lifting method* (based on the familiar notion of elementary row operations in linear algebra). It illustrates through examples and computer explorations how these transforms are used in signal and image processing. Then it develops the general theory of discrete wavelet transforms via the matrix algebra of two-channel filter banks. Finally, it constructs wavelet transforms for analog signals using the filter bank results already established, and sets up the general mathematical framework for *multiresolution analysis* of analog signals, which has already been explained earlier for digital signals.

Linear algebra is the main mathematical tool in the book, instead of calculus and differential equations (the mathematics traditionally used by science and engineering students). Here is some justification for this shift in emphasis:

- The analog model for a signal or image is a real-valued function f of continuous variables: $f(t)$ where t is time for sounds, or $f(x,y)$ where (x,y) are spatial coordinates for images. By sampling the function at a finite number of points, we obtain a digital signal (a column vector or a matrix) as an element of a finite-dimensional vector space.

- Treating digital signals as vectors, we proceed to analyze them by applying suitable *perfect reconstruction linear transforms*. Each such transform uses an invertible matrix A. If $\mathbf{x}$ is the vector encoding the signal, we call the vector $\mathbf{y} = A^{-1}\mathbf{x}$ the *A-transform* of the signal. The original signal can be reconstructed by $\mathbf{x} = A\mathbf{y}$; this expresses $\mathbf{x}$ as a linear combination of the columns of A with the coefficients furnished by the transform vector $\mathbf{y}$.

- The *design problem* is to choose the matrix A such that the entries in the A-transform of the signal give desired information about the signal that is not evident by examining the entries in $\mathbf{x}$. For example, if $\mathbf{x}$ is a sample of an audio signal and A is the *Fourier matrix*, then the entries in $\mathbf{y}$ give the *frequency* content of the signal. For signal compression, A will be a *wavelet matrix*.

- *Signal processing* occurs when we modify some entries in the transformed vector $\mathbf{y}$ to get a new vector $\widetilde{\mathbf{y}}$ before we apply the inverse matrix. For example, to compress the signal for rapid data transmission or compact storage, we fix a threshold level and replace any entry in $\mathbf{y}$ by zero if its absolute value is less than this threshold. We send or store only the nonzero entries in $\widetilde{\mathbf{y}}$ together with their locations in the vector, and we specify the transform A that was used. The receiver can then reconstruct $\widetilde{\mathbf{x}} = A\widetilde{\mathbf{y}}$ as an approximate version of $\mathbf{x}$. The accuracy of the approximation will depend on the threshold and the transform matrix A. The compression step is a nonlinear operation, although the reconstruction is linear. This method is currently used for transmission and storage of fingerprints, for example, with A as a wavelet matrix.

- In typical applications $\mathbf{x}$ has a large number of entries, so it is important to have a fast algorithm (tailored to the matrix A) to calculate $\mathbf{y}$ from $\mathbf{x}$. For Fourier and wavelet transforms, such algorithms arise by factoring the matrix A into a product of suitable *elementary matrices*.

This description of signal processing has emphasized column vectors and matrices because they are familiar to students from an introductory linear algebra course and are the natural language of MATLAB. The function-space point of view, with digital signals considered as functions of a discrete time variable, is equally important. This naturally leads to a more conceptual approach to linear algebra in terms of vector spaces and linear transformations as the book progresses. Mathematical concepts that students first encountered in the linear algebra course (but often did not fully understand) reappear in a new light in the signal processing context. For example, the choice of a square matrix A to be used as a transform matrix is the same as the choice of a basis (the columns of A) and dual basis (the rows of A^{-1}) for the vector space of signals. In the case of the Fourier matrix, the basis consists of sampled complex sinusoidal waves of different frequencies. For a one-scale discrete wavelet transform, the signal space decomposes as the direct sum of the *trend* and *detail* subspaces, which are the column spaces of the partitioned wavelet synthesis matrix. Shifts of the lowpass and highpass filter vectors by an even number of positions furnish bases for these subspaces.

Organization of the book

Each chapter begins with an overview section. Chapter 1 starts with sampling and quantization of signals. Basic concepts from linear algebra are briefly reviewed and the geometric meaning of linear transformations is illustrated through vector graphics. Other linear algebra topics that are important for signal processing (inner product spaces and Fourier series) are treated in somewhat more detail.

Chapter 2 treats the discrete Fourier transform as the digital signal version of the Fourier series. The key signal-processing operation of filtering first appears here in terms of circulant matrices and circular convolution. Filters act on the Fourier transform of the signal by diagonal matrices. Factoring the Fourier matrix yields the *fast Fourier transform*, which is an indispensable tool in signal processing.

Chapter 3 gives an introduction to discrete wavelet transforms and their applications. These transforms separate a finite digital signal into long-term *trend subsignal* and short-term *detail subsignal*, each half the length of the original signal. This separation process can be repeated on the trend subsignal, which leads to the *pyramid algorithm* and the *multiresolution representation* of a signal as a sum of a coarse trend subsignal plus increasingly fine detail subsignals. Beginning with the simple Haar wavelet transform, where the trend subsignal is obtained by averaging adjacent signal values and the detail subsignal by differencing, we construct more effective wavelet transforms by splitting the signal into *even* and *odd* subsignals and applying successive *prediction* and *update* transformations (this is called the *lifting method*). Wavelet transforms of two-dimensional images are obtained by simultaneously transforming the rows and the columns of the matrix that encodes the image. Applications of wavelet transforms to compression of signals and images are given. The philosophy of this chapter is to explore specific wavelet transforms thoroughly as motivation for developing the general mathematical theory in the next chapter.

Chapter 4 shows how to construct wavelet transforms that give effective signal decompositions into trend and detail subsignals. The approach is through *two-channel filter banks* that separate the low-frequency and high-frequency portions of a signal. The ideas and algebraic techniques needed to design these filter banks and the associated discrete wavelet transforms are treated in detail.

Chapter 5, which is at a more advanced mathematical level, is optional for a one-semester course. It gives a brief introduction to wavelet transforms for analog signals, taking advantage of the results already developed in Chapters 3 and 4.

Since the book is written for students with a wide range of mathematical, scientific, and engineering backgrounds and interests, it interweaves motivating discussions and graphics with more precise mathematical statements such as definitions and theorems, and it provides overall strategies and considerable detail in the mathematical arguments. Each chapter has many illustrative examples and computer explorations. There are more than 90 figures in the text. The mathematical concepts are clarified by more than 75 exercises with detailed solutions.

Acknowledgments

The author developed this book from a decade of teaching a one-semester undergraduate course on this topic at Rutgers to a large number of undergraduate mathematics, computer science, physical science, and engineering students. Their enthusiastic comments and course evaluations over the years convinced him to transform and expand the lecture notes, exercises, and computer assignments into a more coherent book form, and to add a chapter on wavelet transforms for analog signals (which was not part of the course). Many students have contributed to improving the exposition and correcting errors, and he thanks them for their help.

The wavelet books [Jensen and la Cour-Harbo (2001)] and [Strang and Nguyen (1997)], together with the classic applied linear algebra book [Strang (2006)], which the author used for many years in a graduate applied linear algebra course, had a strong influence on the presentation here. The books [Broughton and Bryan (2009)] and [Frazier (1999)], also designed for an undergraduate course and emphasizing linear algebra, furnished ideas for the treatment of some of the topics in the present book. The author is indebted to Prof. Charles Dapogny, who made a thorough critical reading of the course lecture notes when he was a colleague at Rutgers and provided a large number of helpful suggestions for improvements. The author would also like to thank Dr. Matthew Thibault, who taught the course at Rutgers in Spring 2015, for many discussions about its goals and content.

He thanks Dr. Enriqueta Rodríguez-Carrington for her editorial help and moral support throughout this project. At World Scientific, executive editor Rochelle Kronzek provided encouragement and desk editor Lai Fun Kwong was of great help in production. Major revisions of the book occurred during the 2014–2015 academic year while the author was supported by a sabbatical leave from Rutgers University.

Contents

Chapter 1

Linear Algebra and Signal Processing

1.1 Overview

In signal processing we sample an *analog signal* of a continuous variable (time or space) to obtain a *digital signal* of a discrete variable. This analog-to-digital conversion lets us manipulate signals using linear algebra: finite-dimensional vector spaces and their bases and dual bases, linear transformations and their matrices, and direct sums of vector spaces and the associated partitioned matrices. We begin with a brief review of these concepts, which will be used throughout the later chapters for signal processing by Fourier and wavelet transforms. As an illustration of the power of the linear algebra approach, we present the *vector graphics* method for efficient encoding of a two-dimensional image. Then we show how to use parametric families of *affine transformations* to animate the image. This application combines geometry and linear algebra in a vivid way.

We next recall the concepts of inner products, norms, orthogonal projections, and unitary matrices from linear algebra. These mathematical tools are important for signal processing because analog to digital conversion creates loss of information (*aliasing* and *quantization error*). Further signal processing, such as *compression* (for rapid transmission and compact storage), noise removal, and emphasizing selected features of a signal (such as the high frequencies in a sound or the edges in an image), also cause more loss of information. The distortion in the original signal caused by these processing steps can be quantified using the *energy* in a signal, which is the square of the norm of the vector representing the signal. When a digital signal is transformed by a unitary matrix, the energy is unchanged.

The vector space of periodic functions of a continuous variable plays an important role in signal processing. The basic idea of Fourier analysis is that the Fourier coefficients of a periodic function give the components of the function relative to the Fourier basis of oscillating waves. Thus the function can be viewed as a column vector with infinitely many components. The function can be well approximated (in the energy norm) by trigonometric polynomials, which correspond to column vectors with only finitely many nonzero components.

1.2   Sampling and Quantization

An analog signal,[1] such as the continuously-varying voltage in an electric current or the gray scale levels in a black-and-white photograph on film, must be converted to digital form in order to be stored as a computer-readable file. This analog-to-digital conversion process is carried out in several steps. The first step is to *sample* the signal. Next the signal is *quantized*. Finally, the quantized signal is *encoded* for efficient storage and transmission.

Example 1.1 (Sampling an Analog Signal). Suppose the sound intensity levels in an audio signal are given by a function $f(t)$ for $a \leq t \leq b$, where t represents time and $f(t)$ is a real number. Choose N equally-spaced *sampling points* $t_0 = a$, $t_1 = a + \Delta t$, $t_2 = a + 2\Delta t$, $\ldots$, $t_{N-1} = a + (N-1)\Delta t$ where $\Delta t = (b-a)/N$ (note that $t_N = b$ is the right end-point of the interval). Here we have arbitrarily chosen the sample points at the *left* end-point of each sample interval; we could also have sampled at the midpoint or right end point of each sample interval[2]. If we assume that $f(t)$ is a continuous function of t, then the value $f(t)$ is approximately the same as the value at the sample point nearest to t. Hence we can approximate the whole function f by a column vector:

$$ f \approx \mathbf{y} = \begin{bmatrix} f(t_0) \\ f(t_1) \\ \cdots \\ f(t_{N-1}) \end{bmatrix}. $$

Each entry in $\mathbf{y}$ is a real number, and so it would need an infinite set of decimal digits to be determined exactly. In practice, it is represented in MATLAB by a specified type of floating-point number, for example double-precision 64-bit. We will return to this topic in Section 2.2.   ■

Example 1.2 (Quantizing a Digital Signal). We can *quantize* the entries in a vector $\mathbf{y} \in \mathbb{R}^N$ representing a digital signal by taking a fixed number of binary digits. For example, suppose each entry in $\mathbf{y}$ is between 0 and 1 (this can always be arranged by rescaling). For *eight bit quantization* we divide the interval $0 \leq y \leq 1$ into $2^8 = 256$ equal subintervals $0 < 1/256 < \cdots < 255/256 < 1$ and replace $\mathbf{y}$ by the integer part of $256\mathbf{y}$. Thus for $N = 4$ the quantization steps change the vector on the left to the (binary) vector on the right:

$$ \begin{bmatrix} 0.814723686393179 \\ 0.905791937075619 \\ 0.126986816293506 \\ 0.913375856139019 \end{bmatrix} \longrightarrow \begin{bmatrix} 208 \\ 231 \\ 32 \\ 233 \end{bmatrix} \longrightarrow \begin{bmatrix} 11010000 \\ 11100111 \\ 00100000 \\ 11101001 \end{bmatrix}. $$

---

[1] The term *analog* means that time and space coordinates are considered as a real variables as in calculus.

[2] A more accurate sampling method, called *preprocessing*, uses an average of several values of $f(t)$ in each sampling interval.

The initial vector **y** in this example was obtained by the MATLAB command `rand(4,1)`, the second vector **z** by `floor(256*y)`, and the third vector **b** by `dec2bin(z)` (the binary representation of the entries in **z**). ∎

Remark 1.1. If the values of a signal $f(t)$ vary over a wide range, it is desirable to take quantization intervals of unequal sizes. The most frequently-occurring ranges of values of $f(t)$ get many small quantization intervals, whereas a few larger intervals are used to include rarely-occurring values. The final step in analog to digital conversion is to encode the blocks of 0 and 1 in the binary vector **b** in an efficient way using methods of *coding theory*. See [Van Fleet (2008), §3.4] to learn about the widely-used *Huffman code*.

Example 1.3 (Sampling an Image). Suppose the gray scale levels in a black and white photograph are given by function $f(x, y)$ in the rectangle $a \leq x \leq b$ and $c \leq y \leq d$. We choose MN equally-spaced sampling points (x_j, y_i) in the rectangle, Here $x_0 = a$, $x_1 = a + \Delta x$, $x_2 = a + 2\Delta x$, ... , $x_N = a + (N-1)\Delta x$ where $\Delta x = (b-a)/N$, and $y_0 = c$, $y_1 = c + \Delta y$, $y_2 = c + 2\Delta y$, ..., $y_{M-1} = c + (M-1)\Delta y$, where $\Delta y = (d-c)/M$. If we assume that $f(x, y)$ is a continuous function of (x, y), then the value $f(x, y)$ is approximately the same as the value at the sample point nearest to (x, y). Hence we can approximate the whole function f on the rectangle by an $M \times N$ matrix:

$$
f \approx \begin{bmatrix} f(x_0, y_0) & f(x_1, y_0) & \cdots & f(x_{N-1}, y_0) \\ f(x_0, y_1) & f(x_1, y_1) & \cdots & f(x_{N-1}, y_1) \\ \vdots & \vdots & \ddots & \vdots \\ f(x_0, y_{M-1}) & f(x_1, y_{M-1}) & \cdots & f(x_{N-1}, y_{M-1}) \end{bmatrix}.
$$

Although the indexing of the matrix entries might look odd, with the column index for the x variable and the row index for the y variable, it corresponds to the usual convention in computer graphics: the origin of coordinates is at the upper left corner, the x-axis points to the right and the y-axis points down. In both cases we start the indexing at 0 rather than at 1. This convention turns out to be natural for purposes of Fourier analysis and wavelet analysis. The quantization and encoding steps are the same as those for audio signals. We will study image processing in Section 3.6. ∎

1.3   Vector Spaces

Introductory linear algebra courses emphasize vector spaces of column vectors and matrices whose entries are real numbers. For Fourier and wavelet analysis we will need more general vector spaces. Sometimes it will be very convenient to use *complex numbers* as scalars.[3] Letting $\mathbb{F}$ denote either the field of real numbers $\mathbb{R}$ or the

---

[3]See Appendix A for a brief review of complex numbers.

field of complex numbers $\mathbb{C}$, we define $\mathbb{F}^n$ to be the set of all column vectors

$$\mathbf{x} = \begin{bmatrix} x_1 \\ \vdots \\ x_n \end{bmatrix} \quad \text{where } x_1, \ldots, x_n \in \mathbb{F}.$$

We add column vectors and multiply column vectors by elements of $\mathbb{F}$ (which we call *scalars*) in terms of their components x_i, using addition and multiplication in $\mathbb{F}$.

More generally, a *vector space* over $\mathbb{F}$ is a set V of elements called *vectors*. The set V has a zero vector, denoted by $\mathbf{0}$, an *addition* operation to combine vectors $\mathbf{u}$ and $\mathbf{v}$ into a new vector $\mathbf{u} + \mathbf{v}$, and a *scalar multiplication* operation to produce a new vector $\lambda\mathbf{u}$ from a scalar $\lambda \in \mathbb{F}$ and a vector $\mathbf{u} \in V$. We assume that these operations satisfy the same commutative, associative, and distributive properties that the vector space $\mathbb{F}^n$ satisfies. Specifically, we assume that the following axioms *A1–A5* are satisfied for all vectors $\mathbf{u}, \mathbf{v}, \mathbf{w}$ in V and all scalars λ, μ in $\mathbb{F}$.

A1 (**zero vector**) $\mathbf{u} + \mathbf{0} = \mathbf{u}$
A2 (**commutative law**) $\mathbf{u} + \mathbf{v} = \mathbf{v} + \mathbf{u}$
A3 (**scalar multiplication**) $1\mathbf{u} = \mathbf{u}$ and $0\mathbf{u} = \mathbf{0}$
A4 (**associative laws**) $(\mathbf{u} + \mathbf{v}) + \mathbf{w} = \mathbf{u} + (\mathbf{v} + \mathbf{w})$ and $\lambda(\mu\mathbf{v}) = (\lambda\mu)\mathbf{v}$
A5 (**distributive laws**) $(\lambda + \mu)\mathbf{u} = \lambda\mathbf{u} + \mu\mathbf{u}$ and $\lambda(\mathbf{u} + \mathbf{v}) = \lambda\mathbf{u} + \lambda\mathbf{v}$

Addition and scalar multiplication of vectors are *binary* operations: they take two inputs (either two vectors or a scalar and a vector) and give a vector as output. The commutative, associative, and distributive laws assert that two different ways of performing operations on scalars and vectors give the same final result. Thus if we take three vectors $\mathbf{u}$, $\mathbf{v}$, $\mathbf{w}$ or two scalars λ, μ and a vector $\mathbf{v}$, the associative laws allow us to omit the parentheses in the expressions $\mathbf{u} + \mathbf{v} + \mathbf{w}$ and $\lambda\mu\mathbf{v}$.

We write $-\mathbf{u}$ for $(-1)\mathbf{u}$. Then $\mathbf{u} - \mathbf{u} = (1 - 1)\mathbf{u} = 0\mathbf{u} = \mathbf{0}$. Hence $-\mathbf{u}$ is the inverse of $\mathbf{u}$ for the operation of vector addition. Likewise, if $\lambda \neq 0$ then the operation of multiplication by λ^{-1} is the inverse of multiplication by λ, since $\lambda^{-1}(\lambda\mathbf{v}) = (\lambda^{-1}\lambda)\mathbf{v} = 1\mathbf{v} = \mathbf{v}$.

Example 1.4. We denote by $\mathbb{F}^{m \times n}$ the set of $m \times n$ matrices whose entries are in $\mathbb{F}$. We add matrices of the same size by adding corresponding elements. Multiplication by a scalar is done on each matrix element. The element $\mathbf{0}$ is the matrix having all entries zero.[4] This definition makes $\mathbb{F}^{m \times n}$ into a vector space over $\mathbb{F}$. Thinking of matrices as vectors might seem confusing, since we are ignoring the possibility of multiplying matrices (when the sizes are compatible). This viewpoint becomes completely natural, however, when we use matrices to encode two-dimensional images as in Example 1.3. ∎

Example 1.5. Let V be the collection of all $\mathbb{F}$-valued functions defined on a set X.

---

[4]The MATLAB command `zeros(m, n)` generates this matrix.

We make V into a vector space by *pointwise* operations:

$$(f+g)(x) = f(x) + g(x), \quad (\alpha f)(x) = \alpha f(x) \quad \text{for } f, g \in V \text{ and } \alpha \in \mathbb{F}.$$

The zero vector $\mathbf{0}$ is the function that has the value 0 everywhere on X. It is easy to check that these operations on V satisfy the vector space axioms. In particular, when $X = \{1, 2, \ldots, n\}$ we can identify V with $\mathbb{F}^n$ by viewing the components of a column vector as a function on X. This is the same idea we used in Example 1.1. Likewise, when

$$X = \{(i, j) : 1 \le i \le m \text{ and } 1 \le j \le n\}$$

we can identify V with $\mathbb{F}^{m \times n}$ by considering a matrix to be a function on the set X, as in Example 1.3. ∎

Definition 1.1. A subset W of a vector space V is called a *subspace* if $\mathbf{0} \in W$ and all the linear combinations $\alpha \mathbf{x} + \beta \mathbf{y}$ are in W for every $\mathbf{x}, \mathbf{y} \in W$ and $\alpha, \beta \in \mathbb{F}$.

If W satisfies Definition 1.1, then Axioms *A1–A5* hold for all vectors in W and all scalars in $\mathbb{F}$ (with the operations of vector addition and scalar multiplication inherited from V). Hence W is a vector space with scalars $\mathbb{F}$. Here are some natural examples of subspaces.

Example 1.6. Let V be a vector space and suppose $S = \{\mathbf{v}_1, \ldots, \mathbf{v}_n\}$ is any collection of vectors in V. Define

$$\text{Span } S = \text{all linear combinations } c_1 \mathbf{v}_1 + \cdots + c_n \mathbf{v}_n \text{ with } c_i \in \mathbb{F}.$$

Then Span S satisfies Definition 1.1, so Span S is a subspace of V. ∎

Example 1.7. Let $V = \mathbb{F}^m$ and take a matrix $A \in \mathbb{F}^{m \times n}$. If S is the set of columns of A, then Span S is the *column space* Col(A). ∎

Example 1.8. Let $V = \mathbb{F}^n$ and let $A \in \mathbb{F}^{m \times n}$. The *null space* Null($A$) is the set of all vectors $\mathbf{x} \in \mathbb{F}^n$ such that $A\mathbf{x} = \mathbf{0}$. Clearly $\mathbf{0} \in$ Null(A). If $\mathbf{x}, \mathbf{y} \in$ Null(A) and $\alpha, \beta \in \mathbb{F}$, then $A(\alpha\mathbf{x} + \beta\mathbf{y}) = \alpha A\mathbf{x} + \beta A\mathbf{y} = \alpha\mathbf{0} + \beta\mathbf{0} = \mathbf{0}$. Hence $\alpha\mathbf{x} + \beta\mathbf{y} \in$ Null(A), so Null(A) is a subspace. ∎

Example 1.9. Let X be an interval of real numbers $a \le x \le b$ or a rectangle $a \le x \le b$, $c \le y \le d$ in two dimensions. Then $C_{\mathbb{F}}(X)$ will denote the set of all continuous $\mathbb{F}$-valued functions on X. Since the sum of two continuous functions and a scalar multiple of a continuous function are continuous, $C_{\mathbb{F}}(X)$ is a subspace of the vector space of all functions on X. ∎

Example 1.10. Let X be the interval $0 \le x \le 1$. Let $W \subset C_{\mathbb{F}}(X)$ consist of all functions $f(x)$ that satisfy the *homogeneous boundary condition* $f(0) = f(1)$. If $f(x)$ and $g(x)$ are in W, then for any scalars α, β the function $h(x) = \alpha f(x) + \beta g(x)$ satisfies

$$h(0) = \alpha f(0) + \beta g(0) = \alpha f(1) + \beta g(1) = h(1).$$

Hence W is a subspace of $C_{\mathbb{F}}(X)$. ∎

1.4   Bases and Dual Bases

Let V be a vector space with scalars $\mathbb{F}$.

Definition 1.2. A set $\mathcal{S} = \{\mathbf{v}_1, \ldots, \mathbf{v}_n\}$ of nonzero vectors in V is *linearly independent* if the only possible linear relation

$$c_1\mathbf{v}_1 + \cdots + c_n\mathbf{v}_n = \mathbf{0} \quad (\text{where } c_j \in \mathbb{F}) \tag{1.1}$$

satisfied by the vectors in $\mathcal{S}$ is the relation with $c_1 = \cdots = c_n = 0$.

Here is another way of understanding Definition 1.2. If we know that two linear combinations of vectors in $\mathcal{S}$ give the same result:

$$a_1\mathbf{v}_1 + \cdots + a_n\mathbf{v}_n = b_1\mathbf{v}_1 + \cdots + b_n\mathbf{v}_n, \tag{1.2}$$

then we can conclude that the corresponding scalars on each side are equal:

$$a_1 = b_1, \; a_2 = b_2, \; \ldots, \; a_n = b_n.$$

To see this, just use the associative and distributive laws to change equation (1.2) into the form (1.1) with $c_i = a_i - b_i$.

Example 1.11. When $V = \mathbb{F}^m$ so that $\mathcal{S} = \{\mathbf{v}_1, \ldots, \mathbf{v}_n\}$ is a set of column vectors, we can form the $m \times n$ matrix $A = \begin{bmatrix} \mathbf{v}_1 & \cdots & \mathbf{v}_n \end{bmatrix}$. Equation (1.1) becomes the system of homogeneous linear equations $A\mathbf{x} = \mathbf{0}$, where $\mathbf{x} = \begin{bmatrix} c_1 & \cdots & c_n \end{bmatrix}^{\mathrm{T}} \in \mathbb{F}^n$. Thus linear independence of $\mathcal{S}$ means that $\mathrm{Null}(A) = \mathbf{0}$. To check whether this is true, recall from linear algebra that elementary row operations can be found to change A into a unique reduced row-echelon form matrix[5] $R = \mathrm{rref}(A)$. An example of a reduced row-echelon form matrix is

$$\begin{bmatrix} 1 & 0 & \star & 0 & \star & \star \\ 0 & 1 & \star & 0 & \star & \star \\ 0 & 0 & 0 & 1 & \star & \star \\ 0 & 0 & 0 & 0 & 0 & 0 \end{bmatrix},$$

where the entries $\star$ can be any scalars; columns 1, 2 and 4 are the *pivot columns* in this example. Since every elementary row operation can be inverted, R and A have the same null space. This space is zero if and only if $R = \begin{bmatrix} \mathbf{e}_1 & \mathbf{e}_2 & \cdots & \mathbf{e}_n \end{bmatrix}$, where $\mathbf{e}_j$ is the jth *standard vector* in $\mathbb{F}^m$ with a 1 in the jth row and 0 elsewhere. In particular, this requires that $n \leq m$. Hence $\mathcal{S}$ *cannot* be linearly independent if $n > m$ (A has more columns than rows). ∎

Definition 1.3. A subset $\mathcal{S} = \{\mathbf{v}_1, \ldots, \mathbf{v}_n\} \subset V$ *spans* V if the equation

$$c_1\mathbf{v}_1 + \cdots + c_n\mathbf{v}_n = \mathbf{x} \quad (\text{where } c_j \in \mathbb{F}) \tag{1.3}$$

has at least one solution for *every* given vector $\mathbf{x} \in V$.

---

[5]It can be obtained by the MATLAB command `rref(A)`.

Example 1.12. When $V = \mathbb{F}^m$, so that $\mathcal{S}$ is a set of column vectors in $\mathbb{F}^m$, we can form the $m \times n$ matrix $A = \begin{bmatrix} \mathbf{v}_1 & \cdots & \mathbf{v}_n \end{bmatrix}$. Then the spanning condition says that the column space of A is $\mathbb{F}^m$. This means that all the standard vectors $\mathbf{e}_1, \ldots, \mathbf{e}_m$ must appear as columns of rref(A). In particular, this forces $n \geq m$. Thus $\mathcal{S}$ *cannot* span $\mathbb{F}^m$ if $n < m$ (A has more rows than columns). ∎

Definition 1.4. A set $\mathcal{S} = \{\mathbf{v}_1, \ldots, \mathbf{v}_n\}$ of vectors in V is a *basis* for V if it is linearly independent and spans V.

When $\mathcal{S}$ is a basis for V then the *number n* of vectors in $\mathcal{S}$ is unique (although there are many choices of the particular vectors). To see this, let $\mathcal{S} = \{\mathbf{v}_1, \ldots, \mathbf{v}_m\}$ be a spanning set for V and $\mathcal{T} = \{\mathbf{w}_1, \ldots, \mathbf{w}_n\}$ an independent set of vectors in V. We claim that $n \leq m$. Indeed, suppose that $n > m$. We can write $\mathbf{w}_j = \sum_{i=1}^m a_{ij} \mathbf{v}_i$ for $j = 1, \ldots, n$. The $m \times n$ matrix A must have a nonzero null space, by Example 1.11, since it has more columns than rows. Let $\mathbf{b} = \begin{bmatrix} b_1 & \cdots & b_n \end{bmatrix}^{\mathrm{T}}$ be a nonzero vector with $A\mathbf{b} = 0$. Then

$$b_1 \mathbf{w}_1 + \cdots + b_n \mathbf{w}_n = \sum_{i=1}^m \left\{ \sum_{j=1}^n a_{ij} b_j \right\} \mathbf{v}_i = \mathbf{0}.$$

Hence $\mathcal{T}$ is not linearly independent, a contradiction. This proves that $n \leq m$. If $\mathcal{S}$ and $\mathcal{T}$ are both bases for V, applying the same argument with the roles of $\mathcal{S}$ and $\mathcal{T}$ interchanged shows that $m = n$.

We call the number of vectors in a basis the *dimension* of V and say that V is a *finite-dimensional* vector space. We call the scalars c_i in (1.3) the *components* of $\mathbf{x}$ relative to the basis $\{\mathbf{v}_1, \ldots, \mathbf{v}_n\}$. These scalars are uniquely determined by $\mathbf{x}$ because of the linear independence of the basis.

Example 1.13. Let A be an $m \times n$ matrix. Set $r = \dim \mathrm{Col}(A)$ (the *rank* of A) and $p = \dim \mathrm{Null}(A)$ (the *nullity* of A). Recall the following facts from elementary linear algebra:

(1) The rank r of A is the number of *pivot columns* in rref(A).
(2) The columns of A in the same positions as the pivot columns of rref(A) give a basis for $\mathrm{Col}(A)$.
(3) The nullity p of A is the number of non-pivot columns in rref(A).
(4) Each non-pivot column of A corresponds to a *free variable* for the equation $A\mathbf{x} = \mathbf{0}$. We obtain p vectors that give a basis for $\mathrm{Null}(A)$ by setting one free variable equal 1 and the other free variables equal 0.

From (1) and (3) we conclude that rank A + nullity $A = n$, the total number of columns in A. In particular, if A is a square matrix and nullity $A = 0$, then A is invertible. ∎

Example 1.14. For any positive integer n, let $\mathcal{P}_n$ be the set of all polynomials $f(x)$ in one variable of degree less than n:

$$f(x) = a_0 + a_1 x + \cdots + a_{n-1} x^{n-1} \quad \text{where} \quad a_j \in \mathbb{F}. \tag{1.4}$$

Then $\mathcal{P}_n$ is a vector space relative to the usual addition and scalar multiplication operations on polynomials. By definition, the monomials $\{1, x, x^2, \ldots, x^{n-1}\}$ span $\mathcal{P}_n$. They are also linearly independent, since the coefficients a_j in (1.4) are uniquely determined by $f(x)$ through Taylor's formula:

$$a_j = \frac{f^{(j)}(0)}{j!} \quad \text{where} \quad f^{(j)}(x) = (d/dx)^j f(x).$$

Thus these monomials give a basis for $\mathcal{P}_n$. Using this basis, we can view the polynomial $f(x) \in \mathcal{P}_n$ as a column vectors under the correspondence

$$f \longleftrightarrow \begin{bmatrix} a_0 \\ a_1 \\ \vdots \\ a_{n-1} \end{bmatrix}.$$

We say that $\mathcal{P}_n$ is *isomorphic* to $\mathbb{F}^n$ as a vector space. ∎

Example 1.15. There are other ways to view polynomials as column vectors. For example, take any n distinct points $x_0, x_1, \ldots, x_{n-1} \in \mathbb{F}$ and evaluate $f(x) \in \mathcal{P}_n$ at these points. Since $f(x)$ is of degree at most $n - 1$, it is uniquely determined by its values at these n points. Conversely, given any set $c_0, \ldots, c_{n-1}$ of n scalars in $\mathbb{F}$, there is an *interpolation polynomial* $f(x)$ of degree at most $n - 1$ such that $f(x_j) = c_j$ for $j = 0, \ldots, n - 1$. This gives an isomorphism

$$f \longleftrightarrow \begin{bmatrix} f(x_0) \\ f(x_1) \\ \vdots \\ f(x_{n-1}) \end{bmatrix}$$

between $\mathcal{P}^n$ and $\mathbb{F}^n$ that is different than the one in Example 1.14. To illustrate this, take $n = 3$ and $x_0 = 0$, $x_1 = 1$, and $x_2 = 2$. For the monomials 1, x, x^2 in $\mathbf{P}_3$ this correspondence is

$$1 \longleftrightarrow \begin{bmatrix} 1 \\ 1 \\ 1 \end{bmatrix}, \quad x \longleftrightarrow \begin{bmatrix} 0 \\ 1 \\ 2 \end{bmatrix}, \quad x^2 \longleftrightarrow \begin{bmatrix} 0 \\ 1 \\ 4 \end{bmatrix}.$$

This set of column vectors is linearly independent and gives a basis for $\mathbb{F}^3$ different from the standard basis $\mathbf{e}_1, \mathbf{e}_2, \mathbf{e}_3$. ∎

Now assume that $V = \mathbb{F}^n$ and $\{\mathbf{v}_1, \ldots, \mathbf{v}_n\}$ is a basis for V. Let $A = [\mathbf{v}_1 \; \mathbf{v}_2 \; \cdots \; \mathbf{v}_n]$ as in Example 1.12. Since the null space of A is zero and the column space is V, we know that A is an invertible matrix by Example 1.13. Let $\mathbf{u}_i \in \mathbb{F}^{1 \times n}$ be the ith row of A^{-1}. If $\mathbf{x} \in \mathbb{F}^n$ then

$$A^{-1}\mathbf{x} = \begin{bmatrix} \mathbf{u}_1\mathbf{x} \\ \vdots \\ \mathbf{u}_n\mathbf{x} \end{bmatrix}.$$

Notice that each entry in the column vector on the right is a scalar (1×1 matrix), since $\mathbf{u}_i$ is a row vector and $\mathbf{x}$ is a column vector. Since $AA^{-1} = I_n$ (the $n \times n$ identity matrix), it follows that we can write every vector $\mathbf{x} \in \mathbb{F}^n$ as

$$\mathbf{x} = A(A^{-1}\mathbf{x}) = \begin{bmatrix} \mathbf{v}_1 & \cdots & \mathbf{v}_n \end{bmatrix} \begin{bmatrix} \mathbf{u}_1\mathbf{x} \\ \vdots \\ \mathbf{u}_n\mathbf{x} \end{bmatrix} = (\mathbf{u}_1\mathbf{x})\mathbf{v}_1 + \cdots + (\mathbf{u}_n\mathbf{x})\mathbf{v}_n .$$

Comparing this formula with (1.3), we conclude that the components c_i of $\mathbf{x}$ relative to the basis $\{\mathbf{v}_1, \ldots, \mathbf{v}_n\}$ are the scalars

$$c_i = \mathbf{u}_i\mathbf{x} \quad \text{(product of a row vector and a column vector).} \tag{1.5}$$

We call the set of row vectors $\{\mathbf{u}_1, \ldots, \mathbf{u}_n\}$ the *dual basis* to the basis of column vectors $\{\mathbf{v}_1, \ldots, \mathbf{v}_n\}$, and we call the vector

$$\tilde{\mathbf{x}} = A^{-1}\mathbf{x} = \begin{bmatrix} \mathbf{u}_1\mathbf{x} \\ \vdots \\ \mathbf{u}_n\mathbf{x} \end{bmatrix}$$

the *A-transform* of $\mathbf{x}$. When we take $\mathbf{x} = \mathbf{v}_j$ in (1.5) we get $c_i = 1$ if $i = j$, and zero otherwise. So the A-transform of $\mathbf{v}_j$ is the standard basis vector $\mathbf{e}_j$. We reconstruct $\mathbf{x}$ from $\tilde{\mathbf{x}}$ as $\mathbf{x} = A\tilde{\mathbf{x}}$.

Example 1.16. Consider the basis $\{\mathbf{v}_1, \mathbf{v}_2, \mathbf{v}_3\}$ for $\mathbb{F}^3$ given by the columns of the upper-triangular matrix

$$A = \begin{bmatrix} 1 & 2 & 4 \\ 0 & 1 & 3 \\ 0 & 0 & 1 \end{bmatrix} .$$

It is easy to check that the inverse matrix[6] in this case is

$$A^{-1} = \begin{bmatrix} 1 & -2 & 2 \\ 0 & 1 & -3 \\ 0 & 0 & 1 \end{bmatrix} .$$

Hence the dual basis is $\mathbf{u}_1 = \begin{bmatrix} 1 & -2 & 2 \end{bmatrix}$, $\mathbf{u}_2 = \begin{bmatrix} 0 & 1 & 3 \end{bmatrix}$, and $\mathbf{u}_3 = \begin{bmatrix} 0 & 0 & 1 \end{bmatrix}$ (the rows of A^{-1}). An arbitrary vector $\mathbf{x} \in \mathbb{F}^3$ has the expansion

$$\begin{bmatrix} x_1 \\ x_2 \\ x_3 \end{bmatrix} = (\mathbf{u}_1\mathbf{x}) \begin{bmatrix} 1 \\ 0 \\ 0 \end{bmatrix} + (\mathbf{u}_2\mathbf{x}) \begin{bmatrix} 2 \\ 1 \\ 0 \end{bmatrix} + (\mathbf{u}_3\mathbf{x}) \begin{bmatrix} 4 \\ 3 \\ 1 \end{bmatrix} ,$$

where the components are $\mathbf{u}_1\mathbf{x} = x_1 - 2x_2 + 2x_3$, $\mathbf{u}_2\mathbf{x} = x_2 - 3x_3$, and $\mathbf{u}_3\mathbf{x} = x_3$. For example, the vector $\mathbf{x} = \begin{bmatrix} 1 & 2 & 3 \end{bmatrix}^{\mathrm{T}}$ has A-transform

$$\tilde{\mathbf{x}} = A^{-1}\mathbf{x} = \begin{bmatrix} 1 & -2 & 2 \\ 0 & 1 & -3 \\ 0 & 0 & 1 \end{bmatrix} \begin{bmatrix} 1 \\ 2 \\ 3 \end{bmatrix} = \begin{bmatrix} 3 \\ -7 \\ 3 \end{bmatrix} .$$

---

[6]You can find the inverse either by row reduction of the augmented matrix $[A : I_3]$ (the Gauss–Jordan method) or by multiplying the elementary matrices that perform the row reduction of A to I_3. The MATLAB command is `inv(A)`.

We reconstruct $\mathbf{x}$ from $\widetilde{\mathbf{x}}$ by $A\widetilde{\mathbf{x}} = 3\begin{bmatrix}1\\0\\0\end{bmatrix} - 7\begin{bmatrix}2\\1\\0\end{bmatrix} + 3\begin{bmatrix}4\\3\\1\end{bmatrix} = \begin{bmatrix}1\\2\\3\end{bmatrix}.$ ■

Example 1.17 (Eigenvector Expansion). Suppose S is an $n \times n$ diagonalizable matrix. Recall that this means we can find a basis $\{\mathbf{v}_1, \ldots, \mathbf{v}_n\}$ for $\mathbb{F}^n$ consisting of *eigenvectors* for S: $S\mathbf{v}_j = \lambda_j \mathbf{v}_j$, where the scalars $\lambda_j \in \mathbb{F}$ are the eigenvalues. Let $A = \begin{bmatrix}\mathbf{v}_1 \cdots \mathbf{v}_n\end{bmatrix}$ be the eigenvector matrix. If $\mathbf{x} \in \mathbb{F}^n$ has A-transform $\widetilde{\mathbf{x}} = \begin{bmatrix}c_1 \cdots c_n\end{bmatrix}^{\mathrm{T}}$, then $\mathbf{x} = c_1\mathbf{v}_1 + \cdots + c_n\mathbf{v}_n$. From this expansion we have

$$S\mathbf{x} = c_1 S\mathbf{v}_1 + \cdots + c_n S\mathbf{v}_n = \lambda_1 c_1 \mathbf{v}_1 + \cdots + \lambda_n c_n \mathbf{v}_n.$$

This means that after we have calculated the A-transform of $\mathbf{x}$, then we can obtain the matrix-vector product $S\mathbf{x}$ very rapidly just using n scalar multiplications.

More generally, given any polynomial $p(z) = a_0 + a_1 z + \cdots + a_k z^k$ we can form the $n \times n$ matrix $p(S) = a_0 I_n + a_1 S + \cdots + a_k S^k$. Since $S^j \mathbf{x} = \lambda_1^j c_1 \mathbf{v}_1 + \cdots + \lambda_n^j c_n \mathbf{v}_n$ for all positive integers j, we obtain a fast algorithm to calculate the matrix-vector product $p(S)\mathbf{x}$ by simply rescaling the coefficients in the A-transform of $\mathbf{x}$:

$$p(S)\mathbf{x} = p(\lambda_1)c_1\mathbf{v}_1 + \cdots + p(\lambda_n)c_n\mathbf{v}_n.$$

Furthermore, if $p(z)$ is a polynomial such that $p(\lambda_j) = 0$, then $p(S)\mathbf{x}$ has zero component in the direction of the vector $\mathbf{v}_j$. ■

Remark 1.2. The *transform method* will be a basic tool in later chapters when S is the *shift matrix* and the A-transform is the *discrete Fourier transform*. In this case the basis vectors $\mathbf{v}_j$ in Example 1.17 will be sampled complex oscillating waves of frequency j. The polynomial $p(z)$ will be chosen so that the matrix $p(S)$ acts as a *filter* to select either the low frequency or the high frequency components in a signal $\mathbf{x}$.

1.5 Linear Transformations and Matrices

Let V and W be vector spaces with scalars $\mathbb{F}$. Let S be a function (*algorithm*) that accepts any input vector $\mathbf{v}$ from V and produces an output vector $\mathbf{w} = S(\mathbf{v})$ in W. Symbolically, we have

$$\mathbf{v} \longrightarrow \boxed{S} \longrightarrow \mathbf{w}$$

(we also write $S : V \to W$ for this input-output relation). We say that S is a *linear transformation* if

$$S(\alpha\mathbf{u} + \beta\mathbf{v}) = \alpha S(\mathbf{v}) + \beta S(\mathbf{u})$$

for all vectors $\mathbf{v}$ and $\mathbf{u}$ in V and all scalars α and β in $\mathbb{F}$.

If $T : V \to W$ is another linear transformation with the same input and output spaces as S, then we define the *sum $S+T$* to be the function from V to W that takes an input vector $\mathbf{v}$ and produces an output vector $\mathbf{w} = S(\mathbf{v}) + T(\mathbf{v})$. Symbolically,

$$
\begin{array}{c}
\boxed{\text{S}} \longrightarrow S(\mathbf{v}) \\
\mathbf{v} \diagup\diagdown \qquad\qquad \diagdown\diagup \boxed{+} \longrightarrow \mathbf{w}\,, \\
\boxed{\text{T}} \longrightarrow T(\mathbf{v})
\end{array}
$$

where $\boxed{+}$ denotes the operation of adding two vectors. Likewise, for $\alpha \in \mathbb{F}$ we define the *scalar multiple αS* of S to be the function from V to W that acts on a vector $\mathbf{v}$ by $(\alpha S)(\mathbf{v}) = \alpha S(\mathbf{v})$. From these definitions we verify the following properties:

- Any linear combination $\alpha S + \beta T$ is a linear transformation.
- The set $\mathcal{L}(V, W)$ of all linear transformations from V to W is a vector space.

Example 1.18. Take $V = \mathbb{F}^{m \times n}$ and $W = \mathbb{F}^{n \times m}$. Let S be the *matrix transpose* operation $S\mathbf{v} = \mathbf{v}^{\mathrm{T}}$ (interchange of rows and columns) for $\mathbf{v} \in V$. Then $S : V \longrightarrow W$ is a linear transformation. Since $(\mathbf{v}^{\mathrm{T}})^{\mathrm{T}} = \mathbf{v}$, the transformation S is invertible and the inverse transformation is again matrix transposition. Note that when $m \neq n$, then $\mathbb{F}^{m \times n}$ and $\mathbb{F}^{n \times m}$ are *different* spaces, although they both have dimension mn. Thus we cannot add elements of V to elements of W. For example, when $n = 1$ then the elements of V are *column vectors* with m components, while the elements of W are *row vectors* with m components. ∎

1.5.1   *Matrix form of a linear transformation*

Suppose that $\dim V = n$ and $\dim W = m$. Fix a basis $\{\mathbf{v}_1, \ldots, \mathbf{v}_m\}$ for V and a basis $\{\mathbf{w}_1, \ldots, \mathbf{w}_m\}$ for W. If $S : V \to W$ is a linear transformation, then S can be represented by a unique $m \times n$ matrix A with entries $a_{ij} \in \mathbb{F}$ in the following way. Write each vector $S\mathbf{v}_j$ as a linear combination of the vectors $\mathbf{w}_i$:

$$
S\mathbf{v}_j = \sum_{i=1}^{m} a_{ij} \mathbf{w}_i \,.
$$

Such a linear combination exists and is unique because $\{\mathbf{w}_1, \ldots, \mathbf{w}_m\}$ is a basis for W. Given a vector $\mathbf{v} = x_1 \mathbf{v}_1 + \cdots + x_n \mathbf{v}_n \in V$, we let $\mathbf{x} \in \mathbb{F}^n$ be the column vector with components x_j and we set $\mathbf{y} = A\mathbf{x} \in \mathbb{F}^m$. Then the linearity of S gives the expansion

$$
S\mathbf{v} = \sum_{j=1}^{n} x_j \, S\mathbf{v}_j = \sum_{j=1}^{n} x_j \left\{ \sum_{i=1}^{m} a_{ij} \mathbf{w}_i \right\}
$$

$$
= \sum_{i=1}^{m} \left\{ \sum_{j=1}^{n} a_{ij} x_j \right\} \mathbf{w}_i = \sum_{i=1}^{m} y_i \mathbf{w}_i \,. \tag{1.6}
$$

Our choice of bases for V and W lets us identify vectors in V and W with column vectors, and (1.6) shows that the action of S becomes multiplication by the matrix A:

$$\mathbf{v} \longleftrightarrow \begin{bmatrix} x_1 \\ \vdots \\ x_n \end{bmatrix} \quad \text{and} \quad S\mathbf{v} \longleftrightarrow \begin{bmatrix} y_1 \\ \vdots \\ y_m \end{bmatrix} = A\mathbf{x}. \tag{1.7}$$

So to each $S \in \mathcal{L}(V,W)$ we have associated a matrix $A \in \mathbb{F}^{m \times n}$. If $T \in \mathcal{L}(V,W)$ is another linear transformation whose matrix is B (relative to the *same* choice of bases for V and W), then from (1.7) we see that $S + T$ corresponds to the matrix $A + B$. Furthermore, for any scalar c the linear transformation cS corresponds to the matrix cA. Thus if we fix bases for V and W, then we can identify V with $\mathbb{F}^n$, W with $\mathbb{F}^m$, and the vector space $\mathcal{L}(V,W)$ with $\mathbb{F}^{m \times n}$.

Example 1.19. Let $V = \mathcal{P}_3$ and $W = \mathcal{P}_2$ (see Example 1.14). Let $T = d/dx$ (differentiation of polynomials). Then $T \in \mathcal{L}(V,W)$ since differentiation is linear and decreases the degree of a polynomial by one. If we take the basis $\{1, x, x^2\}$ for V and the basis $\{1, x\}$ for W as in Example 1.14, then

$$T(1) = 0 \longleftrightarrow \begin{bmatrix} 0 \\ 0 \end{bmatrix}, \qquad T(x) = 1 \longleftrightarrow \begin{bmatrix} 1 \\ 0 \end{bmatrix}, \qquad T(x^2) = 2x \longleftrightarrow \begin{bmatrix} 0 \\ 2 \end{bmatrix}.$$

Thus the matrix for T is

$$B = \begin{bmatrix} 0 & 1 & 0 \\ 0 & 0 & 2 \end{bmatrix}.$$

For the polynomial $f(x) = 2 + 3x + 4x^2$ we have $T(f) = 3 + 8x$. Here

$$T(f) \longleftrightarrow \begin{bmatrix} 3 \\ 8 \end{bmatrix} = \begin{bmatrix} 0 & 1 & 0 \\ 0 & 0 & 2 \end{bmatrix} \begin{bmatrix} 2 \\ 3 \\ 4 \end{bmatrix},$$

in agreement with (1.7). ∎

Let U, V, W be finite-dimensional vector spaces with scalars $\mathbb{F}$. Let $S : V \to W$ and $T : U \to V$ be linear transformations. Use T to transform a vector $\mathbf{u} \in U$ into a vector $\mathbf{v} = T\mathbf{u}$. Then use S to transform $\mathbf{v}$ into $\mathbf{w} = S(T(\mathbf{u})) \in W$. Symbolically[7]

$$\mathbf{u} \longrightarrow \boxed{T} \longrightarrow \mathbf{v} \longrightarrow \boxed{S} \longrightarrow \mathbf{w}.$$

It is easy to check that this formula defines a linear transformation $ST \in \mathcal{L}(U,W)$. Fix bases

$$\begin{aligned} \{\mathbf{u}_1, \ldots, \mathbf{u}_p\} \quad &\text{for } U \quad (\dim U = p), \\ \{\mathbf{v}_1, \ldots, \mathbf{v}_n\} \quad &\text{for } V \quad (\dim V = n), \\ \{\mathbf{w}_1, \ldots, \mathbf{w}_m\} \quad &\text{for } W \quad (\dim W = m). \end{aligned}$$

---

[7]In this diagram the input is on the left and the output is on the right, but in the formula $\mathbf{w} = S(T(\mathbf{u}))$ the input is on the right and the output on the left. This explains why the composite transformation is written as ST.

We want to relate the matrix for ST to the matrices for S and T. Suppose that S has matrix $A = [a_{ij}] \in \mathbb{F}^{m \times n}$ and T has matrix $B = [b_{kl}] \in \mathbb{F}^{n \times p}$ relative to these bases. Recall that the (i, j) element c_{ij} of the *matrix product* $C = AB \in \mathbb{F}^{m \times p}$ is (row i of A) $\times$ (column j of B):

$$c_{ij} = \sum_{k=1}^{n} a_{ik} b_{kj} . \tag{1.8}$$

To find the matrix for ST, we use linearity and equation (1.8) to calculate

$$ST\mathbf{u}_j = S\left\{ \sum_{k=1}^{n} b_{kj} \mathbf{v}_k \right\} = \sum_{k=1}^{n} b_{kj} S\mathbf{v}_k$$

$$= \sum_{k=1}^{n} b_{kj} \left\{ \sum_{i=1}^{m} a_{ik} \mathbf{w}_i \right\} = \sum_{i=1}^{m} \left\{ \sum_{k=1}^{n} a_{ik} b_{kj} \right\} \mathbf{w}_i = \sum_{i=1}^{m} c_{ij} \mathbf{w}_i .$$

This proves the following important property:

($\star$) The matrix for the product (composition) ST of linear transformations is the product AB (in the same order) of the corresponding matrices.

From this property we obtain the *associativity* of matrix multiplication without making any element-by-element calculation: $A(BC) = (AB)C$ when A, B, and C are matrices of compatible sizes, since composition of transformations is always associative.

Example 1.20. Let $U = W = \mathcal{P}_3$ and $V = \mathcal{P}_2$ (see Example 1.14). Let $T = d/dx$ and let S be multiplication by x. Then $T \in \mathcal{L}(W, V)$ and $S \in \mathcal{L}(V, W)$, since S is linear and increases the degree of a polynomial by one. If we take the basis $\{1, x, x^2\}$ for W and the basis $\{1, x\}$ for V, then T has matrix B given in Example 1.20. For S we have

$$S(1) = x \longleftrightarrow \begin{bmatrix} 0 \\ 1 \\ 0 \end{bmatrix} , \qquad S(x) = x^2 \longleftrightarrow \begin{bmatrix} 0 \\ 0 \\ 1 \end{bmatrix} .$$

Thus the matrix for S is

$$A = \begin{bmatrix} 0 & 0 \\ 1 & 0 \\ 0 & 1 \end{bmatrix} .$$

The linear transformation $ST = x(d/dx) : W \to W$ acts on the monomial x^n by $ST(x^n) = nx^n$. Thus the matrix for ST is diagonal with entries $0, 1, 2$. We can also obtain this matrix by multiplying the matrices for S and T:

$$AB = \begin{bmatrix} 0 & 0 \\ 1 & 0 \\ 0 & 1 \end{bmatrix} \begin{bmatrix} 0 & 1 & 0 \\ 0 & 0 & 2 \end{bmatrix} = \begin{bmatrix} 0 & 0 & 0 \\ 0 & 1 & 0 \\ 0 & 0 & 2 \end{bmatrix}$$

by the general property ($\star$) above. ∎

1.5.2 *Direct sums of vector spaces*

Let V and W be vector spaces with scalars $\mathbb{F}$. The *direct sum* vector space $U = V \oplus W$ consists of all pairs of vectors $[\mathbf{v}, \mathbf{w}]$ with $\mathbf{v} \in V$ and $\mathbf{w} \in W$. The vector space operations are carried out in *parallel*:

$$\alpha[\mathbf{v}, \mathbf{w}] + \beta[\mathbf{v}', \mathbf{w}'] = [\alpha\mathbf{v} + \beta\mathbf{v}', \alpha\mathbf{w} + \beta\mathbf{w}']$$

for vectors $\mathbf{v}, \mathbf{v}' \in V$, $\mathbf{w}, \mathbf{w}' \in W$ and scalars $\alpha, \beta \in \mathbb{F}$. If $\{\mathbf{v}_1, \ldots, \mathbf{v}_m\}$ is a basis for V and $\{\mathbf{w}_1, \ldots, \mathbf{w}_n\}$ is a basis for W, then the $m + n$ vectors

$$[\mathbf{v}_1, \mathbf{0}], \ldots, [\mathbf{v}_m, \mathbf{0}], [\mathbf{0}, \mathbf{w}_1], \ldots, [\mathbf{0}, \mathbf{w}_n]$$

comprise a basis for $V \oplus W$, since they span $V \oplus W$ and are linearly independent. Thus

$$\dim(V \oplus W) = \dim V + \dim W.$$

We will identify V with the set of all vectors $[\mathbf{v}, \mathbf{0}]$ and identify W with the set of all vectors $[\mathbf{0}, \mathbf{w}]$ in $V \oplus W$. With this notational convention we can write every vector $\mathbf{x}$ in U uniquely as $\mathbf{x} = \mathbf{v} + \mathbf{w}$, with $\mathbf{v} \in V$ and $\mathbf{w} \in W$.

Example 1.21. When V consists of all row vectors $\mathbf{v}$ with m components and W consists of all row vectors $\mathbf{w}$ with n components, then $U = V \oplus W$ consists of all row vectors with $m + n$ components. In MATLAB $[\mathbf{v}, \mathbf{w}]$ is called the *concatenation* (chaining together) of $\mathbf{v}$ and $\mathbf{w}$. We identify V with the row vectors of the form $[c_1, \ldots, c_m, 0, \ldots, 0]$, and W with the row vectors of the form $[0, \ldots, 0, c_{m+1}, \ldots, c_{m+n}]$, where the entries $c_j \in \mathbb{F}$. For column vectors a similar construction is possible by stacking one column vector on top of the other. Thus from the column vectors $\mathbf{y} \in \mathbb{F}^m$ and $\mathbf{z} \in \mathbb{F}^n$ we obtain a column vector

$$\mathbf{x} = \begin{bmatrix} \mathbf{y} \\ \mathbf{z} \end{bmatrix} = \begin{bmatrix} \mathbf{y} \\ \mathbf{0} \end{bmatrix} + \begin{bmatrix} \mathbf{0} \\ \mathbf{z} \end{bmatrix} \tag{1.9}$$

in $\mathbb{F}^{m+n}$. The MATLAB notation is $[\mathbf{y}; \mathbf{z}]$ in this case. In this way we can view the vector space $\mathbb{F}^{m+n}$ as $\mathbb{F}^m \oplus \mathbb{F}^n$. ∎

Remark 1.3. For the wavelet transforms in Chapters 3 and 4 we start with a vector space U of dimension 2^k and construct subspaces V (trend) and W (detail) of dimension 2^{k-1} such that $U = V \oplus W$. This gives a decomposition $\mathbf{u} = \mathbf{v} + \mathbf{w}$ (sum of trend and detail at level k) of a signal $\mathbf{u} \in U$. We repeat this process by decomposing the level k trend $\mathbf{v}$ into a trend and detail at level $k - 1$, and so forth. This is the *pyramid algorithm* that gives the *multiresolution decomposition* for signal processing by wavelet methods.

1.5.3 *Partitioned matrices and block multiplication*

Let T be a matrix of size $(m + n) \times (m + n)$, where m, n are positive integers. We can write T in partitioned form as

$$T = \begin{bmatrix} A & B \\ C & D \end{bmatrix}, \tag{1.10}$$

where A is of size $m \times m$, B has size $m \times n$, C has size $n \times m$, and D has size $n \times n$. We call A, B, C, D the *blocks* in the partition of T. If $\mathbf{x} \in \mathbb{F}^{m+n}$ is given by (1.9) and T by (1.10), then we can calculate the matrix-vector product $T\mathbf{x}$ using *block multiplication* as if T were a 2×2 matrix and $\mathbf{x}$ were a vector with only two components:

$$T\mathbf{x} = \begin{bmatrix} A & B \\ C & D \end{bmatrix} \begin{bmatrix} \mathbf{y} \\ \mathbf{z} \end{bmatrix} = \begin{bmatrix} A\mathbf{y} + B\mathbf{z} \\ C\mathbf{y} + D\mathbf{z} \end{bmatrix}. \tag{1.11}$$

This follows from the (row)×(column) formula for the entries in a matrix-vector product; notice that the number of columns of A and C is the number of rows of $\mathbf{y}$, while the number of columns of B and D is the number of rows of $\mathbf{z}$, so all the matrix-vector products in (1.11) are defined.

Example 1.22. For 3×3 matrices the partition $3 = 2 + 1$ gives block sizes 2×2, 2×1, 1×2, and 1×1. Thus we can write

$$T = \begin{bmatrix} 1 & 2 & 3 \\ 4 & 5 & 6 \\ 7 & 8 & 9 \end{bmatrix} = \begin{bmatrix} A & B \\ C & D \end{bmatrix} \text{ with } A = \begin{bmatrix} 1 & 2 \\ 4 & 5 \end{bmatrix}, \ B = \begin{bmatrix} 3 \\ 6 \end{bmatrix}, \ C = \begin{bmatrix} 7 & 8 \end{bmatrix}, \text{ and } D = \begin{bmatrix} 9 \end{bmatrix}.$$

Take $\mathbf{y} = \begin{bmatrix} a \\ b \end{bmatrix}$ and $\mathbf{z} = \begin{bmatrix} c \end{bmatrix}$. Then $A\mathbf{y} = \begin{bmatrix} a + 2b \\ 4a + 5b \end{bmatrix}$, $B\mathbf{z} = \begin{bmatrix} 3c \\ 6c \end{bmatrix}$, $C\mathbf{y} = \begin{bmatrix} 7a + 8b \end{bmatrix}$, and $D\mathbf{z} = \begin{bmatrix} 9c \end{bmatrix}$. Thus $A\mathbf{y} + B\mathbf{z} = \begin{bmatrix} a + 2b + 3c \\ 4a + 5b + 6c \end{bmatrix}$ and $C\mathbf{y} + D\mathbf{z} = \begin{bmatrix} 7a + 8b + 9c \end{bmatrix}$.
The block-multiplication formula (1.11) obviously holds in this case. ∎

Let S be another matrix of size $(m + n) \times (m + n)$. Write $S = \begin{bmatrix} A' & B' \\ C' & D' \end{bmatrix}$ in partitioned form with blocks the same sizes as in (1.10). Then we can calculate the matrix product ST by *block multiplication* of 2×2 matrices:

$$ST = \begin{bmatrix} (A'A + B'C) & (A'B + B'D) \\ (C'A + D'C) & (C'B + D'D) \end{bmatrix}. \tag{1.12}$$

To verify equation (1.12) it suffices to take a vector $\mathbf{x} = \begin{bmatrix} \mathbf{y} \\ \mathbf{z} \end{bmatrix}$ and check that $ST\mathbf{x}$ is the vector given by applying the block matrix on the right side of (1.12) to $\mathbf{x}$. This follows by using equation (1.11) twice: first for T multiplying the vector $\mathbf{x}$ and then for S multiplying the vector $T\mathbf{x}$.

Observe that each block in (1.12) is the matrix product of a *row block* of S and a *column block* of T; for example

$$C'A + D'C = \begin{bmatrix} C' & D' \end{bmatrix} \begin{bmatrix} A \\ C \end{bmatrix}.$$

Caution must be used to keep the individual matrix products in the correct order however, since generally $C'A \neq AC'$.

Example 1.23 (Block-Triangular Matrices). The matrix T in partitioned form (1.10) is called *block upper triangular* if $C = \mathbf{0}$. If the partitioned matrix S is also in block upper triangular matrix, then the multiplication formula (1.12) simplifies to

$$ST = \begin{bmatrix} A'A & (A'B + B'D) \\ \mathbf{0} & D'D \end{bmatrix}. \tag{1.13}$$

Hence the product of two block upper triangular matrices is again block upper triangular of the same partitioned form. As a particular case, T is invertible if and only if A and D are invertible; when this is the case, then

$$T^{-1} = \begin{bmatrix} A^{-1} & -A^{-1}BD^{-1} \\ \mathbf{0} & D^{-1} \end{bmatrix}, \tag{1.14}$$

which is also block upper triangular. To verify this formula for T^{-1}, multiply the right side of equation (1.14) by T; using (1.13) we obtain the identity matrix. In the special case $A = I_m$ and $D = I_n$, where I_p denotes the $p \times p$ identity matrix, we call T *block unit upper triangular*. If S and T are such matrices, formulas (1.12) and (1.14) simplify to

$$ST = \begin{bmatrix} I_m & (B + B') \\ \mathbf{0} & I_n \end{bmatrix}, \quad T^{-1} = \begin{bmatrix} I_m & -B \\ \mathbf{0} & I_n \end{bmatrix}. \tag{1.15}$$

Thus the product and inverse remain in block unit upper triangular form. More precisely, the matrix product is calculated by adding the upper right-hand blocks, and the matrix inverse is obtained by replacing the upper right-hand block by its negative.

The matrix T in partitioned form (1.10) is called *block lower triangular* if $B = \mathbf{0}$; thus its transpose is block upper triangular. The matrix is called *block unit lower triangular* if its transpose is block unit upper triangular matrix. All the assertions the previous paragraph about products and inverses of block upper triangular matrices carry over to block lower triangular matrices. ∎

1.6   Vector Graphics and Animation

A two-dimensional image made up of straight line segments can be efficiently encoded by a matrix whose columns give end points (vertices) of the segments.[8] For example, the triangle in $\mathbb{R}^2$ with vertices at $(0,0)$, $(1,2)$, and $(4,3)$ can be encoded by the matrix

$$M = \begin{bmatrix} 0 & 1 & 4 & 0 \\ 0 & 2 & 3 & 0 \end{bmatrix} = \begin{bmatrix} \mathbf{v}_1 & \mathbf{v}_2 & \mathbf{v}_3 & \mathbf{v}_4 \end{bmatrix}. \tag{1.16}$$

---

[8]By contrast, digital images made up of individual pixels (*raster graphics*) are encoded by an $m \times n$ matrix as described in Sections 1.2 and 3.6. See the Wikipedia articles on *vector graphics* and *raster graphics* for comparisons of these methods. The material in this section will not be needed for Fourier and wavelet transforms.

The triangle is constructed from the matrix M by drawing a line segment from the vertex whose coordinates give the vector $\mathbf{v}_j$ to the vertex whose coordinates give the vector $\mathbf{v}_{j+1}$, for $j = 1, 2, 3$. Notice that we need to take $\mathbf{v}_4 = \mathbf{v}_1$ to get all three edges of the triangle.

In general, suppose the line segments making up an image are all connected. In this case we use a matrix whose columns are vectors $\mathbf{v}_1, \ldots, \mathbf{v}_n$ (this list may have repetitions, as in the example of the triangle). Then we draw the image by putting a line segment from the vertex whose coordinates give the vector $\mathbf{v}_j$ to the vertex whose coordinates give the vector $\mathbf{v}_{j+1}$ for $j = 1, \ldots, n$. This gives a *vector graphics* encoding of the image by the $2 \times n$ matrix $B = \begin{bmatrix} \mathbf{v}_1 & \mathbf{v}_2 & \cdots & \mathbf{v}_n \end{bmatrix}$.

There are two advantages of encoding an image by vector graphics:

(1) The image can be transformed geometrically (rescaled, rotated, reflected, sheared, or translated) with no loss of detail. Applying a succession of such transformations of the image creates the visual effect of *animation*.
(2) The computation of transformations is fast, since the number of arithmetic operations is proportional to the number of vertices, not the number of pixels as in raster graphics.

To understand the mathematical basis for these features, recall that if A is an $m \times m$ matrix and B is an $m \times n$ matrix with columns $\mathbf{v}_1, \ldots, \mathbf{v}_n$ in $\mathbb{R}^m$, then

$$AB = \begin{bmatrix} A\mathbf{v}_1 & A\mathbf{v}_2 & \cdots & A\mathbf{v}_n \end{bmatrix}.$$

So if B is the matrix that encodes the image in vector graphics, then AB encodes the transformed image, since each vertex vector $\mathbf{v}_j$ in B becomes $A\mathbf{v}_j$ in the matrix AB. Furthermore, if A is an invertible matrix, then we can recover the original image by another matrix multiplication using A^{-1}, since $A^{-1}(AB) = (A^{-1}A)B = B$ by the associative property of matrix multiplication.

1.6.1 *Geometric transformations of images*

Here are matrix descriptions for some basic types of geometric transformations of two-dimensional images. Each type of transformation depends on one or more real parameters. The parameters can be varied to achieve the animation effect mentioned in (1) above.

Scaling transformations

A diagonal matrix $D_{a,b} = \begin{bmatrix} a & 0 \\ 0 & b \end{bmatrix}$ with parameters $a > 0$ and $b > 0$ acts on a vector $\mathbf{v} = \begin{bmatrix} x_1 \\ x_2 \end{bmatrix}$ as a *scaling transformation*: $D_{a,b}\mathbf{v} = \begin{bmatrix} ax_1 \\ bx_2 \end{bmatrix}$. The parameters for the product of two scaling transformations are the products of the corresponding parameters: $D_{a,b}D_{a',b'} = D_{aa',bb'}$. Also $D_{1,1} = I_2$ (the 2×2 identity matrix), and the transformation inverse to $D_{a,b}$ is $D_{1/a,1/b}$. When $a = b$, so the rescaling is by the same factor in each coordinate, then $D_{a,a} = aI$ is called a *dilation* if $a > 1$ and a *contraction* if $0 < a < 1$. Figure 1.1 shows the triangle encoded by the matrix (1.16), and Figure 1.2 shows the triangle together with its rescaling by the transformation $D_{2,3}$ (horizontal dimensions multiplied by 2 and vertical dimensions multiplied by 3).

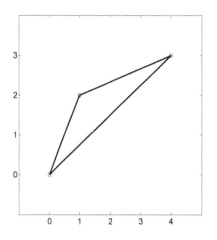

Fig. 1.1   Vector graphics triangle   Fig. 1.2   Scaling transformation of triangle

Rotations

For $\theta \in \mathbb{R}$ let R_θ be a counterclockwise rotation by the angle θ. Then $R_\theta \begin{bmatrix} 1 \\ 0 \end{bmatrix} = \begin{bmatrix} \cos\theta \\ \sin\theta \end{bmatrix}$ and $R_\theta \begin{bmatrix} 0 \\ 1 \end{bmatrix} = \begin{bmatrix} -\sin\theta \\ \cos\theta \end{bmatrix}$. Hence R_θ has matrix $\begin{bmatrix} \cos\theta & -\sin\theta \\ \sin\theta & \cos\theta \end{bmatrix}$. By definition, $R_0 = I_2$. Also, a rotation by angle ϕ followed by rotation by angle θ is a rotation by angle $\theta + \phi$, so we conclude[9] that $R_\theta R_\phi = R_{\theta+\phi}$. Figure 1.3 shows the triangle encoded by the matrix (1.16) and its transformation by $R_{2\pi/3}$ and $R_{4\pi/3}$. We could achieve an animation effect of a rotating triangle by using the transformations R_t with t going from 0 to 2π.

Shearing Transformations

Let $t \in \mathbb{R}$ and consider the linear transformation $\begin{bmatrix} x_1 \\ x_2 \end{bmatrix} \longrightarrow \begin{bmatrix} x_1 + tx_2 \\ x_2 \end{bmatrix}$ on $\mathbb{R}^2$. The matrix for this transformation is $N_t = \begin{bmatrix} 1 & t \\ 0 & 1 \end{bmatrix}$. We see that $N_0 = I_2$ and $N_s N_t = N_{s+t}$ (just as for the rotation matrices R_θ). Thus N_t is invertible with inverse N_{-t}. Figure

---
[9]This matrix multiplication formula yields the *addition formulas* for the sine and cosine functions.

1.4 shows the triangle encoded by the matrix (1.16) and its transformation by N_2 and N_4. We could achieve an animation effect of the original triangle being sheared to the right by using the transformations N_t with t going from 0 to 4.

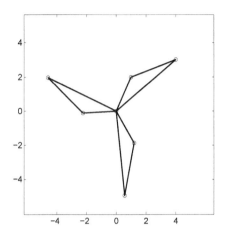

Fig. 1.3 Rotations

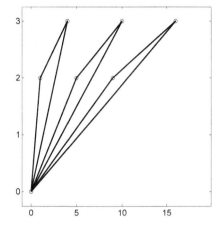

Fig. 1.4 Shearing transformations

Translations

All the transformations described so far are linear. The operation of *translation* by a vector $\mathbf{b} \in \mathbb{R}^2$, which sends $\mathbf{x}$ to $\mathbf{x}+\mathbf{b}$ for each $\mathbf{x} \in \mathbb{R}^2$, is not a linear transformation of $\mathbb{R}^2$ since it moves $\mathbf{0}$ to $\mathbf{b}$. However, we can implement translations by matrix multiplication if we change to a new *homogeneous coordinate system* for $\mathbb{R}^2$ and use 3×3 matrices, as follows.

We can identify $\mathbb{R}^2$ with the plane $\{x_3 = 1\}$ in $\mathbb{R}^3$ by the correspondence

$$\begin{bmatrix} x_1 \\ x_2 \end{bmatrix} \longleftrightarrow \begin{bmatrix} x_1 \\ x_2 \\ 1 \end{bmatrix}.$$

For example, the triangle in $\mathbb{R}^2$ with vertices at $(0,0)$, $(1,2)$, and $(3,4)$ is now encoded in homogeneous coordinates by the 3×4 matrix

$$\widetilde{M} = \begin{bmatrix} 0 & 1 & 4 & 0 \\ 0 & 2 & 3 & 0 \\ 1 & 1 & 1 & 1 \end{bmatrix} = \begin{bmatrix} \mathbf{v}_1 & \mathbf{v}_2 & \mathbf{v}_3 & \mathbf{v}_4 \\ 1 & 1 & 1 & 1 \end{bmatrix}. \tag{1.17}$$

Given a vector $\mathbf{b} = \begin{bmatrix} b_1 \\ b_2 \end{bmatrix} \in \mathbb{R}^2$, we define the matrix $T_{\mathbf{b}} = \begin{bmatrix} 1 & 0 & b_1 \\ 0 & 1 & b_2 \\ 0 & 0 & 1 \end{bmatrix} = \begin{bmatrix} I_2 & \mathbf{b} \\ \mathbf{0} & 1 \end{bmatrix}$,

where the last expression is the block matrix form of $T_{\mathbf{b}}$. When $\mathbf{b} = \mathbf{0}$ then $T_{\mathbf{0}} = I_3$

(the 3×3 identity matrix). We calculate that

$$T_{\mathbf{b}} \begin{bmatrix} x_1 \\ x_2 \\ 1 \end{bmatrix} = \begin{bmatrix} x_1 + b_1 \\ x_2 + b_2 \\ 1 \end{bmatrix},$$

which is a special case of formula (1.11). Thus $T_{\mathbf{b}}$ acts on the homogeneous coordinates of a vector in the same way that translation by $\mathbf{b}$ acts on $\mathbb{R}^2$. By (1.15) we see that $T_{\mathbf{b}} T_{\mathbf{c}} = T_{\mathbf{b}+\mathbf{c}}$ for all vectors $\mathbf{b}, \mathbf{c} \in \mathbb{R}^2$, so multiplication of these translation matrices corresponds to adding the translation vectors.

1.6.2   *Affine transformations*

If A is any invertible 2×2 matrix, let $L_A = \begin{bmatrix} A & \mathbf{0} \\ \mathbf{0} & 1 \end{bmatrix}$ be the 3×3 matrix in block form with A as the upper left-hand block. By formula (1.13) we see that $L_A L_B = L_{AB}$ for any 2×2 matrices A and B. Moreover, by (1.14) we have $\left(L_A \right)^{-1} = L_{A^{-1}}$. Furthermore, if $\mathbf{x} \in \mathbb{R}^2$ then block multiplication gives

$$L_A \begin{bmatrix} \mathbf{x} \\ 1 \end{bmatrix} = \begin{bmatrix} A\mathbf{x} \\ 1 \end{bmatrix}.$$

Thus L_A acts on the plane $\{x_3 = 1\}$ in $\mathbb{R}^3$ in the same way that A acts on $\mathbb{R}^2$. Performing a linear transformation $\mathbf{x} \longrightarrow A\mathbf{x}$ followed by a translation by $\mathbf{b} \in \mathbb{R}^2$, we obtain an *affine transformation* $\mathbf{x} \longrightarrow A\mathbf{x} + \mathbf{b}$ of $\mathbb{R}^2$ (if $\mathbf{b} \neq \mathbf{0}$ then it is not a linear transformation since it moves the zero vector). This transformation acts on the homogeneous coordinate vector for $\mathbf{x}$ by the 3×3 matrix $T_{\mathbf{b}} L_A$, which we can write in block form as

$$\begin{bmatrix} I_2 & \mathbf{b} \\ \mathbf{0} & 1 \end{bmatrix} \begin{bmatrix} A & \mathbf{0} \\ \mathbf{0} & 1 \end{bmatrix} = \begin{bmatrix} A & \mathbf{b} \\ \mathbf{0} & 1 \end{bmatrix}.$$

Thus these affine transformations are given by block upper-triangular matrices. The inverse transformation acts on homogeneous coordinate vectors by the block form matrix

$$\begin{bmatrix} A^{-1} & \mathbf{0} \\ \mathbf{0} & 1 \end{bmatrix} \begin{bmatrix} I_2 & -\mathbf{b} \\ \mathbf{0} & 1 \end{bmatrix} = \begin{bmatrix} A^{-1} & -A^{-1}\mathbf{b} \\ \mathbf{0} & 1 \end{bmatrix},$$

as in formula (1.14).

For example, Figure 1.5 shows the triangle encoded by matrix (1.16) and its translations by the vectors $\mathbf{b} = \begin{bmatrix} -2 \\ 1 \end{bmatrix}$ and $2\mathbf{b}$. Figure 1.6 shows the affine transformation of this triangle obtained by first applying the rotation $R_{\pi/3}$ and then translating by $\mathbf{b}$.

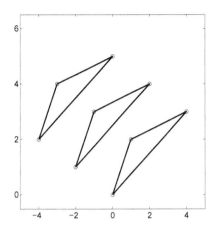

Fig. 1.5   Translations

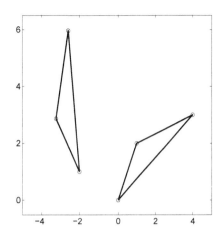

Fig. 1.6   Affine transformation

1.7   Inner Products, Orthogonal Projections, and Unitary Matrices

The *inner product* between vectors $\mathbf{x} = \begin{bmatrix} x_1 \\ \vdots \\ x_n \end{bmatrix}$ and $\mathbf{y} = \begin{bmatrix} y_1 \\ \vdots \\ y_n \end{bmatrix}$ in $\mathbb{R}^n$ is defined as

$$\langle \mathbf{x}, \mathbf{y} \rangle = x_1 y_1 + \cdots + x_n y_n \, .$$

It is also called the *dot product* and often denoted by $\mathbf{x} \cdot \mathbf{y}$, but we will use the bracket notation. We say that $\mathbf{x}$ is *orthogonal* to $\mathbf{y}$ if $\langle \mathbf{x}, \mathbf{y} \rangle = 0$, and we write $\mathbf{x} \perp \mathbf{y}$ in this case. In particular, if $\mathbf{x} \neq \mathbf{0}$ then $\langle \mathbf{x}, \mathbf{x} \rangle = x_1^2 + \cdots + x_n^2 > 0$ since $r^2 > 0$ for any nonzero real number r. Thus a nonzero vector is never orthogonal to itself. The *length* or *norm* of $\mathbf{x}$ is defined as

$$\|\mathbf{x}\| = \sqrt{\langle \mathbf{x}, \mathbf{x} \rangle} = \sqrt{x_1^2 + \cdots + x_n^2}$$

(the Pythagorean formula in n dimensions).

These formulas must be modified when we use vectors in $\mathbb{C}^n$. For example, if $\mathbf{u} = \begin{bmatrix} 1 \\ i \end{bmatrix}$ (where $i = \sqrt{-1}$), then $\mathbf{u}$ is a nonzero vector but $u_1^2 + u_2^2 = 1 - 1 = 0$, since $i^2 = -1$. So if we used the definition above for inner product and norm in this case, we would find that $\mathbf{u}$ is orthogonal to itself and that it has zero length.

The way around this difficulty is to use the fact that $\bar{z}z = (x - iy)(x + iy) = x^2 + y^2 > 0$ if $z = x + iy$ is a nonzero complex number (where x and y are real numbers). Thus when $\mathbb{F}$ is the field of real or complex numbers, we define the *standard inner product* on $\mathbb{F}^n$ to be

$$\langle \mathbf{u}, \mathbf{v} \rangle = \sum_{k=1}^{n} \bar{u}_k v_k \quad \text{for } \mathbf{u} = \begin{bmatrix} u_1 \\ \vdots \\ u_n \end{bmatrix} \text{ and } \mathbf{v} = \begin{bmatrix} v_1 \\ \vdots \\ v_n \end{bmatrix}. \tag{1.18}$$

When $\mathbb{F} = \mathbb{R}$, then $\overline{u}_k = u_k$ and we obtain the inner product on $\mathbb{R}^n$ given above.

Let A be any matrix. We define the *Hermitian transpose* $A^{\mathrm{H}} = \overline{A}^{\mathrm{T}}$, where $\overline{A}$ is the matrix whose entries are the complex conjugates of the entries of A.[10] Using this notation, we can express the inner product between vectors $\mathbf{u}$ and $\mathbf{v}$ in $\mathbb{F}^n$ as $\langle \mathbf{u}, \mathbf{v} \rangle = \mathbf{u}^{\mathrm{H}} \mathbf{v}$. In particular, if $\mathbf{u} \neq \mathbf{0}$ then

$$\langle \mathbf{u}, \mathbf{u} \rangle = \sum_{k=1}^{n} |u_k|^2 > 0.$$

We define the *norm* $\|\mathbf{u}\| = \sqrt{\langle \mathbf{u}, \mathbf{u} \rangle}$. The square of the norm $\|\mathbf{u}\|^2 = \mathbf{u}^{\mathrm{H}} \mathbf{u}$ is often called the *energy* of $\mathbf{u}$.

Remark 1.4. In our definition of inner product we are following the physics convention to make the inner product $\langle \mathbf{u}, \mathbf{v} \rangle$ a *linear* function of $\mathbf{v}$ (the *second* vector) and a *conjugate linear* function of $\mathbf{u}$ (the *first* vector). Most mathematics texts use the opposite convention and define the inner product to be $\mathbf{v}^{\mathrm{H}} \mathbf{u}$ (reversing $\mathbf{u}$ and $\mathbf{v}$). Of course, for vectors with real components there is no difference between these two definitions.

For signal and image processing applications, we will need to use inner products on general vector spaces. Let V be a vector space with scalars $\mathbb{F}$ (real or complex numbers).

Definition 1.5. An *inner product* on V is a $\mathbb{F}$-valued function $\langle \mathbf{u}, \mathbf{v} \rangle$ defined for all $\mathbf{u}, \mathbf{v} \in V$ that satisfies the following three conditions for all vectors $\mathbf{u}, \mathbf{v}, \mathbf{w} \in V$ and scalars α, β:

$$\langle \mathbf{w}, \alpha \mathbf{u} + \beta \mathbf{v} \rangle = \alpha \langle \mathbf{w}, \mathbf{u} \rangle + \beta \langle \mathbf{w}, \mathbf{v} \rangle, \qquad \text{(linearity)}$$

$$\langle \mathbf{u}, \mathbf{v} \rangle = \overline{\langle \mathbf{v}, \mathbf{u} \rangle}, \qquad \text{(conjugate symmetry)}$$

$$\langle \mathbf{u}, \mathbf{u} \rangle > 0 \quad \text{if } \mathbf{u} \neq \mathbf{0}. \qquad \text{(positivity)}$$

The linearity and conjugate symmetry conditions imply that an inner product is *conjugate linear* as a function of its first argument:

$$\langle \alpha \mathbf{u} + \beta \mathbf{v}, \mathbf{w} \rangle = \overline{\langle \mathbf{w}, \alpha \mathbf{u} + \beta \mathbf{v} \rangle} = \overline{\alpha} \, \overline{\langle \mathbf{w}, \mathbf{u} \rangle} + \overline{\beta} \, \overline{\langle \mathbf{w}, \mathbf{v} \rangle}$$
$$= \overline{\alpha} \, \langle \mathbf{u}, \mathbf{w} \rangle + \overline{\beta} \, \langle \mathbf{v}, \mathbf{w} \rangle.$$

When $V = \mathbb{F}^n$, then the standard inner product defined in (1.18) satisfies the linearity, conjugate symmetry, and positivity properties.

Example 1.24. Let $\ell_{\mathbb{F}}^2(\mathbb{Z})$ consist of all $\mathbb{F}$-valued functions f on $\mathbb{Z}$ that are *square-summable*:

$$\sum_{n \in \mathbb{Z}} |f[n]|^2 < \infty$$

---

[10]In MATLAB all matrices are automatically assumed to have complex entries, and the MATLAB expression A' is the Hermitian transpose of the matrix A.

(here we write $f[n]$ for the value of f at $n \in \mathbb{Z}$). For a, b in $\mathbb{F}$ we have $2|ab| \leq |a|^2 + |b|^2$ and thus $|a + b|^2 \leq 2|a|^2 + 2|b|^2$, since $|a + b|^2 \geq 0$. Hence for any functions f and g on $\mathbb{Z}$ and scalars α, β we have the inequalities

$$2|f[n]g[n]| \leq |f[n]|^2 + |g[n]|^2,$$
$$|\alpha f[n] + \beta g[n]|^2 \leq 2\alpha^2|f[n]|^2 + 2\beta^2|g[n]|^2.$$

It follows from these inequalities that if f and g are square-summable then the function $h[n] = \alpha f[n] + \beta g[n]$ is also square-summable. Hence $\ell_{\mathbb{F}}^2(\mathbb{Z})$ is a vector space over $\mathbb{F}$. Furthermore, the infinite series

$$\langle f, g \rangle = \sum_{n \in \mathbb{Z}} \overline{f[n]}g[n] \tag{1.19}$$

converges absolutely and defines an *inner product* on $\ell_{\mathbb{F}}^2(\mathbb{Z})$. For every positive integer N we can identify $\mathbb{F}^N$ (with its usual inner product) with the subspace of $\ell_{\mathbb{F}}^2(\mathbb{Z})$ consisting of all functions that are zero outside the set $\{1, 2, \ldots, N\}$. ∎

Here is another important example of an inner product space in which summation is replaced by integration.

Example 1.25. Consider the vector space V of all $\mathbb{F}$-valued continuous functions on a finite interval $a \leq x \leq b$ in $\mathbb{R}$. Given two functions f and g in V, define

$$\langle f, g \rangle = \int_a^b \overline{f(x)}g(x)\,dx. \tag{1.20}$$

The linearity and conjugate symmetry properties are obviously satisfied by $\langle f, g \rangle$. Since $f(x)\overline{f(x)} = |f(x)|^2 \geq 0$, the *energy* $\|f\|^2$ of a function $f(x)$ is the *total area* under the graph of the function $|f(x)|^2$ over the interval $a \leq x \leq b$. In particular, if $\langle f, f \rangle = 0$ then the nonnegative continuous function $|f(x)|^2$ has zero integral over $a \leq x \leq b$. This implies that $f(x) = 0$ for all $a \leq x \leq b$. Hence the positivity property for the inner product is satisfied. ∎

We now assume that an inner product on V has been specified, and we shall call V an *inner product space*. We define the *norm* of a vector $\mathbf{u}$ in V to be $\|\mathbf{u}\| = \sqrt{\langle \mathbf{u}, \mathbf{u} \rangle}$ (positive square root). If $\alpha \in \mathbb{F}$ then

$$\langle \alpha\mathbf{u}, \alpha\mathbf{u} \rangle = \alpha\bar{\alpha}\langle \mathbf{u}, \mathbf{u} \rangle = |\alpha|^2\langle \mathbf{u}, \mathbf{u} \rangle$$

by the linearity and conjugate symmetry properties of the inner product. Hence the norm satisfies

$$\|\alpha\mathbf{u}\| = |\alpha|\,\|\mathbf{u}\| \quad \text{for all scalars } \alpha \text{ and vectors } \mathbf{u}. \tag{homogeneity}$$

Given $\mathbf{u}$ and $\mathbf{v}$ in V, we can use the linearity and conjugate symmetry properties to expand

$$\|\mathbf{u} + \mathbf{v}\|^2 = \langle \mathbf{u} + \mathbf{v}, \mathbf{u} + \mathbf{v} \rangle = \|\mathbf{u}\|^2 + \langle \mathbf{u}, \mathbf{v} \rangle + \langle \mathbf{v}, \mathbf{u} \rangle + \|\mathbf{v}\|^2$$
$$= \|\mathbf{u}\|^2 + \langle \mathbf{u}, \mathbf{v} \rangle + \overline{\langle \mathbf{u}, \mathbf{v} \rangle} + \|\mathbf{v}\|^2. \tag{1.21}$$

The vectors $\mathbf{u}$ and $\mathbf{v}$ in V are called *orthogonal* if $\langle \mathbf{u}, \mathbf{v} \rangle = 0$, and we write $\mathbf{u} \perp \mathbf{v}$. In this case we see from (1.21) that the norms satisfy

$$\|\mathbf{u} + \mathbf{v}\|^2 = \|\mathbf{u}\|^2 + \|\mathbf{v}\|^2 \quad \text{when } \mathbf{u} \perp \mathbf{v}. \qquad \text{(Pythagorean relation)}$$

Given a pair of vectors $\mathbf{u}$, $\mathbf{v}$ with $\mathbf{v} \neq 0$, we define

$$\mathbf{p} = c\mathbf{v} \quad \text{with} \quad c = \frac{\langle \mathbf{v}, \mathbf{u} \rangle}{\langle \mathbf{v}, \mathbf{v} \rangle}.$$

In this equation we have chosen the scalar c to make $\langle \mathbf{v}, \mathbf{u} - \mathbf{p} \rangle = \langle \mathbf{v}, \mathbf{u} \rangle - c\langle \mathbf{v}, \mathbf{v} \rangle = 0$. Thus $\mathbf{p}$ is a scalar multiple of $\mathbf{v}$ and $(\mathbf{u} - \mathbf{p}) \perp \mathbf{p}$. We call $\mathbf{p}$ the *orthogonal projection* of $\mathbf{u}$ onto the one-dimensional subspace $\mathbb{F}\mathbf{v}$. Since $\mathbf{u} = (\mathbf{u} - \mathbf{p}) + \mathbf{p}$, the Pythagorean relation gives $\|\mathbf{u}\|^2 = \|\mathbf{u} - \mathbf{p}\|^2 + \|\mathbf{p}\|^2$. But $\|\mathbf{u} - \mathbf{p}\|^2 \geq 0$, so by omitting this term we obtain the inequality

$$\|\mathbf{u}\|^2 \geq \|\mathbf{p}\|^2 = |c|^2 \|\mathbf{v}\|^2 = \frac{|\langle \mathbf{v}, \mathbf{u} \rangle|^2}{\|\mathbf{v}\|^4} \|\mathbf{v}\|^2 = \frac{|\langle \mathbf{v}, \mathbf{u} \rangle|^2}{\|\mathbf{v}\|^2}.$$

Here we have used the homogeneity property of the norm to calculate $\|\mathbf{p}\|^2$. Rearranging terms and taking a square root, we obtain the *Cauchy–Schwarz inequality*[11]

$$|\langle \mathbf{u}, \mathbf{v} \rangle| \leq \|\mathbf{u}\| \, \|\mathbf{v}\| \quad \text{for all vectors} \quad \mathbf{u}, \mathbf{v} \in V. \qquad (1.22)$$

Using the Cauchy–Schwarz inequality in (1.21), we obtain

$$\|\mathbf{u} + \mathbf{v}\|^2 \leq \|\mathbf{u}\|^2 + 2\|\mathbf{u}\|\|\mathbf{v}\| + \|\mathbf{v}\|^2 = (\|\mathbf{u}\| + \|\mathbf{v}\|)^2.$$

Taking the square root of the quantities on both sides of this inequality then gives the *triangle inequality*

$$\|\mathbf{u} + \mathbf{v}\| \leq \|\mathbf{u}\| + \|\mathbf{v}\| \quad \text{for all vectors} \quad \mathbf{u}, \mathbf{v} \in V. \qquad (1.23)$$

Definition 1.6. A set of vectors $\mathcal{S} = \{\mathbf{u}_1, \mathbf{u}_2, \ldots\}$ in V is called *orthonormal* if each vector in $\mathcal{S}$ has length one ($\|\mathbf{u}_j\| = 1$) and the vectors are mutually orthogonal ($\mathbf{u}_j \perp \mathbf{u}_k$ for all $j \neq k$).

An orthonormal set $\mathcal{S}$ is always linearly independent. To prove this, suppose we have a linear relation

$$c_1 \mathbf{u}_1 + \cdots + c_n \mathbf{u}_n = \mathbf{0}$$

for some scalars c_j. For each k we take the inner product of $\mathbf{u}_k$ with the left side of this equation. The result is c_k since $\langle \mathbf{u}_k, \mathbf{u}_j \rangle$ is 0 when $k \neq j$ and is 1 when $k = j$. The inner product of $\mathbf{u}_k$ with the right side of the equation is 0, however, so we conclude that $c_k = 0$ for all k.

Now assume that $\mathcal{S} = \{\mathbf{u}_1, \ldots, \mathbf{u}_n\}$ is a finite orthonormal set of vectors in V. Let W be the subspace of V spanned by $\mathcal{S}$. Then $\dim W = n$ since $\mathcal{S}$ is linearly independent. We call $\mathcal{S}$ an *orthonormal basis* for W. Every vector $\mathbf{w} \in W$ can be expressed in terms of this basis as

$$\mathbf{w} = c_1 \mathbf{u}_1 + \cdots + c_n \mathbf{u}_n \quad \text{with} \quad c_k = \langle \mathbf{u}_k, \mathbf{w} \rangle. \qquad (1.24)$$

---

[11]In the argument just given, we assumed $\mathbf{v} \neq 0$; the inequality (1.22) is simply $0 = 0$ when $\mathbf{v} = \mathbf{0}$.

The formula for the coefficient c_k is obtained by taking the inner product of $\mathbf{u}_k$ with $\mathbf{w}$ (the same method used above to prove that $\mathcal{S}$ is linearly independent).

If $\mathbf{v} = d_1\mathbf{u}_1 + \cdots + d_n\mathbf{u}_n$ is another vector in W, then

$$\langle \mathbf{w}, \mathbf{v} \rangle = \sum_{j=1}^{n}\sum_{k=1}^{n} \langle c_j\mathbf{u}_j, d_k\mathbf{u}_k \rangle = \sum_{j=1}^{n}\sum_{k=1}^{n} \bar{c}_j d_k \langle \mathbf{u}_j, \mathbf{u}_k \rangle$$

$$= \bar{c}_1 d_1 + \cdots + \bar{c}_n d_n \,, \tag{1.25}$$

since $\langle \mathbf{u}_j, \mathbf{u}_k \rangle = 0$ when $j \neq k$ and $\langle \mathbf{u}_j, \mathbf{u}_j \rangle = 1$. In particular, when $\mathbf{v} = \mathbf{w}$ we obtain *Parseval's formula*

$$\|\mathbf{w}\|^2 = |c_1|^2 + \cdots + |c_n|^2 \,. \tag{1.26}$$

Notice in this formula that the left side only depends on $\mathbf{w}$, whereas the right side depends on the choice of orthonormal basis.[12]

For any vector $\mathbf{v} \in V$ we define

$$P\mathbf{v} = \langle \mathbf{u}_1, \mathbf{v} \rangle \mathbf{u}_1 + \cdots + \langle \mathbf{u}_n, \mathbf{v} \rangle \mathbf{u}_n \,.$$

Then $P\mathbf{v} \in W$ since it is a linear combination of vectors in $\mathcal{S}$. Furthermore, $\mathbf{v} - P\mathbf{v} \perp W$ since

$$\langle \mathbf{u}_j, \mathbf{v} - P\mathbf{v} \rangle = \langle \mathbf{u}_j, \mathbf{v} \rangle - \sum_{k=1}^{n} \langle \mathbf{u}_k, \mathbf{v} \rangle \langle \mathbf{u}_j, \mathbf{u}_k \rangle$$

$$= \langle \mathbf{u}_j, \mathbf{v} \rangle - \langle \mathbf{u}_j, \mathbf{v} \rangle \langle \mathbf{u}_j, \mathbf{u}_j \rangle = 0 \,,$$

by orthogonality of $\mathcal{S}$ and the normalization condition $\langle \mathbf{u}_j, \mathbf{u}_j \rangle = 1$. We call $P\mathbf{v}$ the *orthogonal projection of* $\mathbf{v}$ *onto the subspace* W. Since each inner product $\langle \mathbf{u}_j, \mathbf{v} \rangle$ is a linear function of $\mathbf{v}$, the transformation $\mathbf{v} \longrightarrow P\mathbf{v}$ is linear.

The vector $P\mathbf{v}$ is uniquely determined as the *best approximation* to $\mathbf{v}$ by vectors in W when we measure the distance between vectors by the norm of their difference:

$$\min_{\mathbf{w} \in W} \|\mathbf{v} - \mathbf{w}\| = \|\mathbf{v} - P\mathbf{v}\| \,. \tag{1.27}$$

To verify this closest vector property, let $\mathbf{w}$ be any vector in W and write

$$\mathbf{v} - \mathbf{w} = (\mathbf{v} - P\mathbf{v}) + (P\mathbf{v} - \mathbf{w}) \,.$$

Since $\mathbf{v} - P\mathbf{v}$ is orthogonal to W and $P\mathbf{v} - \mathbf{w}$ is in W, the Pythagorean relation gives

$$\|\mathbf{v} - \mathbf{w}\|^2 = \|\mathbf{v} - P\mathbf{v}\|^2 + \|P\mathbf{v} - \mathbf{w}\|^2 \geq \|\mathbf{v} - P\mathbf{v}\|^2 \,.$$

This shows that the minimum on the left side of (1.27) is attained by taking $\mathbf{w} = P\mathbf{v}$. Furthermore, any other choice of $\mathbf{w}$ gives a strictly larger value to $\|\mathbf{v} - \mathbf{w}\|$. Thus $P\mathbf{v}$ is the unique minimum-distance vector in W.

Definition 1.7. An $n \times n$ complex matrix U is a *unitary matrix* if the set $\{\mathbf{u}_1, \ldots, \mathbf{u}_n\}$ of columns of U is orthonormal.

---

[12]We will see the significance of this formula later in connection with the discrete Fourier transform and wavelet transforms. The left side gives the *total energy* in $\mathbf{w}$, while the terms on the right give the partition of energy among the components of $\mathbf{w}$ relative to the basis vectors.

An alternate characterization of a unitary matrix is $U^H U = I$, where I is the $n \times n$ identity matrix. This follows because the (i, j) entry in $U^H U$ is $\mathbf{u}_i^H \mathbf{u}_j = \langle \mathbf{u}_i, \mathbf{u}_j \rangle$. Hence U *is unitary if and only if it is invertible and* $U^{-1} = U^H$. It follows that we can express the unitarity condition as

$$\langle U\mathbf{v}, U\mathbf{w} \rangle = \langle \mathbf{v}, \mathbf{w} \rangle \quad \text{for all } \mathbf{v}, \mathbf{w} \in \mathbb{C}^n. \tag{1.28}$$

To see this, write the inner product as a matrix product:

$$\langle U\mathbf{v}, U\mathbf{w} \rangle = (U\mathbf{v})^H (U\mathbf{w}) = \mathbf{v}^H (U^H U)\mathbf{w}. \tag{1.29}$$

If U is unitary, then the right side of (1.29) is $\langle \mathbf{v}, \mathbf{w} \rangle$. Conversely, if the right side of (1.29) is $\langle \mathbf{v}, \mathbf{w} \rangle$ for all $\mathbf{u}$ and $\mathbf{v}$, then by taking $\mathbf{u} = \mathbf{e}_i$ and $\mathbf{v} = \mathbf{e}_j$ we see that the columns of U are orthonormal, since $U\mathbf{e}_i$ is the ith column of U. If an $n \times n$ matrix Q has real entries and is unitary (so $Q^T Q = I$), then we call Q an *orthogonal matrix*.

Now let V and W be finite-dimensional complex inner product spaces of the same dimension, and let T be a linear transformation from V to W. We say that T is a *unitary transformation* if

$$\langle T\mathbf{u}, T\mathbf{v} \rangle = \langle \mathbf{u}, \mathbf{v} \rangle \quad \text{for all vectors } \mathbf{u}, \mathbf{v} \in V. \tag{1.30}$$

Note that in equation (1.30) the inner product on the left is for the space W, while the inner product on the right is for the space V. Taking $\mathbf{u} = \mathbf{v}$, we see that $\|T\mathbf{u}\| = \|\mathbf{u}\|$ for all $\mathbf{u}$. Hence the null space of T is $\mathbf{0}$. Since V and W have the same dimension, T is represented by a square matrix (relative to a choice of bases for V and W). This matrix has nullity zero, so it is invertible (see Example 1.13). Thus every unitary transformation is invertible.

Example 1.26. Let $V = W = \mathbb{C}^n$, and let the linear transformation T have matrix $U = \begin{bmatrix} \mathbf{u}_1 & \cdots & \mathbf{u}_n \end{bmatrix}$ relative to the standard basis $\{\mathbf{e}_1, \ldots, \mathbf{e}_n\}$ of $\mathbb{C}^n$. Then from (1.28) and (1.30) we see that T is a unitary transformation if and only if U is a unitary matrix. ∎

Example 1.27. Let V be an n-dimensional complex vector space with an inner product. Choose an orthonormal basis $\{\mathbf{u}_1, \ldots, \mathbf{u}_n\}$ for V and define a linear transformation $T : V \to \mathbb{C}^n$ by

$$T(c_1 \mathbf{u}_1 + \cdots + c_n \mathbf{u}_n) = \begin{bmatrix} c_1 & \cdots & c_n \end{bmatrix}^T.$$

Since $\dim V = \dim \mathbb{C}^n = n$, formula (1.25) shows that T is a unitary transformation. Thus once we have chosen an orthonormal basis for V, then all calculations in V can be carried over to $\mathbb{C}^n$ by T. The inner product on V becomes the standard inner product on $\mathbb{C}^n$ and the basis $\{\mathbf{u}_1, \ldots, \mathbf{u}_n\}$ becomes the standard orthonormal basis $\{\mathbf{e}_1, \ldots, \mathbf{e}_n\}$. ∎

1.8   Fourier Series

Let V be the complex vector space of piecewise continuous[13] complex-valued functions $f(x)$ that are periodic of period 2π:

$$f(x + 2\pi) = f(x) \quad \text{for all } x \in \mathbb{R}.$$

Because of the periodicity, $f(x)$ is uniquely determined by its restriction to any interval of length 2π, for example the interval $0 \le x \le 2\pi$. The assumption of piecewise continuity lets us integrate $f(x)$ over any interval of finite length.

We define an inner product on V by

$$\langle f, g \rangle = \frac{1}{2\pi} \int_0^{2\pi} \overline{f(x)} g(x)\, dx \,.$$

The square of the norm associated with this inner product is

$$\|f\|^2 = \frac{1}{2\pi} \int_0^{2\pi} |f(x)|^2\, dx \,,$$

and is called the *energy* or the *mean-square average* of f.

The functions $\phi_n(x) = e^{inx} = \cos(nx) + i\sin(nx)$ are in V for $n \in \mathbb{Z}$ (the set of all positive and negative integers), and they are an orthonormal relative to this inner product. To verify this property, take any pair of integers $k \ne n$. Then

$$\langle \phi_k, \phi_n \rangle = \frac{1}{2\pi} \int_0^{2\pi} e^{i(n-k)x}\, dx = \left. \frac{e^{i(n-k)x}}{2\pi i(n-k)} \right|_{x=0}^{x=2\pi} = 0 \,,$$

because $e^{2m\pi i} = 1$ for all integers m. Thus $\phi_n \perp \phi_k$. Furthermore, since $e^0 = 1$, we have

$$\langle \phi_n, \phi_n \rangle = \frac{1}{2\pi} \int_0^{2\pi} 1\, dx = 1 \,.$$

If $f \in V$ then the inner products

$$c_k = \langle \phi_k, f \rangle = \frac{1}{2\pi} \int_0^{2\pi} e^{-ikx} f(x)\, dx \tag{1.31}$$

are called the *Fourier coefficients* of f. When $f(x)$ is a real-valued function, its Fourier coefficients are complex numbers with the *conjugate symmetry* property

$$\overline{c_k} = c_{-k} \,, \tag{1.32}$$

where the bar denotes complex conjugation. This follows because $\overline{\phi_k(x)} = \phi_{-k}(x)$, $\overline{f(x)} = f(x)$, and hence $\overline{\langle \phi_k, f \rangle} = \langle \overline{\phi_k}, \overline{f} \rangle = \langle \phi_{-k}, f \rangle$.

Definition 1.8. For an integer $n \ge 0$ the space $\mathcal{TP}_n$ of *trigonometric polynomials* of degree at most n is the span of the set of functions $\{\phi_k(x) : |k| \le n\}$.

---

[13]We allow $f(x)$ to have a finite number of jump discontinuities for $0 \le x \le 2\pi$.

The functions ϕ_k for $|k| \leq n$ give an orthonormal basis for $\mathcal{TP}_n$, so $\dim \mathcal{TP}_n = 2n + 1$. We use the term *polynomial* since after the change of variable $z = \mathrm{e}^{\mathrm{i}x}$ the function $\phi_n(x)$ becomes z^n. Hence we can write elements of $\mathcal{TP}_n$ as polynomials in positive and negative powers of z:

$$c_{-n} z^{-n} + c_{-n+1} z^{-n+1} + \cdots + c_0 + \cdots + c_{n-1} z^{n-1} + c_n z^n. \qquad (1.33)$$

These polynomials (usually called *Laurent polynomials* because they can have both positive and negative powers of z) will be used extensively in connection with the discrete Fourier transform and wavelet transforms.

If $f(x) \in \mathcal{TP}_n$, then its Fourier coefficients satisfy $c_k = 0$ for $|k| > n$. Thus

$$f(x) = \sum_{-n \leq k \leq n} c_k \, \phi_k(x).$$

For example, the formulas

$$\sin(nx) = \frac{1}{2\mathrm{i}} \mathrm{e}^{\mathrm{i}nx} - \frac{1}{2\mathrm{i}} \mathrm{e}^{-\mathrm{i}nx}, \qquad \cos(nx) = \frac{1}{2} \mathrm{e}^{\mathrm{i}nx} + \frac{1}{2} \mathrm{e}^{-\mathrm{i}nx} \qquad (1.34)$$

show that the real-valued functions $f_n(x) = \sin(nx)$ and $g_n(x) = \cos(nx)$ are in $\mathcal{TP}_n$ and their Fourier coefficients satisfy the conjugate symmetry property (1.32).

Example 1.28. The set of functions $\{\phi_{-2}, \phi_{-1}, \phi_0, \phi_1, \phi_2\}$ is an orthonormal basis for $\mathcal{TP}_2$. If $\psi(x) \in \mathcal{TP}_2$, define

$$U\psi = \begin{bmatrix} c_{-2} \; c_{-1} \; c_0 \; c_1 \; c_2 \end{bmatrix}^{\mathrm{T}}, \qquad (1.35)$$

where c_k are the Fourier coefficients of $\psi(x)$. Since the Fourier coefficients depend linearly on $\psi(x)$, it is clear that U is a linear transformation from $\mathcal{TP}_2$ to $\mathbb{C}^5$. The basis functions $\phi_k(x)$ for $\mathcal{TP}_2$ are transformed by U into the standard basis vectors $\mathbf{e}_{3+k}$ for $\mathbb{C}^5$ (see Example 1.27). Hence U is unitary.

For example, by equation (1.34) we see that the functions $f_2(x) = \sin(2x)$ and $g_2(x) = \cos(2x)$ have transforms

$$Uf_2 = \frac{1}{2\mathrm{i}} \begin{bmatrix} -1 \; 0 \; 0 \; 0 \; 1 \end{bmatrix}^{\mathrm{T}}, \qquad Ug_2 = \frac{1}{2} \begin{bmatrix} 1 \; 0 \; 0 \; 0 \; 1 \end{bmatrix}^{\mathrm{T}}.$$

Notice that there is no variable x displayed in formula (1.35) or these last equations; $U\psi$, Uf_2, and Ug_2 are vectors in $\mathbb{C}^5$ with numerical components, not functions of x. Going the other way, however, we obtain functions of x by using the inverse transform. For example $U^{-1}\left(\begin{bmatrix} 0 \; 1 \; 1 \; 1 \; 0 \end{bmatrix}^{\mathrm{T}} \right)$ is the function $\mathrm{e}^{\mathrm{i}x} + 1 + \mathrm{e}^{-\mathrm{i}x} = 2\cos(x) + 1$.

Since U is unitary, it follows that

$$\frac{1}{2\pi} \int_0^{2\pi} \sin(2x) \cos(2x) \, dx = \langle f_2, g_2 \rangle = \langle Uf_2, Ug_2 \rangle = 0,$$

$$\frac{1}{2\pi} \int_0^{2\pi} \sin^2(2x) \, dx = \langle f_2, f_2 \rangle = \langle Uf_2, Uf_2 \rangle = \frac{1}{2}.$$

The two integrals on the left could be evaluated by double-angle formulas but this is not necessary; it is easier to calculate the inner products directly using the corresponding vectors in $\mathbb{C}^5$.   ∎

For any function $f(x) \in V$ and integer $n \geq 0$, the trigonometric polynomial

$$\psi_n(x) = \sum_{-n \leq k \leq n} c_k \phi_k(x), \quad \text{where } c_k = \langle \phi_k, f \rangle, \qquad (1.36)$$

is the projection of $f(x)$ onto the subspace $\mathcal{TP}_n$. From the closest vector property (1.27) the function $\psi_n(x)$ is the unique *best approximation* to $f(x)$ among all trigonometric polynomials of degree at most n:

$$\|f - \psi_n\| = \min_{g \in \mathcal{TP}_n} \|f - g\|.$$

It is an important result of Fourier analysis that $f(x)$ can be represented by its *Fourier series*:

$$f(x) \sim \sum_{k \in \mathbb{Z}} c_k \, \phi_k(x). \qquad (1.37)$$

This is analogous to the representation of a vector in $\mathbb{C}^n$ in terms of an orthonormal basis for $\mathbb{C}^n$, but now we must sum over an infinite orthonormal set. The meaning of the symbol $\sim$ in (1.37) is that the distance between $f(x)$ and the projection $\psi_n(x)$ of $f(x)$ onto the finite-dimensional space $\mathcal{TP}_n$ goes to zero as n goes to infinity:

$$\lim_{n \to \infty} \|f - \psi_n\| = 0. \qquad (1.38)$$

This approximation property by trigonometric polynomials follows[14] from *Parseval's formula*:

$$\|f\|^2 = \frac{1}{2\pi} \int_0^{2\pi} |f(x)|^2 \, dx = \sum_{k \in \mathbb{Z}} |c_k|^2. \qquad (1.39)$$

To see this connection, note that the function $g_n(x) = f(x) - \psi_n(x)$ has Fourier coefficients $d_k = \langle \phi_k, f - \psi_n \rangle$ that are zero for $|k| \leq n$, whereas $d_k = c_k$ for $|k| > n$. If we use (1.39) with $f(x)$ replaced by $g_n(x)$, we find that the difference in energy between $f(x)$ and $\psi_n(x)$ is

$$\|f - \psi_n\|^2 = \frac{1}{2\pi} \int_0^{2\pi} |f(x) - \psi_n(x)|^2 \, dx = \sum_{|k| > n} |c_k|^2. \qquad (1.40)$$

The infinite series on the right side of (1.39) is convergent. Hence the sum in (1.40) goes to zero as n goes to infinity since it is the tail of a convergent series.[15]

Remark 1.5. We will use the results from this section in several ways in later chapters. Here is a preview of some of these applications.

(1) We can approximate a periodic analog signal by a finite trigonometric polynomial of sufficiently high degree. In Chapter 2 we will find the coefficients of the approximating polynomial by sampling the polynomial at a sufficiently high rate and taking the discrete Fourier transform of the sample.

---

[14]To prove (1.39) or to investigate the pointwise convergence of the Fourier series (1.37) would require a long detour into real analysis that we omit.

[15]We will use similar arguments in Section 5.2.2 in connection with wavelet transforms for analog signals.

(2) In Chapter 4 we will use *finite impulse response filters* (lowpass and highpass) to construct discrete wavelet transforms. The frequency characteristics of such filters are expressed by the trigonometric polynomials obtained from the filter coefficients.

(3) In Chapter 5 we will construct an orthonormal basis for analog signals using dilations by powers of 2 and integer shifts of a *scaling function* and *wavelet function*. The coefficients in the wavelet expansion of an analog signal will be inner products (now given by an integral over $\mathbb{R}$) with the basis functions.

1.9   Computer Explorations

If you are unfamiliar with MATLAB see Appendix A.3.1.

1.9.1   *Sampling and quantizing an audio signal*

You will explore analog-digital conversion and quantization of digital signals following Section 1.2.

(a) Sampling

Consider the function $f(t) = \sin(2\pi(440)t)$ on the interval $0 < t < 1$, where t is time measured in seconds. Sample this function at $8192 = 2^{13}$ points (this is the default sampling rate in MATLAB) to obtain a vector $\mathbf{f} \in \mathbb{R}^{8192}$ with components $\mathbf{f}[k] = f(k\Delta T)$, where $\Delta T = 1/8192$. You can do this in MATLAB by the command

```
f = sin(2*pi*440*(1/8192)*(0:8191));
```

Don't forget to put a semicolon at the end of the command; otherwise, MATLAB *will send the 8192 values to the computer screen.* In this command (1/8192)*(0:8191) generates a row vector $\mathbf{t}$ in $\mathbb{R}^{8192}$ with components $0, \Delta T, 2\Delta T, \dots, 8191\Delta T$.

The sample vector $\mathbf{f}$ is stored in double precision floating point (about 15 significant figures), so the components of $\mathbf{f}$ can be considered as real numbers that can vary continuously. Type sound(f) to play the sound through the computer. By default, MATLAB plays all sound files at 8192 samples per second, and assumes that the sampled audio signal is in the range -1 to 1. Your signal satisfies these conditions, and you should hear the *standard tuning pitch* A (440 Hertz) for one second.

The graph of $f(t)$ goes up and down 440 times in the range $0 < t < 1$ corresponding to the frequency of the sound; consequently, plotting the entire signal on one page would give a almost solid black band. To show the sine-wave shape clearly plot just the first 100 samples by the command

```
plot(f(1:100)), hold on
```

(b) 2-bit quantization

Construct a 2-bit version of the audio signal $f(t)$ with 2^2 quantization levels by dividing the range $-1 \leq f(t) \leq 1$ into four equal intervals. Define the *two-bit quantized vector* **q2f** by

```
q2f = min(floor(2*(f+1)), 3) ;
```

Look at the first 16 entries in **q2f** by q2f(1:16). Then prove the following:

when $-1 \leq f(t) < -0.5$, the corresponding entry in **q2f** is 0;
when $-0.5 \leq f(t) < 0$, the corresponding entry in **q2f** is 1;
when $0 \leq f(t) < 0.5$, the corresponding entry in **q2f** is 2;
when $0.5 \leq f(t) \leq 1$, the corresponding entry in **q2f** is 3.

Thus all the entries in the vector **q2f** are unsigned 2-bit integers $0, 1, 2, 3$ (in binary notation: $00, 01, 10, 11$). A double-precision floating point number requires 64 bits, so the 2-bit quantized signal requires only $2/64$ (about 3%) as many data bits to store or transmit as the original signal.

To construct a 2-bit approximation $\widetilde{\mathbf{f2}}$ to the signal **f** from the quantized signal, change the values $0, 1, 2, 3$ in **q2f** to $-0.75, -0.25, 0.25, 0.75$ (the midpoints of the four quantization intervals). Define

```
f2tilde = -0.75 + 0.5*q2f ;
```

Plot the first hundred values of $\widetilde{\mathbf{f2}}$ on the MATLAB figure from *(a)* that shows **f** by the command

```
plot(f2tilde(1:100))
```

Click on Insert and title in the Figure toolbar and insert the title Sine Wave and 2-bit Quantization. Play the quantized signal with the command sound(f2tilde). It will sound very harsh compared to f.

(c) 2-bit quantization error

The vertical distances in the MATLAB figure between the two graphs generated in *(a)* and *(b)* indicate the distortion in the quantized signal. This distortion corresponds to the vector $\mathbf{f} - \widetilde{\mathbf{f2}}$. You can measure the distortion (as a percentage) by the ratio of the energy in the distortion vector to the energy in the original signal:

$$100\|\mathbf{f} - \widetilde{\mathbf{f2}}\|^2 / \|\mathbf{f}\|^2 .$$

Calculate this ratio in MATLAB by the command

```
distortion2bit = 100*norm(f - f2tilde)^2/norm(f)^2
```

The bad sound generated by $\widetilde{\mathbf{f2}}$ corresponds to the relatively large distortion energy ratio (around 5%).

(d) 4-bit quantization

Now construct a 4-bit version of the audio signal $f(t)$ with 2^4 quantization levels by dividing the range $-1 \leq f(t) \leq 1$ into 16 equal intervals. Define the *four-bit quantized vector* **q4f** by

```
q4f = min(floor(8*(f+1)), 15) ;
```

Look at the first 16 entries in **q4f** by the command `q4f(1:16)`. Observe that they are unsigned 4-bit integers $0, 1, \ldots, 15$ (in binary notation: $0000, 0001, \ldots, 1111$). A double-precision floating point number requires 64 bits, so the 4-bit quantized signal requires only 4/64 (about 6%) as many data bits to store or transmit as the original signal.

To construct a 4-bit approximation $\widetilde{\mathbf{f}4}$ to the signal **f** from the quantized signal, change the values $0, 1, \ldots, 14, 15$ in **q4f** to $-0.9375, -0.8125, \ldots, 0.8125, 0.9375$ (the midpoints of the sixteen quantization intervals). Define

```
f4tilde = -0.9375 + 0.125*q4f ;
```

Generate a new copy of the graph of the first 100 samples of **f** by

```
figure, plot(f(1:100)), hold on
```

Then plot the first hundred values of $\widetilde{\mathbf{f}4}$ on the same MATLAB figure by the command

```
plot(f4tilde(1:100))
```

Click on Insert and title in the Figure toolbar and insert the title Sine Wave and 4-bit Quantization. Play the quantized signal with the command `sound(f4tilde)`. It will sound harsh compared to **f**, but not as harsh as **f2tilde**. The sound quality might be acceptable for telephone communication but not for music.

(e) 4-bit quantization error

Repeat the calculations in *(c)* with $\widetilde{\mathbf{f}2}$ replaced by $\widetilde{\mathbf{f}4}$:

```
distortion4bit = 100*norm(f - f4tilde)^2/norm(f)^2
```

The improved sound generated by $\widetilde{\mathbf{f}4}$ corresponds to the much smaller distortion energy ratio. Calculate the ratio of the two distortion measurements:

```
distortion4bit/distortion2bit
```

It should be about 0.05.

1.9.2 *Vector graphics*

You will explore the use of matrices and vectors in two-dimensional computer vector graphics following Section 1.6. Read the Wikipedia article on *Vector Graphics* for more information on this topic. You will need to create some MATLAB *function* m-files. Use the text editor in MATLAB to create and save the following files.

(i) plot2d.m Create a *function* m-file with the commands

```
function plot2d(X)
x = X(1,:)'; y = X(2,:)';
plot(x, y, 'ro', x, y, 'g-');
axis([-10 10 -10 10])
axis('square')
```

(note the semicolons at the end of the lines–these suppress output to the screen). Save this file under the name `plot2d.m`. The input to this program is any matrix X with at least two rows and any number of columns. The output of `plot2d(X)` is the plot of the line figure whose vertices have x-y coordinates given by the first and second entries in each column of X (the vertices of consecutive columns of X are connected by a line). If the first and last columns of X are the same the figure will be a closed polygon. The `axis` command sets the scale to $[-10, 10]$ in both directions. Test the file by clicking on the MATLAB window and typing

```
X = [0 1 1 0; 0 1 -1 0]
plot2d(X)
```

at the MATLAB prompt. You should get a triangle with vertices at $(0,0)$, $(1,1)$, and $(1,-1)$ (if MATLAB can't find the file `plot2d.m`, check the *Path* settings).

(ii) randtri.m Create a function m-file to generate random triangles with one vertex at $(0,0)$ and scaling factor s by using the commands

```
function Y = randtri(s)
X = zeros(2,4);
X(:, 2:3) = rand(2,2)
Y = s*X
```

Save this m-file under the name `randtri.m`. Test it by typing

```
Y = randtri(5); plot2d(Y)
```

(a) Rotations

Create a function m-file to generate 2×2 rotation matrices by using the commands

```
function R = rot(t)
R = [cos(t), -sin(t); sin(t), cos(t)]
```

Save this m-file under the name `rot.m`. Test it by typing

```
R = rot(pi/3); Y = randtri(5);
figure, plot2d(Y); hold on
```

at the MATLAB prompt to generate a random triangle. Then type

```
Y = R*Y; plot2d(Y)
```

The triangle encoded by Y should appear in the figure rotated counterclockwise by $60°$ ($\pi/3$ radians). Repeat this command four more times using the up-arrow key ↑ to generate six triangles that make a symmetric figure. Be sure that you typed the hold on command so that all the triangles appear in the same graphic window. Click on Insert and title in the Figure toolbar and insert the title Figure with 6-fold Symmetry.

(b) Homogeneous coordinates

Create the following function m-file to transform a two-rowed vector or matrix A into a three-rowed vector or matrix with every entry in the third row equal 1:

```
function B = homcoord(A)
n = size(A, 2);
B = [A; ones(1,n)];
```

Save this m-file under the name homcoord.m. Test it by typing

```
u = rand(2,1), v = homcoord(u)
```

at the MATLAB prompt. You should get a random vector $\mathbf{u} \in \mathbb{R}^2$ and a vector $\mathbf{v} \in \mathbb{R}^3$. The first two components of $\mathbf{v}$ are the same as those of $\mathbf{u}$ and third component is 1.

(c) Translations

Create the following function m-file to implement translation by a vector $\mathbf{u} \in \mathbb{R}^2$ as a 3×3 matrix:

```
function T = translate(u)
A = [eye(2) u];
T =[A; 0 0 1];
```

Save this m-file under the name translate.m. Test it by typing

```
u = 2*rand(2,1), T = translate(u)
```

at the MATLAB prompt. You should get a random vector $\mathbf{u} \in \mathbb{R}^2$ and a 3×3 matrix T, with the third column of T being the vector of homogeneous coordinates of $\mathbf{u}$.

Generate a random triangle and plot it using the commands

```
Y = randtri(3); Z = homcoord(Y);
figure, plot2d(Z); hold on;
```

Now translate the triangle by **u**:

```
Z = T*Z; plot2d(Z)
```

The graphic window should show your random triangle and the translation of your triangle in the direction of your random vector **u** (notice that the function `plot2d(Z)` only uses rows one and two of the matrix Z, which are the homogeneous coordinates of the vertices of the triangle). Use the up-arrow key to repeat the command `Z = T*Z; plot2d(Z)` three times. This should give five triangles in the same graph, each one a translate of the original triangle. Put the title Translations of a Triangle on the figure.

(d) Affine transformations

The action of a 2×2 matrix A on $\mathbb{R}^2$ is given in homogeneous coordinates by a 3×3 matrix C (see Section 1.6). Create the following function m-file to convert from A to C.

```
function C = affine(A)
B = [A zeros(2,1)];
C = [B; 0 0 1];
```

Save this m-file under the name `affine.m`. Now type

```
S = randtri(3); H = homcoord(S);
figure, plot2d(H), hold on
```

at the MATLAB prompt. Click on `Insert` in the tool bar of the Figure window, and then use the `Arrow` and `TextBox` tools to label the triangle S (you may have to experiment to put the arrow and label in a good position). Now apply translations and rotations to S:

```
u = 7*rand(2,1); T = translate(u);
R = rot(2*pi/3); C = affine(R);
plot2d(T*H); plot2d(C*H); plot2d(T*C*H)
```

Create arrows and text labels for the other three triangles in the figure. Identify the different triangles with the correct labels: Translated Triangle, Rotated Triangle, and Translated and Rotated Triangle. Insert the title Translated and Rotated Triangles.

(e) Matrices and affine transformations

Generate a random vector `u = rand(2,1)` and form the translation matrix `T = translate(u)`. Then generate a random 2×2 matrix `A = rand(2,2)` and form the affine version `B = affine(A)`.

(i) Use MATLAB to calculate `A*u` and the matrix product `B*T`. How are the entries in `B*T` related to the matrix A and the vector **u**?

(ii) Use MATLAB to calculate the matrix product B*T*B^(-1). How are the entries in this matrix related to the entries in $A\mathbf{u}$?

1.10   Exercises

(1) Let $x(t) = 7t$ and $y(t) = 5\sin(3\pi t)$ be analog signals on the interval $0 \le t \le 1$ and set $z(t) = x(t) + 2y(t)$.

 (a) Sample $x(t)$ and $y(t)$ at times $t = 0, 0.25, 0.5, 0.75$ to produce a sample vectors $\mathbf{x}, \mathbf{y} \in \mathbb{R}^4$. Verify that $\mathbf{x} + 2\mathbf{y}$ is the vector $\mathbf{z}$ obtained by sampling $z(t)$ at these same times. *Sampling is linear.*
 (b) For $\mathbf{u} \in \mathbb{R}^4$ let $q(\mathbf{u})$ denote the vector whose components are obtained by applying the `floor` function to each component of $\mathbf{u}$. If $\mathbf{z}$, $\mathbf{x}$, and $\mathbf{y}$ are the vectors in (a), show that $q(\mathbf{z}) \ne q(\mathbf{x}) + 2q(\mathbf{y})$. *Quantization is nonlinear.*

(2) Let $\mathbb{F}$ be $\mathbb{R}$ or $\mathbb{C}$. Show that the following sets V with the given operations of vector addition and scalar multiplication satisfy the vector space axioms A1-A5 in Section 1.3.

 (a) $V = \mathbb{F}^{m \times n}$, with the usual addition and scalar multiplication of matrices.
 (b) Let X be the interval of real numbers $a \le x \le b$ and $V = C_{\mathbb{F}}(X)$ (all continuous $\mathbb{F}$-valued functions on X) with the usual addition and scalar multiplication of functions. (HINT: From calculus $cf(x)$ and $f(x) + g(x)$ are continuous if $f(x)$ and $g(x)$ are continuous and c is a constant.)
 (c) Let V be the set of all polynomials $f(x) = a_0 + a_1 x + \cdots + a_n x^n$ with $a_j \in \mathbb{F}$ and n any nonnegative integer, with the usual addition and scalar multiplication of functions.

(3) Determine whether the following sets M of matrices are subspaces of $\mathbb{R}^{2 \times 2}$. When this is the case, find a basis for M.

 (a) M consists of all 2×2 diagonal matrices.
 (b) M consists of all 2×2 upper triangular matrices (zero below diagonal).
 (c) M consists of all symmetric 2×2 matrices.
 (d) M consists of all 2×2 matrices A with $\det A = 0$.

(4) Let V be the real vector space of the continuous real-valued functions on the interval $-1 \le x \le 1$. Let U be the set of all functions $f \in V$ such that $f(-1) = 2f(1)$. Prove that U is a subspace of V.

(5) Let $\mathbf{v}_1, \mathbf{v}_2, \mathbf{v}_3$ be the columns of the matrix $A = \begin{bmatrix} 2 & -1 & 0 \\ -1 & 2 & -1 \\ 0 & -1 & 2 \end{bmatrix}$.

 (a) Prove that these three vectors form a basis for $\mathbb{R}^3$.
 (b) Find the dual basis $\mathbf{u}_1, \mathbf{u}_2, \mathbf{u}_3$ of row vectors.

(c) Use the dual basis to find c_1, c_2, c_3 such that $\begin{bmatrix} 1 \\ 2 \\ 3 \end{bmatrix} = c_1\mathbf{v}_1 + c_2\mathbf{v}_2 + c_3\mathbf{v}_3$.

(6) Let $\mathcal{P}_3$ be the space of all polynomials $f(x)$ of degree less than 3 with real coefficients. Let $f_1(x) = x + 2$, $f_2(x) = x + 3$, and $f_3(x) = x^2 + x$.

(a) Show that $\{f_1, f_2, f_3\}$ spans $\mathcal{P}_3$.
(b) Show that $\{f_1, f_2, f_3\}$ is a linearly independent set.

(7) Let $\mathcal{P}_3$ be the space of all polynomials $f(x)$ of degree less than 3 with real coefficients. In each of the following, find a basis and the dimension for the subspace V of $\mathcal{P}_3$ spanned by the given polynomials.

(a) $f(x) = x - 1$, $g(x) = x^2 - 1$, $h(x) = x^2 - x$.
(b) $f(x) = x - 1$, $g(x) = x^2 - 1$, $h(x) = x^2 + 1$.

(8) Let $\mathcal{P}_3$ be the space of all polynomials $f(x)$ of degree less than 3 with real coefficients. Let V be the subspace of $\mathcal{P}_3$ consisting of all polynomials $f(x) = ax^2 + 3bx + 4a + 5b$, with a and b arbitrary real numbers. Find a basis for V. (HINT: First find a spanning set for V. Then show that your set is linearly independent.)

(9) Let $\mathcal{P}_4$ be the space of all polynomials $f(x)$ of degree less than 4 with real coefficients. Let V be the subspace of $\mathcal{P}_4$ consisting of all polynomials such that $f(0) = 0$ and $f(1) = 0$.

(a) Show that every function in V is of the form $x(x - 1)(ax + b)$ for real numbers a, b.
(b) Use the result of (a) to find a basis for V.

(10) Let $L : \mathbb{R}^2 \to \mathbb{R}^2$ be a linear transformation. Suppose $L\begin{bmatrix} 1 \\ 1 \end{bmatrix} = \begin{bmatrix} 2 \\ 3 \end{bmatrix}$ and $L\begin{bmatrix} 1 \\ -1 \end{bmatrix} = \begin{bmatrix} 4 \\ 5 \end{bmatrix}$.

(a) Find $L\mathbf{e}_1$ and $L\mathbf{e}_2$, where $\mathbf{e}_1$, $\mathbf{e}_2$ are the standard basis vectors for $\mathbb{R}^2$.
(b) Find the matrix A of L with respect to the standard basis $\mathbf{e}_1, \mathbf{e}_2$ for $\mathbb{R}^2$. Check your answer by calculating $A\begin{bmatrix} 1 \\ 1 \end{bmatrix}$ and $A\begin{bmatrix} 1 \\ -1 \end{bmatrix}$.

(11) Let V be the vector space of all polynomials $f(x) = a_0 + a_1x + a_2x^2$ of degree at most 2 with real coefficients. Given $f \in V$ define $T(f) = \begin{bmatrix} 2f(-1) \\ 3f(0) \\ 4f(1) \end{bmatrix}$.

(a) Let $\mathbf{v}_1(x) = 1$, $\mathbf{v}_2(x) = x$, and $\mathbf{v}_3(x) = x^2$. Calculate $T\mathbf{v}_1$, $T\mathbf{v}_2$, and $T\mathbf{v}_3$.
(b) Prove that T is a linear transformation from V to $\mathbb{R}^3$.

(c) Use your results in (a) to find the 3×3 matrix A for T relative to the basis $\mathbf{v}_1$, $\mathbf{v}_2$, $\mathbf{v}_3$ for V and the standard basis $\mathbf{e}_1$, $\mathbf{e}_2$, $\mathbf{e}_3$ for $\mathbb{R}^3$.

(12) Let $\mathcal{P}_2$ be the space of all polynomials of degree less than 2 and $\mathcal{P}_3$ the space of all polynomials of degree less than 3. Let L be the linear transformation from $\mathcal{P}_2$ to $\mathcal{P}_3$ given by $Lf(x) = 2f'(x) + (3x + 4)f(x)$, where $f'(x) = (d/dx)f(x)$.

(a) Calculate the action of L on the ordered basis $\{x, 1\}$ for $\mathcal{P}_2$ in terms of the ordered basis $\{x^2, x, 1\}$ for $\mathcal{P}_3$.

(b) Use the result in (a) to find the matrix A for L relative to these ordered bases.

(c) Let $f(x) = 6x + 5$. Calculate $Lf(x)$ first using calculus and then using the result in (b) and matrix-vector multiplication.

(13) Let $V = \mathcal{P}_3$ be the space of all polynomials $f(x)$ of degree less than 3. Let L be the linear transformation from V to V given by $Lf(x) = (3x + 4)f'(x) + \int_0^1 12tf(t)\,dt$, where $f'(x) = (d/dx)f(x)$; note that the definite integral gives a constant function. Write $\mathbf{v}_0(x) = 1$, $\mathbf{v}_1(x) = x$, and $\mathbf{v}_2(x) = x^2$.

(a) Calculate $L\mathbf{v}_0$, $L\mathbf{v}_1$, and $L\mathbf{v}_2$ as linear combinations of $\mathbf{v}_0$, $\mathbf{v}_1$, and $\mathbf{v}_2$.

(b) Find the matrix A for L relative to the ordered basis $\{\mathbf{v}_0, \mathbf{v}_1, \mathbf{v}_2\}$ for V.

(c) Check your answer to (b) when $f(x) = 6x^2 + 5x + 4$ by calculating $Lf(x)$ first using calculus and then using matrix-vector multiplication.

(14) Let $M = \begin{bmatrix} 0 & 0 & 1 & 2 & 1 & 0 \\ 0 & 1 & 2 & 1 & 0 & 0 \\ 1 & 1 & 1 & 1 & 1 & 1 \end{bmatrix}$.

(a) Use the vector graphics method to draw the figure in $\mathbb{R}^2$ whose vertices have homogeneous coordinates given by the columns of M.

(b) For each of the following choices of the matrix A, sketch the graph of the figure in $\mathbb{R}^2$ represented by AM in vector graphics. Describe geometrically the effect of the transformation given by A.

i. $A = \begin{bmatrix} 2 & 0 & 0 \\ 0 & 1/2 & 0 \\ 0 & 0 & 1 \end{bmatrix}$.   iii. $A = \begin{bmatrix} 1 & -3 & 0 \\ 0 & 1 & 0 \\ 0 & 0 & 1 \end{bmatrix}$.

ii. $A = \begin{bmatrix} 1/2 & -\sqrt{3}/2 & 0 \\ \sqrt{3}/2 & 1/2 & 0 \\ 0 & 0 & 1 \end{bmatrix}$.   iv. $A = \begin{bmatrix} 1 & 0 & -2 \\ 0 & 1 & 3 \\ 0 & 0 & 1 \end{bmatrix}$.

(15) For each of the following transformations from $\mathbb{R}^2$ to $\mathbb{R}^2$, find the 3×3 matrix representing the transformation in the homogeneous coordinate system.

(a) The transformation R that rotates each vector by 120 degrees in the counterclockwise direction.

(b) The transformation C that contracts each vector by a factor of one-third.

(c) The transformation T that translates each point 3 units to the left and 5 units up.

(d) The transformation L that contracts each vector by a factor of one-third and then translates each point 3 units to the left and 5 units up.

(16) Let $\mathbf{u}_1 = \dfrac{1}{3\sqrt{2}} \begin{bmatrix} 1 \\ 1 \\ -4 \end{bmatrix}$, $\mathbf{u}_2 = \dfrac{1}{3} \begin{bmatrix} 2 \\ 2 \\ 1 \end{bmatrix}$, $\mathbf{u}_3 = \dfrac{1}{\sqrt{2}} \begin{bmatrix} 1 \\ -1 \\ 0 \end{bmatrix}$.

(a) Show that $\{\mathbf{u}_1, \mathbf{u}_2, \mathbf{u}_3\}$ is an orthonormal basis for $\mathbb{R}^3$.

(b) Let $\mathbf{x} = \begin{bmatrix} 1 & 1 & 1 \end{bmatrix}^{\mathrm{T}}$. Write $\mathbf{x}$ as a linear combination of $\mathbf{u}_1$, $\mathbf{u}_2$, and $\mathbf{u}_3$.

(c) Calculate $\|\mathbf{x}\|^2$ using Parseval's formula (1.26) for this orthonormal basis, and compare with the direct calculation using the components of $\mathbf{x}$ relative to the standard basis.

(17) Let $\{\mathbf{u}_1, \mathbf{u}_2, \mathbf{u}_3\}$ be an orthonormal set of vectors in an inner product space V. Consider the vectors $\mathbf{u} = \mathbf{u}_1 + 2\mathbf{u}_2 + 3\mathbf{u}_3$ and $\mathbf{v} = 4\mathbf{u}_1 + 5\mathbf{u}_2$ in V. Calculate $\langle \mathbf{u}, \mathbf{v} \rangle$, $\|\mathbf{u}\|$, and $\|\mathbf{v}\|$ using equations (1.25) and (1.26).

(18) Consider the complex vector space $\mathbb{C}^2$ with the standard inner product $\langle \mathbf{u}, \mathbf{v} \rangle = \mathbf{u}^{\mathrm{H}}\mathbf{v}$.

(a) Let $\mathbf{u} = \begin{bmatrix} 2 \\ 3 + 4i \end{bmatrix}$ and $\mathbf{v} = \begin{bmatrix} 5i \\ 1 + i \end{bmatrix}$. Calculate $\|\mathbf{u}\|$, $\|\mathbf{v}\|$, and $\langle \mathbf{u}, \mathbf{v} \rangle$.

(b) Let $\mathbf{u}, \mathbf{v} \in \mathbb{C}^2$ be vectors such that $\|\mathbf{u}\| = 3$, $\|\mathbf{v}\| = 4$, and $\langle \mathbf{u}, \mathbf{v} \rangle = 1 + 2i$. Calculate $\|\mathbf{u} + i\mathbf{v}\|^2$.

(19) Consider the matrix $U = \dfrac{1}{5} \begin{bmatrix} x & 3i \\ 3i & x \end{bmatrix}$, where x is a real number.

(a) Calculate the matrix product $U^{\mathrm{H}}U$, where U^{H} denotes the conjugate transpose matrix.

(b) Use your calculation in (a) to find all real values of x for which U is a unitary matrix.

(20) Consider the 2×2 matrix $U = \begin{bmatrix} z & -\overline{w} \\ w & \overline{z} \end{bmatrix}$, where z and w are complex numbers.

(a) Calculate the matrix product $U^{\mathrm{H}}U$, where U^{H} denotes the conjugate transpose matrix. Use your answer to find the equations on z and w that make U a unitary matrix.

(b) If W is any 2×2 unitary matrix, show that there exists $\theta \in \mathbb{R}$ such that the matrix $e^{-i\theta}W$ has determinant 1 (HINT: Use the equation $W^{\mathrm{H}}W = I$ to show that $|\det W| = 1$.)

(c) Use your answers to (a) and (b) to write down the most general 2×2 unitary matrix W.

(21) Let V be the vector space of continuous real-valued functions defined on $0 \le x \le 1$, with inner product $\langle f, g \rangle = \int_0^1 f(x)g(x)\,dx$. Let $f(x) = x^2$ and $g(x) = x$.

(a) Calculate $\langle f, g \rangle$.

(b) Calculate $\|f\|$ and $\|g\|$.

(c) Show that the results in (a) and (b) are consistent with the Cauchy-Schwarz inequality.

(d) Calculate the orthogonal projection $p(x)$ of $f(x)$ onto the subspace spanned by $g(x)$.

(e) Verify that the function $h(x) = f(x) - p(x)$ is orthogonal to $g(x)$.

(22) Let V be the vector space of continuous real-valued functions defined on $0 \leq x \leq 1$. Give V the inner product $\langle f, g \rangle = \int_0^1 f(x)g(x)\,dx$. Let U be the two-dimensional subspace of V with basis $\{u_1, u_2\}$, where $u_1(x) = 1$ and $u_2(x) = \sqrt{3}\,(2x - 1)$.

(a) Show that $\{u_1, u_2\}$ is an orthonormal set of functions.

(b) Let $f(x) = x^2$ and let $p(x)$ be the projection of $f(x)$ onto U. Calculate the appropriate integrals to find the scalars c_1 and c_2 such that $p(x) = c_1\,u_1(x) + c_2\,u_2(x)$.

(c) Let $f(x)$ and $p(x)$ be as in (b). Use the fact that $\{u_1, u_2\}$ is an orthonormal set to calculate $\|p\|^2$ and $\|f - p\|^2$.

(23) Let V be the vector space of continuous complex-valued functions defined on $0 \leq x \leq 2\pi$, with inner product $\langle f, g \rangle = \frac{1}{2\pi} \int_0^{2\pi} \overline{f(x)}g(x)\,dx$ and norm $\|f\| = \sqrt{\langle f, f \rangle}$. Let $\varphi_k(x) = e^{ikx}$.

(a) Let $f(x) = 2\varphi_{-1}(x) + 3\varphi_1(x)$ and $g(x) = 4\varphi_{-1}(x) + (5 + 6i)\varphi_1(x) + 7\varphi_4(x)$. Find the numerical values of $\langle f, g \rangle$ and $\|g\|^2$ *without calculating any integrals*.

(b) Let $h(x)$ be the orthogonal projection of $g(x)$ onto the subspace spanned by $\varphi_{-1}(x), \varphi_0(x), \varphi_1(x)$. Then $h(x) = $ _____ $\varphi_{-1}(x) + $ _____ $\varphi_0(x) + $ _____ $\varphi_1(x)$ and $\langle \varphi_4, h \rangle = $ _____. Fill in the blanks with the correct scalars *without calculating any integrals*.

(24) Let $f(x) = x$ for $0 \leq x \leq 2\pi$.

(a) Calculate the Fourier coefficients $c_k = \langle \phi_k, f \rangle$. (HINT: Calculate c_0 directly. For $k \neq 0$ use an integration by parts.)

(b) Calculate $\|f\|^2$.

(c) Use the Parseval formula (1.39) together with the results of (a) and (b) to obtain the remarkable identity $\sum_{k=1}^{\infty} \frac{1}{k^2} = \pi^2/6$.

Chapter 2

Discrete Fourier Transform

2.1 Overview

We examine more closely the sampling process for analog signals. We illustrate the *Nyquist sampling criterion* for avoiding the *aliasing* that occurs when the sampling rate is too low relative to the frequencies of the sinusoidal waves contained in the signal.

The main tool for determining the frequency content of a digital signal is the *discrete Fourier transform* (DFT). We introduce this transform first using the $N \times N$ *Fourier matrix* applied to the vector in $\mathbb{C}^N$ representing the signal. This gives the *Fourier basis* representation of the signal in terms of sampled sinusoidal waves of frequencies $0, 1, \ldots, N-1$. These Fourier basis vectors are eigenvectors for the $N \times N$ shift matrix S_N; hence they are eigenvectors for any *circulant matrix* $C = p(S_N)$ that is a polynomial in S_N. The eigenvalues of C are values of the polynomial $p(z)$ at Nth roots of unity. We can also view digital signals with N components as N-periodic functions on the integers. In this picture circulant matrices correspond to *shift-invariant filters*, and multiplication of a vector by a circulant matrix becomes *circular convolution*.

Direct numerical calculation of the DFT using the Fourier matrix F_N is slow because every entry in the matrix is nonzero. However, when N is even the matrix F_N has a simple factorization in 2×2 block form involving $F_{N/2}$ and a permutation matrix that performs *downsampling*. Assuming that N is a power of 2, we can continue this matrix factorization with $N \to N/2 \to \cdots \to 2$. By this means we obtain the *fast Fourier transform* (FFT) method for calculating the DFT, which only requires $N \log N$ arithmetic operations (scalar multiplications and additions) instead of the N^2 operations required by direct matrix-vector multiplication. This dramatic speed-up is essential in signal processing applications.

2.2 Sampling and Aliasing

Suppose we have a function $f(t)$ that measures the sound level at time t of an analog audio signal, and we want to analyze the signal on the time interval $a \le t \le b$. We obtain a *digital signal* ϕ by sampling $f(t)$ at the N equal-spaced values $t_j = a + j\Delta t$ for $j = 0, 1, \ldots, N-1$ with $\Delta t = (b-a)/N$, as in Example 1.1. We write $\phi[j] = f(t_j)$ for the sampled values of the signal[1] and we call the integer N the *sampling rate*. We can view ϕ either as a function on the set $\{0, 1, \ldots, N-1\}$ or as a column vector[2]

$$\mathbf{y} = \begin{bmatrix} \phi[0] & \phi[1] & \cdots & \phi[N-1] \end{bmatrix}^{\mathsf{T}}. \tag{2.1}$$

The function viewpoint emphasizes the relation to analog signals and is well adapted to signal processing operations, whereas the column vector form is natural for linear algebra calculations with matrices. We will use both representations.

Example 2.1. Suppose the signal is

$$f_k(t) = e^{2\pi i k t} = \cos(2\pi k t) + i \sin(2\pi k t) \tag{2.2}$$

where k is a positive integer. This is an oscillating wave, and if $T = 1/k$ then

$$f_k(t + T) = e^{2\pi i k(t + 1/k)} = e^{2\pi i k t} e^{2\pi i} = f_k(t)$$

because $e^{2\pi i} = 1$. Hence $f_k(t)$ has *period* T (measured in units of time) and *frequency* $1/T = k$. In particular, $f_k(t + 1) = f_k(t + kT) = f_k(t)$ so $f_k(t)$ is determined by its values in the interval $0 \le t < 1$. Furthermore,

$$\int_0^1 \overline{f_k(t)}\, f_m(t)\, dt = \begin{cases} 1 & \text{if } k = m, \\ 0 & \text{if } k \ne m, \end{cases}$$

by the change of variable $x = 2\pi t$ and the calculations in Section 1.8. Thus the set of functions $\{f_k(t) : k \in \mathbb{Z}\}$ is orthonormal, relative to the inner product

$$\langle f, g \rangle = \int_0^1 \overline{f(t)}\, g(t)\, dt.$$

Sampling $f_k(t)$ at the N equally-spaced points $t_j = j/N$ gives the values

$$f_k(t_j) = e^{2\pi i k j / N} = \omega^{kj} \quad \text{for } j = 0, 1, \ldots, N-1, \quad \text{with } \omega = e^{2\pi i / N}.$$

Thus by sampling an oscillating wave of frequency k at rate N, we obtain the function $\phi_{k,N}[j] = \omega^{kj}$ on $\mathbb{Z}$ and the corresponding column vector

$$\mathbf{E}_k = \begin{bmatrix} 1 & \omega^k & \omega^{2k} & \cdots & \omega^{(N-1)k} \end{bmatrix}^{\mathsf{T}} \tag{2.3}$$

in $\mathbb{C}^N$. Note that $\phi_{k,N}[j + N] = \omega^{kj}\omega^{kN} = \omega^{kj}$. Hence $\phi_{k,N}[j]$ is periodic of period N as a function of j.   ∎

The oscillating functions $f_k(t)$ in Example 2.1 are the building blocks for the Fourier analysis of analog signals (see Section 1.8). Their sampled versions $\phi_{k,N}$ and the corresponding vectors $\mathbf{E}_k$ play a similar role for digital signals. However,

---

[1]Throughout this book $\phi[j]$ denotes the value of a digital signal ϕ at a discrete time j.
[2]The indexing of the components is different than MATLAB indexing going from 1 to N.

the conversion from analog to digital loses information if the sampling rate is too low. This occurs because

$$\phi_{k+N,N} = \phi_{k,N} \quad \text{and} \quad \mathbf{E}_{k+N} = \mathbf{E}_k \,,$$

since $\omega^{j(k+N)} = \omega^{jk}\omega^{jN} = \omega^{jk}(\omega^N)^j = \omega^{jk}$. Thus the analog signals $f_k(t)$ and $f_{k+N}(t)$ yield the same digital signal when sampled on the interval $0 \leq t \leq 1$ at rate N. This phenomenon is called *aliasing*. On the other hand, if $0 < |k - p| < N$, then $\omega^{k-p} \neq 1$ and hence $\omega^k \neq \omega^p$. So $\mathbf{E}_k \neq \mathbf{E}_p$ in this case and the functions $f_k(t)$ and $f_p(t)$ give two different (non-proportional) vectors in $\mathbb{C}^N$ when sampled on the interval $0 \leq t \leq 1$ at rate N. Now there is no aliasing. The following example illustrates this phenomenon.

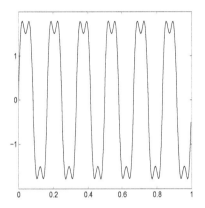

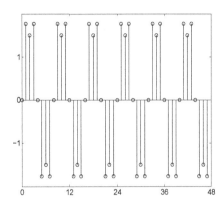

Fig. 2.1   Analog signal   Fig. 2.2   Digital sample of signal $(N = 48)$

Example 2.2. Suppose that time t is measured in seconds and the signal is a superposition of a fundamental low-frequency oscillation $2\sin(6\pi t)$ with frequency 3 Hertz (cycles per second) and maximum amplitude 2, together with a weaker high-frequency oscillation $0.5\sin(18\pi t)$ with frequency 9 Hertz (which is three times the fundamental frequency) and maximum amplitude 0.5:

$$s(t) = 2\sin(6\pi t) + 0.5\sin(18\pi t) \,.$$

If we want to sample the signal over the time interval $0 \leq t \leq 2$, we can rescale the interval to $0 \leq x \leq 1$ by setting $x = t/2$ and

$$f(x) = s(2x) = 2\sin(12\pi x) + 0.5\sin(36\pi x)$$
$$= -\mathrm{i}\left(\mathrm{e}^{12\pi\mathrm{i}x} - \mathrm{e}^{-12\pi\mathrm{i}x} + 0.25\mathrm{e}^{36\pi\mathrm{i}x} - 0.25\mathrm{e}^{-36\pi\mathrm{i}x}\right) \,.$$

We have plotted $f(x)$ for $0 \leq x \leq 1$ in Figure 2.1; notice the *modulation* of the low-frequency wave by the high-frequency wave.

If we sample $f(x)$ on the interval $0 \leq x \leq 1$ at rate N, we get the vector

$$-\mathrm{i}\left(\mathbf{E}_6 - \mathbf{E}_{-6} + 0.25\mathbf{E}_{18} - 0.25\mathbf{E}_{-18}\right) = -\mathrm{i}\left(\mathbf{E}_6 - \mathbf{E}_{N-6} + 0.25\mathbf{E}_{18} - 0.25\mathbf{E}_{N-18}\right) \,.$$

If $N > 36$, then there is no aliasing; in fact, we will prove in Proposition 2.1 that the vectors $\mathbf{E}_6$, $\mathbf{E}_{N-6}$, $\mathbf{E}_{18}$, and $\mathbf{E}_{N-18}$ are mutually orthogonal in this case. Thus the digital signal contains all the information (frequencies and amplitudes of the waves at each frequency) in the original analog signal. For example, if we sample at rate $N = 48$, we get the vector plotted as a stem graph in Figure 2.2. Notice how the graph faithfully shows the modulation of the low-frequency wave by the high-frequency wave as in Figure 2.1.

If we use a smaller value of N, say $N = 24$, then $\mathbf{E}_{N-6} = \mathbf{E}_{-6}$ and $\mathbf{E}_{N-18} = \mathbf{E}_6$. Thus the sampled vector becomes $(3/4i)\left(\mathbf{E}_6 - \mathbf{E}_{-6}\right) \in \mathbb{C}^{24}$. This is the same vector that we would get by sampling the function $g(x) = 1.5\sin(12\pi x)$ at this rate (see Figure 2.3, where we have plotted $g(x)$ as a dashed line). Notice that the high frequency component in the signal disappears, while the amplitude of the low frequency component is increased. ■

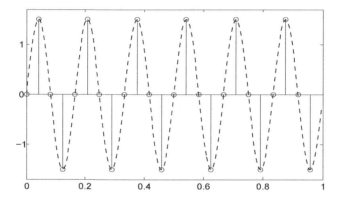

Fig. 2.3   Under-sampled digital signal with aliasing ($N = 24$)

In Example 2.2 the highest frequency was 18 cycles (when the time interval was rescaled to be $0 \le x \le 1$). Sampling without aliasing required a sample rate $N > 36$, larger than twice the highest frequency present in the signal. This is a special case of the following general fact (see Exercises 2.8 #2).

Nyquist sampling criterion: *If a digital sample of an analog signal is to reproduce frequencies up to m Hertz without aliasing, then the sampling rate must be greater than 2m Hertz.*

For example, in audio recording the frequency of 20,000 Hertz is taken as an upper limit of human hearing. The commercial compact disk (CD) digital recording standard with a sampling rate $N = 44,100$ Hertz, which replaced the long-playing (LP) analog recording technology in the 1980's, satisfies the Nyquist criterion.

2.3 Discrete Fourier Transform and Fourier Matrix

Fix a positive integer N. The complex number $\omega = e^{2\pi i/N}$ satisfies $\omega^N = 1$. Its powers $1 = \omega^0, \omega, \omega^2, \ldots, \omega^{N-1}$ subdivide the unit circle into N equal segments. We call ω a *primitive Nth root of unity* because every root of the equation $z^N = 1$ is of the form ω^m for some integer m (see Figure 2.4 for the case $N = 8$). We will

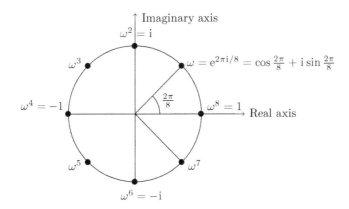

Fig. 2.4 Solutions to $z^8 = 1$ are $z = 1, \omega, \omega^2, \ldots, \omega^7$ with $\omega = (1 + i)/\sqrt{2}$

denote ω as ω_N when the value of N needs to be indicated. When $N = 2m$ is an even integer, then

$$(\omega_N)^{2k} = e^{4k\pi i/2m} = e^{2k\pi i/m} = (\omega_{N/2})^k. \tag{2.4}$$

Furthermore, $(\omega_N)^{N/2} = e^{2m\pi i/2m} = e^{\pi i} = -1$. Relation (2.4) will play a basic role in the *fast Fourier transform* algorithm in Section 2.6.

Here is another important property of Nth roots of unity. We illustrate it first with the case $N = 4$ and $\omega = \cos(\pi/2) + i \sin(\pi/2) = i$. Then $\omega^2 = -1$ and $\omega^3 = -i$; thus $1 + \omega + \omega^2 + \omega^3 = 1 + i - 1 - i = 0$. In general for any positive integer N, if $u \neq 1$ is any complex number such that $u^N = 1$, then a similar equation holds:

$$1 + u + u^2 + \cdots + u^{N-1} = 0. \tag{2.5}$$

To prove (2.5), call the sum on the left s. Then $us = u + u^2 + \cdots + u^N = s$, because $u^N = 1$. Hence $(u - 1)s = 0$. Since $u \neq 1$, we conclude that $s = 0$.

Proposition 2.1. *The vectors* $\mathbf{E}_k \in \mathbb{C}^N$ *in equation (2.3), with* $\omega = e^{2\pi i/N}$, *have the following properties (here* $\langle \mathbf{u}, \mathbf{v} \rangle = \mathbf{u}^H \mathbf{v}$ *is the inner product on* $\mathbb{C}^N$ *):*

(1) $\langle \mathbf{E}_k, \mathbf{E}_k \rangle = N$ *and* $\langle \mathbf{E}_m, \mathbf{E}_k \rangle = 0$ *if* $0 < |k - m| < N$.
(2) The set $\{\mathbf{E}_0, \mathbf{E}_1, \ldots, \mathbf{E}_{N-1}\}$ *is an orthogonal basis for* $\mathbb{C}^N$.
(3) If $\mathbf{y} \in \mathbb{C}^N$, *then* $\mathbf{y} = d_0 \mathbf{E}_0 + d_1 \mathbf{E}_1 + \cdots + d_{N-1} \mathbf{E}_{N-1}$ *with* $d_k = \frac{1}{N} \langle \mathbf{E}_k, \mathbf{y} \rangle$.

Proof. Since $\bar{\omega} = \omega^{-1}$, the inner product between $\mathbf{E}_m$ and $\mathbf{E}_k$ is

$$(\mathbf{E}_m)^H \mathbf{E}_k = 1 + \omega^{k-m} + (\omega^{k-m})^2 + \cdots + (\omega^{k-m})^{N-1}. \tag{2.6}$$

Set $u = \omega^{k-m}$. Then the right side of equation (2.6) is the sum $1 + u + \cdots + u^{N-1}$ of powers of u. If $k = m$, then $u = 1$ and this sum is N. If $0 < |k - m| < N$, then $u \neq 1$ and this sum is 0 by equation (2.5), proving statement (1). Since the N vectors in the set $\{\mathbf{E}_0, \mathbf{E}_1, \ldots, \mathbf{E}_{N-1}\}$ are nonzero and mutually orthogonal, this set is linearly independent. Hence it spans $\mathbb{C}^N$ because $\dim \mathbb{C}^N = N$. This proves statement (2). Statement (3) follows from statements (1) and (2) as a general fact about an orthogonal basis for a vector space. Recall that the formula for d_k comes from using the expansion of $\mathbf{y}$ to evaluate the inner product of $\mathbf{E}_k$ with $\mathbf{y}$. $\qquad\square$

Remark 2.1. The expansion of vectors in $\mathbb{C}^N$ in Proposition 2.1(3) is the digital signal version of the Fourier series expansion (1.37) for analog signals, with the Fourier coefficients c_k replaced by d_k. The vector $\mathbf{E}_k$ is the sampled digital waveform of frequency k and $|d_k|^2$ is the amount of energy in the signal $\mathbf{y}$ at this frequency.[3]

We now express the results in Proposition 2.1 in matrix form.

Definition 2.1. The *Fourier matrix* F_N is the $N \times N$ matrix whose rows are $\mathbf{E}_0^H, \ldots, \mathbf{E}_{N-1}^H$. Thus the (j, k) entry of F_N is $\omega^{-(j-1)(k-1)}$ for $j, k = 1, \ldots, N$, where $\omega = e^{2\pi i/N}$.

For $N = 2$ the Fourier matrix is

$$F_2 = \begin{bmatrix} 1 & 1 \\ 1 & -1 \end{bmatrix}, \tag{2.7}$$

since $e^{2\pi i/2} = -1$. For $N = 4$ we have $w = e^{2\pi i/4} = i$ and $\omega^{-1} = -i$. Hence

$$F_4 = \begin{bmatrix} 1 & 1 & 1 & 1 \\ 1 & -i & (-i)^2 & (-i)^3 \\ 1 & (-i)^2 & (i)^4 & (-i)^6 \\ 1 & (-i)^3 & (-i)^6 & (-i)^9 \end{bmatrix} = \begin{bmatrix} 1 & 1 & 1 & 1 \\ 1 & -i & -1 & i \\ 1 & -1 & 1 & -1 \\ 1 & i & -1 & -i \end{bmatrix}. \tag{2.8}$$

For any positive integer N the matrix F_N is symmetric and the entries in the first column (or row) are all 1. The second column (or row) consists of the powers of ω^{-1} from 0 to $N-1$, the third column (or row) consists of the powers of ω^{-2} from 0 to $N-1$, and so on.

Definition 2.2. Given $\mathbf{y} \in \mathbb{C}^N$, we set $\mathbf{Y} = F_N \mathbf{y}$ and call $\mathbf{Y}$ the *discrete Fourier transform* (DFT) of $\mathbf{y}$.

From the definition of the Fourier matrix we can write the DFT of $\mathbf{y}$ as

$$\mathbf{Y} = \begin{bmatrix} \mathbf{E}_0^H \mathbf{y} & \mathbf{E}_1^H \mathbf{y} & \cdots & \mathbf{E}_{N-1}^H \mathbf{y} \end{bmatrix}^T = \begin{bmatrix} d_0 & d_1 & \cdots & d_{N-1} \end{bmatrix}^T,$$

where the components $d_0, d_1, \ldots, d_{N-1}$ are the inner products in Proposition 2.1(3). In particular, taking $\mathbf{y} = \mathbf{E}_k$, we have

$$F_N \mathbf{E}_k = N \mathbf{e}_{k+1} \quad \text{for } k = 0, 1, \ldots, N-1, \tag{2.9}$$

---

[3]The division by N in the formula for d_k is for normalization purposes, just as the division by 2π in equation (1.31).

where $\mathbf{e}_1, \ldots, \mathbf{e}_N$ are the standard basis vectors for $\mathbb{C}^N$. Since F_N is a symmetric matrix, the complex conjugate matrix $\overline{F}_N$ has columns $\mathbf{E}_0, \mathbf{E}_1, \ldots, \mathbf{E}_{N-1}$. Hence for any vector $\mathbf{y} \in \mathbb{C}^N$

$$\overline{F}_N \mathbf{Y} = d_0 \mathbf{E}_0 + d_1 \mathbf{E}_1 + \cdots + d_{N-1} \mathbf{E}_{N-1} = N\mathbf{y}, \tag{2.10}$$

where we have used Proposition 2.1(3) to obtain the last equality.

Theorem 2.1. *The inverse of the Fourier matrix is $(1/N)\overline{F}_N$. Furthermore, the normalized matrix $(1/\sqrt{N})F_N$ is unitary.*

Proof. The first assertion follows from (2.10). By Proposition 2.1 the columns of $(1/\sqrt{N})F_N$ are an orthonormal set. This proves the second assertion. $\qquad\square$

Corollary 2.1. *Let $\{\mathbf{e}_1, \ldots, \mathbf{e}_N\}$ be the standard basis for $\mathbb{C}^N$. Set $\mathbf{u}_j = (1/\sqrt{N})F_N \mathbf{e}_j$. Then $\{\mathbf{u}_1, \ldots, \mathbf{u}_N\}$ is an orthonormal basis for $\mathbb{C}^N$, called the Fourier basis. Hence if $\mathbf{y} \in \mathbb{C}^N$ and $\mathbf{Y} = F_N\mathbf{y}$, then $\|\mathbf{y}\|^2 = (1/N)\|\mathbf{Y}\|^2$.*

Proof. For the first assertion, note that $\mathbf{u}_j$ is the jth column of the unitary matrix $(1/\sqrt{N})F_N$. Since $(1/\sqrt{N})\mathbf{Y} = (1/\sqrt{N})F_N\mathbf{y}$ and $(1/\sqrt{N})F_N$ is a unitary matrix, the vectors $(1/\sqrt{N})\mathbf{Y}$ and $\mathbf{y}$ have the same norm. $\qquad\square$

Example 2.3. Suppose $N = 4$ and $\mathbf{y} = \begin{bmatrix} 1 & 2 & -1 & 0 \end{bmatrix}^{\mathrm{T}}$. Then

$$\mathbf{Y} = F_4\mathbf{y} = \begin{bmatrix} 1 & 1 & 1 & 1 \\ 1 & -i & -1 & i \\ 1 & -1 & 1 & -1 \\ 1 & i & -1 & -i \end{bmatrix} \begin{bmatrix} 1 \\ 2 \\ -1 \\ 0 \end{bmatrix} = \begin{bmatrix} 2 \\ 2-2i \\ -2 \\ 2+2i \end{bmatrix}.$$

In this case $\|\mathbf{y}\|^2 = 1 + 2^2 + (-1)^2 = 6$, while we have

$$\frac{1}{4}\|\mathbf{Y}\|^2 = \frac{1}{4}[2^2 + |2-2i|^2 + (-2)^2 + |2+2i|^2] = \frac{1}{4}[4+8+4+8] = 6,$$

as predicted by Corollary 2.1. $\qquad\blacksquare$

If we think of the standard basis $\mathbf{e}_j$ as a sampled version of a signal, then this signal is localized in (discrete) *time*, since only one component of $\mathbf{e}_j$ is nonzero. By contrast, all the entries in $\mathbf{u}_j$ are nonzero, so the Fourier matrix removes the time localization. In the opposite direction, the vector $\mathbf{E}_k$ is the sampled version of a wave having only one frequency. This digital signal is completely unlocalized in time since all the entries in $\mathbf{E}_k$ have absolute value one. But the DFT of $\mathbf{E}_k$ is $N\mathbf{e}_{k+1}$, which has only one nonzero component and hence is localized in (discrete) *frequency*. We illustrate this phenomenon in the following example.

Example 2.4. Suppose we sample the signal $f(x) = 2\sin(12\pi x) + 0.5\sin(36\pi x)$ from Example 2.2 at a rate satisfying the Nyquist criterion, say $N = 48$. Since $-6 \equiv 42 \mod (48)$ and $-18 \equiv 30 \mod (48)$, the sampled vector is

$$\mathbf{y} = -i\left(\mathbf{E}_6 + 0.25\mathbf{E}_{18} - 0.25\mathbf{E}_{30} - \mathbf{E}_{42}\right) \in \mathbb{C}^{48}.$$

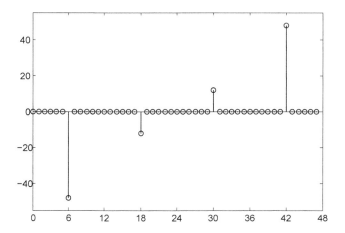

Fig. 2.5   Discrete Fourier transform in Example 2.4

By (2.9) we see that

$$\mathbf{Y} = -48\mathrm{i}\Big(\mathbf{e}_7 + 0.25\mathbf{e}_{19} - 0.25\mathbf{e}_{31} - \mathbf{e}_{43}\Big) \in \mathbb{C}^{48}. \qquad (2.11)$$

The signal vector $\mathbf{y}$ (which has real components $\mathbf{y}[k]$ for $k = 0, \ldots, 47$) was already plotted in Figure 2.2. Figure 2.5 shows the imaginary part of its discrete Fourier transform vector $\mathbf{Y} = F_{48}\mathbf{y}$ (which has purely imaginary components $\mathbf{Y}[k]$ for $k = 0, \ldots, 47$). The horizontal axis in Figure 2.2 represents *time*, whereas the horizontal axis in Figure 2.5 represents *frequency*. Notice that the vector $\mathbf{y}$ is spread out in time. By contrast, the only nonzero entries in $\mathbf{Y}$ are $\mathbf{Y}[6]$, $\mathbf{Y}[18]$, $\mathbf{Y}[30]$ and $\mathbf{Y}[42]$. The Fourier transform vector $\mathbf{Y}$ is skew-symmetric around the midpoint of the frequency range (this is called the *Nyquist point*). Figure 2.5 shows the two frequencies present in the original signal. The values $\mathbf{Y}[6]$, $\mathbf{Y}[42]$ correspond to the low frequency sine wave, whereas the values $\mathbf{Y}[18]$, $\mathbf{Y}[30]$ correspond to the high frequency sine wave.   ■

2.4   Shift-Invariant Transformations and Circulant Matrices

2.4.1   *Moving averages and shift operator*

Consider a finite digital signal $\mathbf{y}$ with N values by $\mathbf{y}[0], \mathbf{y}[1], \ldots, \mathbf{y}[N-1]$. As in Section 2.2, we can view $\mathbf{y}$ either as a column vector

$$\begin{bmatrix} \mathbf{y}[0] \\ \mathbf{y}[1] \\ \vdots \\ \mathbf{y}[N-1] \end{bmatrix} \in \mathbb{C}^N, \qquad (2.12)$$

or as a function $j \longrightarrow \mathbf{y}[j]$ of a discrete time variable $j = 0, 1, \ldots, N-1$.

A basic operation in signal processing is to take a *moving average* of the signal. For example, we can replace each value $\mathbf{y}[j]$ by the average of the values $\mathbf{y}[j-1]$ and $\mathbf{y}[j+1]$. This gives a new signal $\mathbf{z}$ with

$$\mathbf{z}[j] = (\mathbf{y}[j-1] + \mathbf{y}[j+1])/2. \tag{2.13}$$

There is a bug in formula (2.13), however. To calculate $\mathbf{z}[0]$ or $\mathbf{z}[N-1]$ we need the values $\mathbf{y}[-1]$ and $\mathbf{y}[N]$, which aren't available. We will solve this problem by defining $\mathbf{y}[k]$ for all integers k (positive and negative) using the *periodic extension* of $\mathbf{y}$:

$$\mathbf{y}[j+kN] = \mathbf{y}[j] \quad \text{for } j = 0, 1, \ldots, N-1 \text{ and all integers } k. \tag{2.14}$$

In terms of *modular arithmetic*, we can write $\mathbf{y}[m] = \mathbf{y}[j]$ when $m \equiv j \pmod{N}$. For the extended function $\mathbf{y}$ we have $\mathbf{y}[-1] = \mathbf{y}[N-1]$ and $\mathbf{y}[N] = \mathbf{y}[0]$ since $-1 = N - 1 + N$ and $N = 0 + N$. Now when we write $\mathbf{y}$ we will mean this N-periodic function on the integers that is uniquely determined by the corresponding vector (2.12) in $\mathbb{C}^N$.

With the function $\mathbf{y}$ defined to be N-periodic, formula (2.13) makes sense for all integers j. It can also be written in a case-by-case way just using the values $\mathbf{y}[j]$ for $j = 0, 1, \ldots, N-1$:

$$\mathbf{z}[j] = \begin{cases} (\mathbf{y}[N-1] + \mathbf{y}[1])/2 & \text{for } j = 0, \\ (\mathbf{y}[j-1] + \mathbf{y}[j+1])/2 & \text{for } j = 1, \ldots, N-2, \\ (\mathbf{y}[N-2] + \mathbf{y}[0])/2 & \text{for } j = N-1. \end{cases}$$

For example, if $N = 4$ and $\mathbf{y} = [\, 1,\, 2,\, -1,\, 0\,]^T$ as in Example 2.3, then

$$\mathbf{z}[0] = (0+2)/2, \quad \mathbf{z}[1] = (1-1)/2, \quad \mathbf{z}[2] = (2+0)/2, \quad \mathbf{z}[3] = (-1+1)/2.$$

We will use the notation $\ell_{\mathbb{C}}[\mathbb{Z}/N\mathbb{Z}]$ for the set of all complex-valued functions ϕ defined on the integers $\mathbb{Z}$ that are periodic of period N:

$$\phi[k] = \phi[k+N] \quad \text{for all integers } k.$$

Any linear combination of N-periodic functions is again N-periodic, so $\ell_{\mathbb{C}}[\mathbb{Z}/N\mathbb{Z}]$ is a complex vector space. Every such function ϕ is uniquely determined by the vector

$$\mathbf{y} = \begin{bmatrix} \phi[0] \\ \phi[1] \\ \vdots \\ \phi[N-1] \end{bmatrix} \tag{2.15}$$

in $\mathbb{C}^N$. We recover the function ϕ by taking the N-periodic extension of $\mathbf{y}$ from equation (2.14).

We define the inner product and the norm of N-periodic functions ϕ and ψ as

$$\langle \phi, \psi \rangle = \sum_{j=0}^{N-1} \overline{\phi[j]}\, \psi[j] \quad \text{and} \quad \|\phi\| = \sqrt{\langle \phi, \phi \rangle}.$$

Then equation (2.15) gives a unitary transformation from $\ell_{\mathbb{C}}[\mathbb{Z}/N\mathbb{Z}]$ to $\mathbb{C}^N$. If $\phi \longleftrightarrow \mathbf{x} \in \mathbb{C}^N$ and $\psi \longleftrightarrow \mathbf{y} \in \mathbb{C}^N$, then $\langle \phi, \psi \rangle = \mathbf{x}^H \mathbf{y} = \langle \mathbf{x}, \mathbf{y} \rangle$. So for all calculations with N-periodic functions we can use the corresponding vectors in $\mathbb{C}^N$. However, for many signal processing operations the function point of view is more natural, as we have just seen in formula (2.13). For example, using the periodicity of the functions we can calculate the inner product by summing over any N consecutive integers:

$$\langle \phi, \psi \rangle = \sum_{j=k}^{k+N-1} \overline{\phi[j]}\, \psi[j] \,.$$

Definition 2.3. The *shift operator* S acts on a complex-valued function ϕ defined on $\mathbb{Z}$ by $(S\phi)[j] = \phi[j-1]$ for all $j \in \mathbb{Z}$. Thus the graph of $S\phi$ is obtained by shifting the graph of ϕ one unit to the right.

It is clear that S is a linear transformation on the vector space of all complex-valued functions on $\mathbb{Z}$. Also S is invertible, with inverse transformation $S^{-1}\phi[j] = \phi[j+1]$. Thus the operator $\mathbf{y} \longrightarrow \mathbf{z}$ given by (2.13) is the average of a forward shift and a backward shift:

$$\mathbf{z} = \frac{1}{2}\left(S\mathbf{y} + S^{-1}\mathbf{y}\right). \tag{2.16}$$

If ϕ is periodic of period N, then

$$(S\phi)[j+N] = \phi[j+N-1] = \phi[j-1] = (S\phi)[j] \quad \text{for all } j \in \mathbb{Z}\,.$$

Hence $S\phi$ also has period N (we can also see this from the action of S on the graph of ϕ). Thus S gives a linear transformation on $\ell_{\mathbb{C}}[\mathbb{Z}/N\mathbb{Z}]$ that we will continue to denote by S. Note that the periodicity condition implies that

$$(S^N\phi)[j] = \phi[j-N] = \phi[j] \quad \text{for all } \phi \in \ell_{\mathbb{C}}[\mathbb{Z}/N\mathbb{Z}]\,.$$

Hence S^N acts as the identity linear transformation on N-periodic functions.

2.4.2 *Shift-invariant transformations*

We call a linear transformation T on N-periodic signals *shift-invariant* if changing the input signal $\mathbf{y}$ to $S\mathbf{y}$ changes the output signal $\mathbf{z} = T\mathbf{y}$ to $S\mathbf{z}$. This condition should hold for all inputs $\mathbf{y}$, so it the same as the commutativity relation $TS = ST$ between the linear transformations. Any linear transformation of the form

$$c_k S^k + c_{k+1} S^{k+1} + \cdots + c_m S^m \tag{2.17}$$

(a linear combination of positive or negative powers of S, where each coefficient c_j is a complex number) is shift-invariant, since $SS^j = S^j S$ for all integers j. For example, the transformation defined by formula (2.13) is shift invariant, and equation (2.16) shows that it is a linear combination of powers of S. We will prove that every shift-invariant linear transformation is of the form (2.17).

We can identify the inner product space $\ell_{\mathbb{C}}[\mathbb{Z}/N\mathbb{Z}]$ with $\mathbb{C}^N$ by the unitary transformation (2.15). This allows us to describe shift-invariant linear transformations in terms of their matrices relative to the standard basis for $\mathbb{C}^N$. It turns out that these matrices have a very special form.

We begin by finding the $N \times N$ matrix S_N for the shift transformation. The standard basis vector $\mathbf{e}_k$ (with 1 in row k and zeros elsewhere) corresponds to the N-periodic function $\mathbf{y}_{k-1}$ with $\mathbf{y}_k[j] = 1$ if $j \equiv k \pmod{N}$, and otherwise $\mathbf{y}_k[j] = 0$. Here the index shift comes from the relation (2.15). Since $S\mathbf{y}_k = \mathbf{y}_{k+1}$, we have

$$S_N \mathbf{e}_k = \mathbf{e}_{k+1} \quad \text{for } k = 1, 2, \ldots, N. \tag{2.18}$$

For this formula to be valid we must label the standard basis vectors circularly modulo N: $\mathbf{e}_{N+1} = \mathbf{e}_1$, $\mathbf{e}_{N+2} = \mathbf{e}_2$ and so on. This means that S_N acts as a *circular permutation* of the standard basis vectors.

Example 2.5. Suppose $N = 3$. Then $S_3\mathbf{e}_1 = \mathbf{e}_2$, $S_3\mathbf{e}_2 = \mathbf{e}_3$, and $S_3\mathbf{e}_3 = \mathbf{e}_1$. Hence $(S_3)^2\mathbf{e}_1 = \mathbf{e}_3$, $(S_3)^2\mathbf{e}_2 = \mathbf{e}_1$, and $(S_3)^2\mathbf{e}_3 = \mathbf{e}_2$. Also $(S_3)^3 = I$, so $(S_3)^{-1} = (S_3)^2$. Thus the matrix of the shift operator and its inverse relative to the standard basis for $\mathbb{C}^3$ are

$$S_3 = \begin{bmatrix} 0 & 0 & 1 \\ 1 & 0 & 0 \\ 0 & 1 & 0 \end{bmatrix} \quad \text{and} \quad S_3^{-1} = \begin{bmatrix} 0 & 1 & 0 \\ 0 & 0 & 1 \\ 1 & 0 & 0 \end{bmatrix}.$$

By direct matrix multiplication we can check that $(S_3)^3 = I$. ∎

The general features of Example 2.5 are valid for the shift matrix for any value of N. From (2.18) we see that $(S_N)^N = I_N$, so $(S_N)^{-1} = (S_N)^{N-1}$. Since S_N is a permutation matrix, it is real and orthogonal. Hence we also have $(S_N)^{-1} = (S_N)^T = (S_N)^{N-1}$, as illustrated in Example 2.5.

Theorem 2.2. *Suppose C is an $N \times N$ matrix that satisfies $CS_N = S_NC$. Let the first column of C be $[c_0, c_1, \ldots, c_{N-1}]^T$. Then*

$$C = c_0 I_N + c_1 S_N + c_2 (S_N)^2 + \cdots + c_{N-1}(S_N)^{N-1}, \tag{2.19}$$

where I_N denotes the $N \times N$ identity matrix.

Proof. The first column of C is the vector $C\mathbf{e}_1$, so this vector can be written in terms of the standard basis as

$$C\mathbf{e}_1 = c_0\mathbf{e}_1 + c_1\mathbf{e}_2 + \cdots + c_{N-1}\mathbf{e}_N. \tag{2.20}$$

Now we calculate the columns $C\mathbf{e}_k$ of C for $k = 2, \ldots, N$. The period N is fixed, so we shall denote the matrix S_N simply as S and I_N as I to avoid clutter in the formulas. Since C is shift-invariant we have $S^{k-1}C = CS^{k-1}$. Thus if we multiply both sides of (2.20) by S^{k-1} and use the property $S^{k-1}\mathbf{e}_1 = \mathbf{e}_k$, we obtain

$$C\mathbf{e}_k = CS^{k-1}\mathbf{e}_1 = S^{k-1}C\mathbf{e}_1$$
$$= c_0 S^{k-1}\mathbf{e}_1 + c_1 S^{k-1}\mathbf{e}_2 + c_2 S^{k-1}\mathbf{e}_3 + \cdots + c_{N-1}S^{k-1}\mathbf{e}_N$$
$$= c_0\mathbf{e}_k + c_1 S\mathbf{e}_k + c_2 S^2\mathbf{e}_k + \cdots + c_{N-1}S^{N-1}\mathbf{e}_k.$$

So the kth column of the matrix C is the same as the kth column of the matrix $c_0 I + c_1 S + c_2 S^2 + \cdots + c_{N-1} S^{N-1}$ for $k = 1, \ldots, N$. $\square$

Example 2.6. Suppose $N = 3$ and $C = c_0 I_3 + c_1 S_3 + c_2 (S_3)^2$ is a 3×3 shift-invariant matrix. From Example 2.5 we have

$$C = c_0 \begin{bmatrix} 1\,0\,0 \\ 0\,1\,0 \\ 0\,0\,1 \end{bmatrix} + c_1 \begin{bmatrix} 0\,0\,1 \\ 1\,0\,0 \\ 0\,1\,0 \end{bmatrix} + c_2 \begin{bmatrix} 0\,1\,0 \\ 0\,0\,1 \\ 1\,0\,0 \end{bmatrix} = \begin{bmatrix} c_0\ c_2\ c_1 \\ c_1\ c_0\ c_2 \\ c_2\ c_1\ c_0 \end{bmatrix} .$$

Hence the successive columns of C are obtained by circular permutation of the first column. Matrices of this form are called *circulant matrices*. For example, when $N = 4$ the averaging operation from equation (2.13) is given by the circulant matrix

$$C = \frac{1}{2} \left(S_4 + (S_4)^3 \right) = \frac{1}{2} \begin{bmatrix} 0\,1\,0\,1 \\ 1\,0\,1\,0 \\ 0\,1\,0\,1 \\ 1\,0\,1\,0 \end{bmatrix} .$$

Here we have used the relation $(S_4)^{-1} = (S_4)^3$.   ■

2.4.3 *Eigenvectors and eigenvalues of circulant matrices*

We now use the Fourier matrix to find the eigenvectors and eigenvalues of a circulant matrix.[4] The key step is to find the eigenvectors and eigenvalues of the shift matrix S_N. Let $\omega = e^{2\pi i/N}$ and let $\mathbf{E}_k$ be the vector in $\mathbb{C}^N$ in equation (2.3). Recall that we obtained this vector by sampling the analog signal $f_k(x) = e^{2\pi i k x}$ of frequency k at N equally-spaced points $x = 0, 1/N, \ldots, (N-1)/N$ in the interval $0 \leq x \leq 1$.

Theorem 2.3. *Each vector in the basis* $\{\mathbf{E}_0, \mathbf{E}_1, \ldots, \mathbf{E}_{N-1}\}$ *for* $\mathbb{C}^N$ *is an eigenvector of the shift matrix* S_N. *The corresponding eigenvalues are the* N *complex numbers* ω^{-k}, *where* $k = 0, 1, \ldots, N - 1$ *(the* Nth *roots of unity enumerated around the unit circle in clockwise order).*

Proof. Since S_N shifts the entries in $\mathbf{E}_k$ down one place with the last entry moved to the top, we have

$$S_N \mathbf{E}_k = \begin{bmatrix} \omega^{(N-1)k} \\ 1 \\ \omega^k \\ \vdots \\ \omega^{(N-2)k} \end{bmatrix} = \omega^{-k} \begin{bmatrix} \omega^{Nk} \\ \omega^k \\ \omega^{2k} \\ \vdots \\ \omega^{(N-1)k} \end{bmatrix} = \omega^{-k} \mathbf{E}_k . \tag{2.21}$$

This equation shows that $\mathbf{E}_k$ is an eigenvector for S_N with eigenvalue ω^{-k}. $\square$

---

[4]It is quite remarkable that we can do this in explicit symbolic form; this is not possible for a general $N \times N$ matrix.

Now that we have found eigenvalues and eigenvectors of the shift matrix, we can do the same for any circulant matrix.

Theorem 2.4. *Suppose that $C = c_0 I_N + c_1 S_N + \cdots + c_{N-1}(S_N)^{N-1}$ is an $N \times N$ circulant matrix. Define the polynomial $p(z) = c_0 + c_1 z + c_2 z^2 + \cdots + c_{N-1} z^{N-1}$. Then $\mathbf{E}_k$ is an eigenvector for C with eigenvalue $p(\omega^{-k})$. Hence for every vector*

$$\mathbf{y} = d_0 \mathbf{E}_0 + d_1 \mathbf{E}_1 + \cdots + d_{N-1} \mathbf{E}_{N-1} \tag{2.22}$$

in $\mathbb{C}^N$, the vector $C\mathbf{y}$ is obtained by multiplying the coefficient d_k by the eigenvalue $p(\omega^{-k})$:

$$C\mathbf{y} = p(\omega^0) \, d_0 \mathbf{E}_0 + p(\omega^{-1}) \, d_1 \mathbf{E}_1 + \cdots + p(\omega^{-N+1}) \, d_{N-1} \mathbf{E}_{N-1}. \tag{2.23}$$

Proof. Using equation (2.21) repeatedly and denoting S_N as S, we find that $S^j \mathbf{E}_k = \omega^{-jk} \mathbf{E}_k$ for all integers j. Hence

$$
\begin{aligned}
C\mathbf{E}_k &= c_0 \mathbf{E}_k + c_1 S \mathbf{E}_k + c_2 S^2 \mathbf{E}_k + \cdots + c_{N-1} S^{N-1} \mathbf{E}_k \\
&= \left(c_0 + c_1 \omega^{-k} + c_2 \omega^{-2k} + \cdots + c_{N-1} \omega^{-(N-1)k} \right) \mathbf{E}_k \\
&= p(\omega^{-k}) \mathbf{E}_k \ .
\end{aligned}
$$

Equation (2.23) now follows by applying C to each term in the expansion of $\mathbf{y}$. $\quad\square$

Remark 2.2. The coefficients d_k in (2.22) are the entries in the (normalized) DFT vector $(1/N) F_N \mathbf{y}$. Furthermore, if we view the circulant matrix C as a linear transformation T of $\mathbb{C}^N$, then Theorem 2.4 asserts that the matrix of T relative to the basis $\{\mathbf{E}_0, \ldots, \mathbf{E}_{N-1}\}$ of eigenvectors for T is diagonal.

Example 2.7. Consider the 4×4 circulant matrix $C = \left(S_4 + (S_4)^{-1} \right)/2$, as in Example 2.6. Since $(S_4)^{-1} = (S_4)^3$, we can write $C = \left(S_4 + (S_4)^3 \right)/2$ and $p(z) = (z + z^3)/2$. The fourth roots of 1 are $1, \mathrm{i}, -1, -\mathrm{i}$, so the eigenvalues of C are

$$
\begin{aligned}
&p(1) = 1, && p(-\mathrm{i}) = (-\mathrm{i} + (-\mathrm{i})^3)/2 = 0 \ , \\
&p(-1) = (-1 + (-1)^3)/2 = -1 \ , && p(\mathrm{i}) = (\mathrm{i} + (\mathrm{i})^3)/2 = 0 \ ,
\end{aligned}
$$

and C acts on the eigenvectors by $C\mathbf{E}_0 = \mathbf{E}_0$, $C\mathbf{E}_1 = \mathbf{0}$, $C\mathbf{E}_2 = -\mathbf{E}_2$, $C\mathbf{E}_3 = \mathbf{0}$. Thus if we write a vector $\mathbf{y} = d_0 \mathbf{E}_0 + d_1 \mathbf{E}_1 + d_2 \mathbf{E}_2 + d_3 \mathbf{E}_3$ as a linear combination of eigenvectors, then $C\mathbf{y} = d_0 \mathbf{E}_0 - d_2 \mathbf{E}_2$. $\quad\blacksquare$

2.5 Circular Convolution and Filters

Now we translate the results about circulant matrices into the language of digital signal processing. Let T be a shift-invariant linear transformation on the vector space $\ell_{\mathbb{C}}(\mathbb{Z}/N\mathbb{Z})$. If we identify N-periodic functions with vectors in $\mathbb{C}^N$ by the correspondence (2.15), then T becomes a linear transformation on $\mathbb{C}^N$ and the shift operator S becomes the $N \times N$ shift matrix S_N. Let C be the matrix for T relative to the standard basis for $\mathbb{C}^N$. Then $CS_N = S_N C$, so by Theorem 2.2 there are complex numbers $c_0, \ldots, c_{N-1}$ such that

$$C = c_0 I_N + c_1 S_N + \cdots + c_{N-1} S_N^{N-1}. \qquad (2.24)$$

We can write this equation as $C = p(S_N)$, where $p(z) = c_0 + c_1 z + \cdots + c_{N-1} z^{N-1}$, as in Theorem 2.4. Since we know from Section 1.5 that the correspondence between linear transformations and their matrices preserves sums and products, it follows that $T = p(S)$ as a linear transformation on $\ell_{\mathbb{C}}[\mathbb{Z}/N\mathbb{Z}]$. Thus the action of T on an N-periodic function $\mathbf{y}$ is

$$T\mathbf{y}[j] = c_0\mathbf{y}[j] + c_1\mathbf{y}[j-1] + c_2\mathbf{y}[j-2] + \cdots + c_{N-1}\mathbf{y}[j-N+1]. \qquad (2.25)$$

This formula shows that $T\mathbf{y}$ is a kind of *moving average*[5] of the original function $\mathbf{y}$, since the value of $T\mathbf{y}$ at time j is a linear combination of the values of $\mathbf{y}$ at times $j, j-1, \ldots, j-N+1$.

Define the function $\mathbf{f}[k] = c_k$ for $k = 0, 1, \ldots, N-1$. Then (2.25) becomes

$$T\mathbf{y}[j] = \sum_{k=0}^{N-1} \mathbf{f}[k]\,\mathbf{y}[j-k]\,. \qquad (2.26)$$

Since S^{k+N} has the same action on N-periodic functions as does S^k, it is natural to extend the function $\mathbf{f}$ to be an N-periodic function on $\mathbb{Z}$. We can then describe formula (2.26) in the following way.

Definition 2.4. The *circular convolution* of N-periodic functions $\mathbf{f}$ and $\mathbf{y}$ is the function

$$\mathbf{z}[j] = \sum_{k=0}^{N-1} \mathbf{f}[k]\,\mathbf{y}[j-k] \quad \text{for } j \in \mathbb{Z}\,,$$

denoted by $\mathbf{z} = \mathbf{f} \star \mathbf{y}$.

It is evident that $\mathbf{z}[j+N] = \mathbf{z}[j]$, so the function $\mathbf{f} \star \mathbf{y}$ is N-periodic. Thus circular convolution is a new way of multiplying N-periodic functions that is different from pointwise multiplication. This new product operation is exactly what is needed for shift-invariant transformations, since equation (2.26) can be written as $T\mathbf{y} = \mathbf{f} \star \mathbf{y}$. Conversely, if we take any N-periodic function $\mathbf{f}$ on the integers, we can define $T\mathbf{y} = \mathbf{f} \star \mathbf{y}$ for $\mathbf{y} \in \ell_{\mathbb{C}}[\mathbb{Z}/N\mathbb{Z}]$. Then T is a linear transformation on $\ell_{\mathbb{C}}[\mathbb{Z}/N\mathbb{Z}]$, since the convolution product $\mathbf{f} \star \mathbf{y}$ is a linear function of $\mathbf{y}$. Furthermore,

$$(ST\mathbf{y})[j] = \sum_{k=0}^{N-1} \mathbf{f}[k]\,\mathbf{y}[j-1-k] = \sum_{k=0}^{N-1} \mathbf{f}[k]\,S\mathbf{y}[j-k] = (TS\mathbf{y})[j]\,.$$

This shows that $ST = TS$, so T is shift invariant. If we take the N-periodic function $\mathbf{y}$ with $\mathbf{y}[0] = 1$ and $\mathbf{y}[k] = 0$ for $0 < k < N$, then $T\mathbf{y}[j] = \mathbf{f}[j]$. Hence we can

---

[5]It is an average in the usual probability sense if all the coefficients c_k are nonnegative and add up to 1; we don't impose that restriction here.

recover **f** from the linear transformation T. We can thus restate Theorem 2.2 in terms of circular convolution as follows.

Corollary 2.2. *Every linear transformation T of N-periodic functions that is shift invariant is given by the circular convolution operation $T\mathbf{y} = \mathbf{f} \star \mathbf{y}$ for a unique function $\mathbf{f} \in \ell_{\mathbb{C}}[\mathbb{Z}/N\mathbb{Z}]$. This function is called the* filter *corresponding to T.*

Our next goal is to obtain the linear filter version of Theorem 2.4. We first describe the discrete Fourier transform in terms of N-periodic functions.

Definition 2.5. *If $\mathbf{y} \in \ell_{\mathbb{C}}[\mathbb{Z}/N\mathbb{Z}]$ then the* discrete Fourier transform *(DFT) of $\mathbf{y}$ is the function*

$$\widehat{\mathbf{y}}[k] = \sum_{j=0}^{N-1} \mathbf{y}[j]\, \omega^{-jk} \quad \text{for } k \in \mathbb{Z}, \text{ where } \omega = e^{2\pi i/N} . \tag{2.27}$$

We note that $\widehat{\mathbf{y}}$ is periodic of period N since $\omega^{-j(k+N)} = \omega^{-jk} \omega^{-jN} = \omega^{-jk}$. Thus $\widehat{\mathbf{y}}$ is determined by its values at $k = 0, \ldots, N - 1$. When we identify the function $\mathbf{y}$ with a column vector in $\mathbb{C}^N$ by (2.15) and multiply this vector by the Fourier matrix F_N we obtain a vector whose entries are given by (2.27). Hence Definition 2.5 is consistent with the previous definition of the DFT on $\mathbb{C}^N$.

Remark 2.3. When $\mathbf{y}$ is obtained by sampling an analog signal $f(x)$ on $0 \le x \le 2\pi$ at rate N, then $(1/N)\widehat{\mathbf{y}}[k]$ is a discrete approximation to the integral in formula (1.31) for the Fourier coefficient c_k of $f(x)$. Using the functions $\phi_{k,N}$ in Example 2.1 we can write the formula for the DFT of $\mathbf{y}$ as an inner product: $\mathbf{y}[k] = \langle \phi_{k,N} , \mathbf{y} \rangle$. This corresponds to the formula $c_k = \langle \phi_k, f \rangle$.

Let $\mathbf{y}$ be a real-valued digital signal of period N. Its DFT has the symmetry[6] $\overline{\widehat{\mathbf{y}}[k]} = \widehat{\mathbf{y}}[-k]$ for all integers k. This follows from (2.27) since $\overline{\omega} = \omega^{-1}$. Since $\widehat{\mathbf{y}}[k + (N/2)] = \widehat{\mathbf{y}}[k - (N/2)]$ by periodicity, we can express this symmetry as

$$\overline{\widehat{\mathbf{y}}[(N/2) + k]} = \widehat{\mathbf{y}}[(N/2) - k] . \tag{2.28}$$

We call $N/2$ the *Nyquist point*. Equation (2.28) shows that the DFT of $\mathbf{y}$ is completely determined by its values at frequencies k from zero up to the Nyquist point; this was already illustrated in Figure 2.5.

We can now restate the result of Theorem 2.3 in terms of the Fourier transform and circular convolution as follows.

Theorem 2.5. *Let $T\mathbf{y} = \mathbf{f} \star \mathbf{y}$ be the circular convolution operator (2.26) on N-periodic functions $\mathbf{y}$ associated with the filter $\mathbf{f}$. Then the discrete Fourier transform of $T\mathbf{y}$ is the pointwise product*

$$\widehat{T\mathbf{y}}[k] = \widehat{\mathbf{f}}[k]\,\widehat{\mathbf{y}}[k] \tag{2.29}$$

of the DFT of $\mathbf{f}$ and the DFT of $\mathbf{y}$.

---

[6]The Fourier coefficients of a periodic real analog signal have a corresponding symmetry (1.32).

Proof. Let $C = p(S_N)$ be the circulant matrix associated with T from equation (2.24). By Theorem 2.3 the eigenvectors for C are the vectors $\mathbf{E}_k$ for $k = 0, 1, \ldots, N - 1$. The corresponding eigenvalues are $p(\omega^{-k}) = \widehat{\mathbf{f}}[k]$. When $\mathbf{y}$ and $T\mathbf{y}$ are written as linear combinations of the eigenvectors, the coefficients of $\mathbf{E}_k$ are $(1/N)\widehat{\mathbf{y}}[k]$ and $(1/N)\widehat{T\mathbf{y}}[k]$, respectively. But T acts on $\mathbf{y}$ by multiplying the coefficient of $\mathbf{E}_k$ by the eigenvalue $\widehat{\mathbf{f}}[k]$. This proves (2.29). $\qquad\square$

Remark 2.4. We can give an alternate proof of Theorem 2.5 without using the eigenvector result of Theorem 2.3, as follows. By the definitions of the discrete Fourier transform and circular convolution, we have

$$\widehat{\mathbf{f} \star \mathbf{y}}[k] = \sum_{j=0}^{N-1} (\mathbf{f} \star \mathbf{y})[j]\, \omega^{-jk} = \sum_{j=0}^{N-1}\sum_{\ell=0}^{N-1} \mathbf{f}[j - \ell]\, \mathbf{y}[\ell]\, \omega^{-jk}\,.$$

Make the substitution $m = j - \ell$ and use the periodicity of $\mathbf{f}$ and $\mathbf{y}$ to shift the range of summation to $0 \leq m < N$ and $0 \leq \ell < N$. This gives

$$\widehat{\mathbf{f} \star \mathbf{y}}[k] = \sum_{m=0}^{N-1}\sum_{\ell=0}^{N-1} \mathbf{f}[m]\, \mathbf{y}[l]\, \omega^{-(m+\ell)k}$$

$$= \left\{ \sum_{m=0}^{N-1} \mathbf{f}[m]\, \omega^{-mk} \right\}\left\{ \sum_{\ell=0}^{N-1} \mathbf{y}[\ell]\, \omega^{-\ell k} \right\} = \widehat{\mathbf{f}}[k]\,\widehat{\mathbf{y}}[k]\,.$$

The key property used in this calculation is the *law of exponents* $\omega^{a+b} = \omega^a\,\omega^b$ for integers a, b. This is the same property making $\mathbf{E}_k$ an eigenvector for S_N.

Example 2.8. Consider the shift-invariant operator

$$T\mathbf{y}[j] = \frac{1}{4}\mathbf{y}[j - 1] + \frac{1}{2}\mathbf{y}[j] + \frac{1}{4}\mathbf{y}[j + 1]$$

acting on an N-periodic function $\mathbf{y}$ with $N \geq 3$. This formula calculates a genuine *moving average* of the values of $\mathbf{y}$, since the coefficients $1/4$, $1/2$, $1/4$ are positive and add to 1.

In terms of the shift operator, $T = (S + 2I + S^{-1})/4$. We can write $T\mathbf{y} = \mathbf{f} \star \mathbf{y}$, where the N-periodic filter $\mathbf{f}$ is defined by

$$\mathbf{f}[1] = 1/4, \quad \mathbf{f}[0] = 1/2, \quad \mathbf{f}[-1] = 1/4,$$

and $\mathbf{f}[j] = 0$ for all j that are not congruent to $1, 0, -1$ modulo N. Since $\widehat{\mathbf{f}}[k] = (\omega^k + 2 + \omega^{-k})/4$ with $\omega = e^{2\pi i/N}$, Theorem 2.5 gives

$$\widehat{T\mathbf{y}}[k] = \widehat{\mathbf{f}}[k]\widehat{\mathbf{y}}[k] = \frac{1}{4}(\omega^k + 2 + \omega^{-k})\widehat{\mathbf{y}}[k]\,.$$

To understand the effect of T on the DFT of $\mathbf{y}$, write

$$\widehat{\mathbf{f}}[k] = \frac{1}{4}\left(e^{k\pi i/N} + e^{-k\pi i/N}\right)^2 = \cos^2(k\pi/N)\,.$$

This formula shows that $\widehat{\mathbf{f}}$ is obtained by sampling the function $\varphi(\theta) = \cos^2(\theta/2)$ of the continuous *frequency variable* θ on the interval $0 \leq \theta \leq 2\pi$. The sample points are at $0, \Delta\theta, 2\Delta\theta, \ldots, (N - 1)\Delta\theta$, where $\Delta\theta = 2\pi/N$. The function $\varphi(\theta)$ is 2π periodic and satisfies $\varphi(0) = 1$ and $\varphi(\pi) = 0$. In Example 2.9 we will see that T acts as a *lowpass filter*. ■

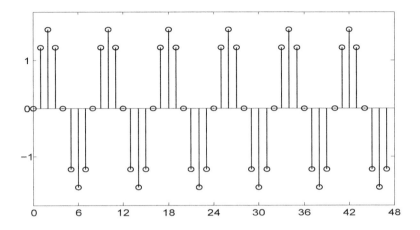

Fig. 2.6   Low-pass filtering of digital signal ($N = 48$)

Example 2.9. Take the filter **f** in Example 2.8 and consider its action on the digital signal corresponding to the vector $\mathbf{y} \in \mathbb{C}^{48}$ from Example 2.4—a sample of the sum of low frequency (3 Hertz) sine wave with amplitude 2 and a higher frequency sine wave (9 Hertz) with amplitude $1/2$. The DFT **Y** in equation (2.11) corresponds to the function $\widehat{\mathbf{y}}$ given by

$$\widehat{\mathbf{y}}[k] = -48\mathrm{i} \begin{cases} 1 & \text{if } k = 6\,, \\ 1/4 & \text{if } k = 18\,, \\ -1/4 & \text{if } k = 30\,, \\ -1 & \text{if } k = 42\,, \\ 0 & \text{otherwise}\,, \end{cases}$$

for $k = 0, 1, \ldots, 47$ and extended periodically of period 48. Since $\widehat{\mathbf{y}}$ has purely imaginary values, it is skew-symmetric around the Nyquist point $k = 24$, in accordance with (2.28). The filtered signal $\mathbf{f} \star \mathbf{y}$ has DFT

$$\widehat{\mathbf{f}}[k]\,\widehat{\mathbf{y}}[k] = -48\mathrm{i} \begin{cases} c^2 & \text{if } k = 6\,, \\ (1 - c^2)/4 & \text{if } k = 18\,, \\ -(1 - c^2)/4 & \text{if } k = 30\,, \\ -c^2 & \text{if } k = 42\,, \\ 0 & \text{otherwise}\,, \end{cases}$$

for $k = 0, 1, \ldots, 47$, where $c^2 = \cos^2\left(\frac{6\pi}{48}\right) = 0.8536$. The inverse DFT of $\widehat{\mathbf{f}}[k]\widehat{\mathbf{y}}[k]$ is the vector $T\mathbf{y} = -\mathrm{i}c^2\left(\mathbf{E}_6 - \mathbf{E}_{42}\right) + \mathrm{i}(1 - c^2)/4\left(\mathbf{E}_{18} - \mathbf{E}_{30}\right)$ in $\mathbb{C}^{48}$. The digital signal $\mathbf{z}$ of period 48 corresponding to $T\mathbf{y}$ is

$$\mathbf{z}[k] = 2c^2 \sin(3k\pi/4) + \frac{1 - c^2}{2} \sin(9k\pi/4)\,.$$

Thus in the filtered signal $\mathbf{z}$ the ratio of the amplitude of the low frequency part $2c^2 \sin(3k\pi/4)$ to the amplitude of the high frequency part $\left((1-c^2)/2\right)\sin(9k\pi/4)$ is $4c^2/(1-c^2) \approx 23{:}1$. By contrast, in the unfiltered signal $\mathbf{y}$ this ratio is $4{:}1$. Thus the filtered signal consists almost entirely of samples of the 3 Hertz sine wave, but with slightly smaller amplitude. This is confirmed by the graph of $T\mathbf{y}$ shown in Figure 2.6 (compare this graph to Figure 2.2). This illustrates why $\mathbf{f}$ is called a *lowpass filter*. The construction of lowpass (and highpass) filters is one of the central topics in Chapter 4.   ■

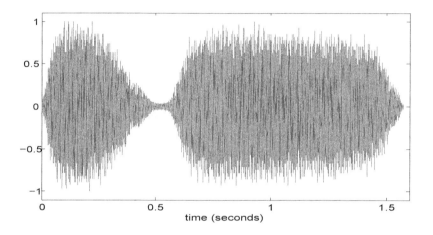

Fig. 2.7   Train whistle

Example 2.10. Figure 2.7 shows a digital signal $\mathbf{y}$ of length 12,880 that records a train whistle at rate $8192 = 2^{13}$ samples per second (the signal lasts $12{,}880/8{,}192 \approx 1.57$ seconds). Figure 2.8 is the graph of the absolute value of the DFT $\mathbf{Y} = F_{12,880}\mathbf{y}$, with the frequency range rescaled to 0–8192 Hertz. The Nyquist point 4,092 in this graph now represents a frequency in Hertz, and we will call it the Nyquist frequency; the graph is symmetric about this frequency, as we expect from equation (2.28). Furthermore, we know from the Nyquist sampling criterion that with this sampling rate we can only detect sinusoidal oscillations of frequency less than 4,096 Hertz in the whistle.

From the plot in the time domain we see that the whistle consists of a short toot followed by a long toot, but we get no frequency information. In the frequency domain we see that there are three large spikes in the range 750 to 1,250 Hertz (which are mirrored around the Nyquist frequency to the range 7,000 to 7,500 Hertz), and three much smaller spikes in the range 1,500 to 3,500 Hertz (which are mirrored around the Nyquist frequency to the range 4,500 to 6,500 Hertz). Zooming in on this plot, we can determine these dominant frequencies quite precisely.[7]

---

[7]The three between 750 and 1,250 Hertz give the pitches (f, a, d) of an inverted d-minor chord.

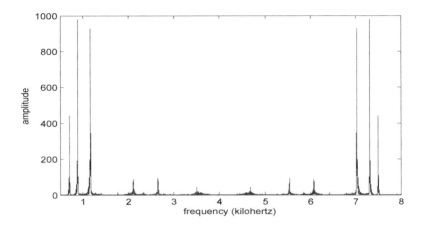

Fig. 2.8   Discrete Fourier transform of whistle (absolute values)

In Figure 2.8 the spikes at the dominant frequencies are slightly spread out. This is due to the starting and stopping of the signal in the time domain, which in the frequency domain becomes convolution with the DFT of the signal envelope.[8] But from the frequency plot alone we cannot know that the whistle signal code is *short–long*, which meant to inspect the train brake line for leaks. This illustrates the fundamental *time localization vs. frequency localization* constraint in signal processing. This shortcoming of the Fourier transform can be partially overcome by using the *short-time Fourier transform* with an appropriate window, and plotting the results as a *spectrogram* in the time-frequency plane—see [Jensen and la Cour-Harbo (2001), §9.5] and [Walker (2008), §5.7].   ∎

2.6   Downsampling and Fast Fourier Transform

The effectiveness of the DFT as a computational tool depends on a remarkable *fast algorithm* for calculating the matrix-vector product $F_N \mathbf{v}$ when $N = 2^k$ is a power of 2. This so-called *fast Fourier transform* (FFT) algorithm[9] is possible because when $N = 2m$ is an even integer, then the Fourier matrix F_N can be factored as the product of a *permutation matrix* (which has no arithmetic computational cost) and a 2×2 *block matrix* whose blocks are F_m or a diagonal matrix multiplying F_m.

Example 2.11. Consider the Fourier matrix when $N = 4$. Recall that

$$F_2 = \begin{bmatrix} 1 & 1 \\ 1 & -1 \end{bmatrix}, \quad F_4 = \begin{bmatrix} 1 & 1 & 1 & 1 \\ 1 & -i & -1 & i \\ 1 & -1 & 1 & -1 \\ 1 & i & -1 & -i \end{bmatrix} = \begin{bmatrix} \mathbf{h}_0 & \mathbf{h}_1 & \mathbf{h}_2 & \mathbf{h}_3 \end{bmatrix}.$$

---

[8]This follows from Theorem 2.5 with time and frequency variables interchanged.

[9]Similar fast algorithms exist for every *highly composite* number N, such as $N = 2^p 3^q$ with p and q large integers.

Let $\mathbf{y} \in \mathbb{C}^4$. By the definition of matrix-vector multiplication we can write

$$F_4 \mathbf{y} = \mathbf{y}[0]\, \mathbf{h}_0 + \mathbf{y}[1]\, \mathbf{h}_1 + \mathbf{y}[2]\, \mathbf{h}_2 + \mathbf{y}[3]\, \mathbf{h}_3$$

as a linear combination of the columns of the Fourier matrix. Rearrange this sum by grouping the even and odd indices of the components of $\mathbf{y}$:

$$\mathbf{y}[0]\, \mathbf{h}_0 + \mathbf{y}[2]\, \mathbf{h}_2 = \begin{bmatrix} 1 & 1 \\ 1 & -1 \\ 1 & 1 \\ 1 & -1 \end{bmatrix} \begin{bmatrix} \mathbf{y}[0] \\ \mathbf{y}[2] \end{bmatrix},$$

$$\mathbf{y}[1]\, \mathbf{h}_1 + \mathbf{y}[3]\, \mathbf{h}_3 = \begin{bmatrix} 1 & 1 \\ -i & i \\ -1 & -1 \\ i & -i \end{bmatrix} \begin{bmatrix} \mathbf{y}[1] \\ \mathbf{y}[3] \end{bmatrix}.$$

(2.30)

Define

$$\mathbf{y}_{\text{even}} = \begin{bmatrix} \mathbf{y}[0] \\ \mathbf{y}[2] \end{bmatrix}, \qquad \mathbf{y}_{\text{odd}} = \begin{bmatrix} \mathbf{y}[1] \\ \mathbf{y}[3] \end{bmatrix}, \qquad D_2 = \begin{bmatrix} 1 & 0 \\ 0 & -i \end{bmatrix}.$$

Since $\begin{bmatrix} 1 & 1 \\ -i & i \end{bmatrix} = D_2 F_2$, we can use block multiplication as in (1.11) to write formulas (2.30) as

$$\mathbf{y}[0]\, \mathbf{h}_0 + \mathbf{y}[2]\, \mathbf{h}_2 = \begin{bmatrix} F_2 \\ F_2 \end{bmatrix} \mathbf{y}_{\text{even}}, \qquad \mathbf{y}[1]\, \mathbf{h}_1 + \mathbf{y}[3]\, \mathbf{h}_3 = \begin{bmatrix} D_2 F_2 \\ -D_2 F_2 \end{bmatrix} \mathbf{y}_{\text{odd}}.$$

The splitting of $\mathbf{y}$ into even/odd vectors of two components can be accomplished by the permutation matrix

$$P_4 = \begin{bmatrix} \mathbf{e}_1 & \mathbf{e}_3 & \mathbf{e}_2 & \mathbf{e}_4 \end{bmatrix} = \begin{bmatrix} 1 & 0 & 0 & 0 \\ 0 & 0 & 1 & 0 \\ 0 & 1 & 0 & 0 \\ 0 & 0 & 0 & 1 \end{bmatrix}, \qquad P_4\, \mathbf{y} = \begin{bmatrix} \mathbf{y}_{\text{even}} \\ \mathbf{y}_{\text{odd}} \end{bmatrix}.$$

We can write the calculations above in block form as

$$F_4\, \mathbf{y} = \begin{bmatrix} F_2 \mathbf{y}_{\text{even}} + D_2 F_2 \mathbf{y}_{\text{odd}} \\ F_2 \mathbf{y}_{\text{even}} - D_2 F_2 \mathbf{y}_{\text{odd}} \end{bmatrix} = \begin{bmatrix} I_2 & D_2 \\ I_2 & -D_2 \end{bmatrix} \begin{bmatrix} F_2 & \mathbf{0} \\ \mathbf{0} & F_2 \end{bmatrix} P_4\, \mathbf{y}.$$

This shows that F_4 has the factorization $\begin{bmatrix} I_2 & D_2 \\ I_2 & -D_2 \end{bmatrix} \begin{bmatrix} F_2 & \mathbf{0} \\ \mathbf{0} & F_2 \end{bmatrix} P_4$ in terms of F_2 and the diagonal matrix D_2. ∎

The same splitting into even and odd components that we used in Example 2.11 will give us a factorization of the DFT of any signal

$$\mathbf{y} = \begin{bmatrix} \mathbf{y}[0], \mathbf{y}[1], \dots, \mathbf{y}[2m-2], \mathbf{y}[2m-1] \end{bmatrix}^{\mathrm{T}}$$

of even length $N = 2m$. Set

$$\mathbf{y}_{\text{even}} = \begin{bmatrix} \mathbf{y}[0], \dots, \mathbf{y}[2m-2] \end{bmatrix}^{\mathrm{T}} \quad \text{and} \quad \mathbf{y}_{\text{odd}} = \begin{bmatrix} \mathbf{y}[1], \dots, \mathbf{y}[2m-1] \end{bmatrix}^{\mathrm{T}}.$$

Here we are using the terms *even* and *odd* because we view $\mathbf{y}$ as a function on $\{0, 1, \ldots, 2m - 1\}$. Thus $\mathbf{y}_{\text{even}}$ uses the values of $\mathbf{y}$ at the even integers, while $\mathbf{y}_{\text{odd}}$ uses the values of $\mathbf{y}$ at the odd integers.[10] This splitting of $\mathbf{y}$ into vectors $\mathbf{y}_{\text{even}}$ and $\mathbf{y}_{\text{odd}}$ with half the number of components is called *downsampling*; it will play an important role for wavelet transforms.

Write $\omega = e^{2\pi i/N} = e^{\pi i/m}$ and $\zeta = \omega^2 = e^{2\pi i/m}$. Then

$$F_{2m}\mathbf{y}[j] = \sum_{k=0}^{m-1} \omega^{-j(2k)}\mathbf{y}[2k] + \sum_{k=0}^{m-1} \omega^{-j(2k+1)}\mathbf{y}[2k+1]$$

$$= \sum_{k=0}^{m-1} \zeta^{-jk}\mathbf{y}_{\text{even}}[k] + \omega^{-j} \sum_{k=0}^{m-1} \zeta^{-jk}\mathbf{y}_{\text{odd}}[k].$$

Since ζ is the mth root of unity used in F_m, we see that

$$F_{2m}\mathbf{y}[j] = F_m\mathbf{y}_{\text{even}}[j] + \omega^{-j}F_m\mathbf{y}_{\text{odd}}[j] \quad \text{for } j = 0, 1, \ldots, 2m - 1. \tag{2.31}$$

The key insight for the fast Fourier transform[11] is that in formula (2.31) the functions $F_m\mathbf{y}_{\text{even}}$ and $F_m\mathbf{y}_{\text{odd}}$ are periodic in j of period m. Thus it suffices to calculate $F_m\mathbf{y}_{\text{even}}[j]$ and $F_m\mathbf{y}_{\text{odd}}[j]$ for $j = 0, 1, \ldots, m - 1$, and to observe that $\omega^{-(m+j)} = -\omega^{-j}$ and $\zeta^{-(m+j)k} = \zeta^{-jk}$. Then we obtain the rest of the entries as

$$F_{2m}\mathbf{y}[m + j] = F_m\mathbf{y}_{\text{even}}[j] - \omega^{-j}F_m\mathbf{y}_{\text{odd}}[j] \quad \text{for } j = 0, 1, \ldots, m - 1. \tag{2.32}$$

We now express these formulas in concise block-matrix form, just as in the case $N = 4$ that we worked out in Example 2.11. Define the $m \times m$ diagonal matrix

$$D_m = \begin{bmatrix} 1 & 0 & 0 & \cdots & 0 \\ 0 & \omega^{-1} & 0 & \cdots & 0 \\ 0 & 0 & \omega^{-2} & \cdots & 0 \\ \vdots & \vdots & \vdots & \ddots & \vdots \\ 0 & 0 & 0 & \cdots & \omega^{-(m-1)} \end{bmatrix} \quad \text{(caution: } \omega^m = -1\text{)}. \tag{2.33}$$

Note that the diagonal of D_m only contains half of the $2m$th roots of 1. Let P_{2m} be the permutation matrix that splits $\mathbf{y}$ into its even and odd components:

$$P_{2m} = \begin{bmatrix} \mathbf{e}_1 \ \mathbf{e}_3 \ \cdots \ \mathbf{e}_{2m-1} \ \mathbf{e}_2 \ \mathbf{e}_4 \ \cdots \ \mathbf{e}_{2m} \end{bmatrix}^{\mathrm{T}}, \quad P_{2m}\mathbf{y} = \begin{bmatrix} \mathbf{y}_{\text{even}} \\ \mathbf{y}_{\text{odd}} \end{bmatrix}.$$

Since P_{2m} only rearranges the components of a vector, calculating $P_{2m}\mathbf{y}$ is free of computational cost. Then, just as in the case $N = 4$, the equations (2.31) and (2.32) for $F_{2m}\mathbf{y}$ can be written as a single vector equation

$$\begin{bmatrix} F_m\mathbf{y}_{\text{even}} + D_m F_m\mathbf{y}_{\text{odd}} \\ F_m\mathbf{y}_{\text{even}} - D_m F_m\mathbf{y}_{\text{odd}} \end{bmatrix} = G_{2m} \begin{bmatrix} F_m & \mathbf{0} \\ \mathbf{0} & F_m \end{bmatrix} P_{2m}\mathbf{y}, \tag{2.34}$$

---

[10]Note that in the MATLAB indexing convention the vector $\mathbf{y}_{\text{even}}$ uses components $1, 3, \ldots, 2m - 1$ of the vector $\mathbf{y}$.

[11]This was already noted by Gauss in the early nineteenth century.

where $G_{2m} = \begin{bmatrix} I_m & D_m \\ I_m & -D_m \end{bmatrix}$. The key point in the matrix factorization (2.34) is that because D_m is a diagonal matrix, applying the matrix G_{2m} to a vector only requires m scalar multiplications (on the last m components of a vector with $2m$ components) followed by $2m$ scalar additions, rather than the $(2m)^2$ scalar multiplications and $2m(2m-1)$ scalar additions that are necessary for a general $2m \times 2m$ matrix.[12] The *fast Fourier transform* (FFT) algorithm calculates F_N when N is a power of 2 by iterating formula (2.34).

Example 2.12. Take $N = 1024 = 2^{10}$. Then by the factorization (2.34)

$$F_{1024}\,\mathbf{y} = G_{1024} \begin{bmatrix} F_{512} & \mathbf{0} \\ \mathbf{0} & F_{512} \end{bmatrix} P_{1024}\,\mathbf{y}\,.$$

The product with G_{1024} needs 512 scalar multiplications. We can use the factorization (2.34) again, but now with $m = 256$, to express each copy of F_{512} in terms of G_{512} and F_{256} applied to signals with 256 components. But now we have two copies of G_{512}, so we need $2 \cdot 256 = 512$ more scalar multiplications. At the next stage, there are four copies of G_{256}, requiring $4 \cdot 128 = 512$ more scalar multiplications. Thus at each stage of the factorization, the number of scalar multiplications remains $512 = 2^9$, and there are 10 stages to get down to F_1. So the total scalar multiplication count to calculate $F_{1024}\,\mathbf{y}$ by this factorization method is $10 \cdot 2^9$. By contrast, direct evaluation of $F_{1024}\,\mathbf{y}$ as a matrix-vector product requires 2^{20} scalar multiplications, so the FFT method gives a speedup for multiplications by a factor of $2^{11}/10$ (more than 200 times faster). ∎

In general, the factorization argument given in Example 2.12 shows that to calculate $F_N\mathbf{y}$ by the FFT matrix factorization method when $N = 2^k$ needs at most

$$k2^{k-1} = \frac{1}{2}N \log_2 N \tag{2.35}$$

scalar multiplications. The speedup compared to the N^2 scalar multiplications needed in a direct matrix-vector product is by a factor of $2N/\log_2 N$. For example, when $N = 2^{20}$ this speedup factor is more than 100,000. The same sort of counting of the number of scalar addition operations needed in the FFT gives an upper bound of $N \log_2 N$. Thus the total number of scalar arithmetic operations in the FFT algorithm is bounded by $(3/2)N \log_2 N$, yielding a comparable speedup factor. Without the FFT algorithm digital signal processing would be impractical.

2.7   Computer Explorations

2.7.1   *Fourier matrix and sampling*

This project uses MATLAB to explore the results in Sections 2.2 and 2.3.

---

[12]Multiplying a row vector and column vector, each with $2m$ components, requires $2m$ scalar multiplications followed by $2m - 1$ scalar additions.

(a) Fourier matrix

The MATLAB function `fft` takes the discrete Fourier transform of each column of a matrix argument. Hence for any positive integer N, the command `F = fft(eye(N))` generates the Fourier matrix F_N, since `eye(N)` generates the $N \times N$ identity matrix (whose columns are the standard basis vectors $\mathbf{e}_k$). Check this by the following commands (use the semicolon ; so that the large matrix `F8` doesn't appear on screen).

```
F2 = fft(eye(2)), F4 = fft(eye(4)), F8 = fft(eye(8));
```

Compare the matrices `F2` and `F4` with the examples of Fourier matrices in the Section 2.3. Then use MATLAB to check that the normalized matrix $U = (1/\sqrt{8})F_8$ is unitary by calculating the distance between the matrix UU^H and the identity matrix:

```
norm((1/8)*F8*F8' - eye(8))
```

Remember that the MATLAB notation `A'` means the *conjugate transpose* matrix A^H. The computed norm should be less than 10^{-15}, which is considered as zero for numerical purposes.

(b) Sampling and the Nyquist point

This exploration is based on [Moler (2004), §8.3]. To carry it out you will need the m-file `fftgui.m` (*finite Fourier transform graphic user interface*). This file (and the entire collection of m-files used in Moler's book) can be downloaded from

www.mathworks.com/moler/ncmfilelist.html

or from the publisher's web page for this book:

www.worldscientific.com/worldscibooks/10.1142/9835#t=suppl

If `y` is a column vector with complex entries, then the MATLAB command `fftgui(y)` generates a window with four subplots: The top two plots indicate the real and imaginary parts of `y`, and the bottom two indicate the real and imaginary parts of the discrete Fourier transform `Y = fft(y)`, all plotted as stem graphs (lollipops). Try this with a vector of length 16 with a single 1 in the first entry and the other entries zero:

```
y = zeros(16, 1); y(1) = 1; fftgui(y)
```

Use the uparrow to repeat this, taking `y(2) = 1` and the other entries zero, and then taking `y(3) = 1` and the other entries zero. Describe the plots of `fft(y)` in each case in terms of sampled sine and cosine waves (see Section 2.3).

In these plots consider the points on the horizontal axis numbered from 0 to $N - 1$, where N is the length of `y`. The point numbered $N/2$ on the frequency axis is called the *Nyquist point*. Hence when $N = 16$, then the Nyquist point is the

9th dot from the left. What symmetries around the Nyquist point do the graphs of
`real(fft(y))` and `imag(fft(y))` have when `y` is a real signal? Check that equation
(2.28) holds for these examples.

Now close the `FFTgui` figure window and enter the command `fftgui` at the
MATLAB prompt. The figure window will reappear with all plots initialized to `y
= zeros(32, 1)`. You can use the mouse to move any of the points in the upper
(or lower) two plot windows, and the points in the other plot windows will change
accordingly. Press the `reset` button to return to the initial plot. Experiment by
using the mouse to repeat the three transforms of vectors with only one nonzero
entry.

Reset the `FFTgui` window. Use the mouse in the `real(y)` plot window to draw
a random waveform with large and small oscillations. Since the period is now
$N = 32$, the Nyquist point is the 17th dot from the left. Check that the graphs of
`real(fft(y))` and `imag(fft(y))` have the correct symmetries around this point.

2.7.2 *Applications of the discrete Fourier transform*

These applications are based on [Moler (2004), §8.1 and §8.10]. See Section 1.9 for
details about writing MATLAB function m-files.

(a) Touch-tone dialing

Telephone dialing is an example of everyday use of Fourier analysis. The basis for
touch-tone dialing is the Dual Tone Multi-Frequency (DMF) system. The telephone
dialing pad acts as a 4–by–3 matrix. Associated with each row and column is
a frequency. Each digit in a telephone number is encoded by a signal which is
the sum of two sine waves whose frequencies are the row and column frequencies
associated with the digit. Here is the matrix, with the row frequencies indicated on
the side and the column frequencies on the bottom:

$$
\begin{array}{cccc}
697 & \boxed{1} & \boxed{2} & \boxed{3} \\
770 & \boxed{4} & \boxed{5} & \boxed{6} \\
852 & \boxed{7} & \boxed{8} & \boxed{9} \\
941 & \boxed{*} & \boxed{0} & \boxed{\#} \\
& 1209 & 1336 & 1477
\end{array}
$$

For example, $\boxed{1}$ is encoded by the pair of frequencies 697, 1209, while $\boxed{3}$ is encoded
by the pair 697, 1477. Create the following function m-file to generate dial tones:

```
function y = f(j,k)        % dial tone for button
                           % in row j and column k
fr = [697  770  852  941]; % row frequencies
fc = [1209  1336  1477];   % column frequencies
```

```
Fs = 32768;              % sampling rate 2^15
t = 0:1/Fs:0.25;         % vector of sample points
                         % in time interval (0, 0.25)
y1 = sin(2*pi*fr(j)*t);  % sine wave of
                         % row frequency for button
y2 = sin(2*pi*fc(k)*t);  % sine wave of
                         % column frequency for button
y = (y1 + y2)/2;         % tone generated by button
```

(be sure to put semicolons after each MATLAB command, so that these long vectors are not displayed on the screen). Save this function file under the name tone.m. Test it by typing

```
y = tone(1,1); Fs = 2^15;   sound(y, Fs)
```

If the sound on your computer is turned on, you should hear a 1/4 second touch-tone for the $\boxed{1}$ button. If you get an error message, check your typing and also check that you have set the MATLAB path.

Now use MATLAB to generate a random integer r between 0 and 9 by

```
r = fix(10*rand(1))
```

Determine the row j and column k of the telephone dial pad for your number r, and use MATLAB to generate the corresponding tone

```
y = tone(j,k);   sound(y, Fs)
```

(here you must replace j and k by the appropriate row/column numbers). Plot 16 milliseconds of the tone waveform by

```
t = (1/Fs)*[1:512];
figure, plot(t, y(1:512))
```

The waveform will be a superposition of the sine waves of two different frequencies; however, you cannot determine the frequencies just by looking at the graph. Click on **Insert** in the Figure toolbar, label the horizontal axis as time (seconds) and insert the title DMF Tone for r on your graph (replace r by your particular randomly-generated number).

Now you will use the Discrete Fourier Transform to find the pair of frequencies in your dial tone, and hence the number r that the tone encodes. Create the following MATLAB m-file that will plot the absolute value of the Fourier transform Y of a signal y as a function of frequency over a specified range of frequencies:

```
function powergraph(y, Fs)
n = length(y);
Y = fft(y);
pow = abs(Y);
pmax = norm(pow, inf);
```

```
freq = (0:n-1)*(Fs/n);
figure, plot(freq,pow); axis([500 1700 0 pmax])
```

Save this function file under the name `powergraph.m`. Here `pmax` is the *maximum* of the absolute values of the entries in the vector `pow` (this number is called the ℓ^∞ norm of `pow`).

At the MATLAB prompt type

```
powergraph(y, Fs)
```

This command uses the variables `y` (the signal) and `Fs` (the sampling frequency) that you already defined. The horizontal axis in the graph gives the values of `freq` (the frequency), and the range is the DTMF frequency range. The vertical axis shows the *intensity* (absolute value) of the Fourier transform at each frequency (the square of the intensity is usually called the *power*). The graph should show two sharp peaks, corresponding to the two frequencies used to encode your number. Use the zoom feature (the magnifying glass in the toolbar) to determine the two frequency peaks. Confirm that the pair of frequencies are the correct ones for your original random number. Then restore the graph to the original version and put labels on the graph indicating the frequency peaks, label the horizontal axis frequency (Hertz), and title the graph as Powergraph of DMT Signal for r (replace r by your particular randomly-generated number).

(b) Analyzing a train whistle

Type   `load train`   at the MATLAB prompt. This will give you a long vector `y` and a scalar `Fs` whose value is the number of samples per second. The time increment is `1/Fs` seconds. The command

```
sound(y, Fs)
```

will play the signal—one short and one long pulse of a train whistle. Generate the time vector `t` and plot the signal as a function of time by the commands

```
n = length(y); t = (1/Fs)*[1:n];
figure, plot(t,y)
```

The graph clearly shows the two short-long whistle pulses, but it gives no information about the frequencies of the tones in the whistle. This is the graph in Figure 2.7.

Now use the function file `powergraph` that you wrote in part *(a)* to obtain part of the graph in Figure 2.8. You can use your graph to find the dominant frequencies in the whistle. Notice that the powergraph gives detailed frequency information about the whistle, but it does not show that there were two pulses (short followed by long).

Zoom in on the peaks in the powergraph to find approximate values of the three main frequencies less than 1,250 Hertz in the whistle. The characteristic mournful sound of the whistle is due to the fact that the ratio of each pair of frequencies is very close to a rational number p/q, where $p < q$ are small positive integers (5 or less), and when sounded together these frequencies give a minor triad. Find these ratios (one of them is 3/5; in musical terminology a major sixth).

2.7.3   Circulant matrices and circular convolution

This section uses MATLAB to explore the results in Sections 2.4 and 2.5.

(a) Shift operator

Create the following MATLAB function m-file that generates the $n \times n$ *shift matrix* S:

```
function S = shift(n)
S = zeros(n,n); S(1,n) = 1;
for k = 1:n-1
S(k+1,k) = 1;
end
```

Save this function file as `shift.m`. Then test it by typing `S = shift(4)`. You should get the 4×4 shift matrix (the 3×3 shift matrix is shown in Example 2.5). Generate a random integer vector **v** by

```
v = fix(10*rand(4,1))
```

Then calculate `S*v` to see that S shifts the components of **v** cyclically.

By Theorem 2.3 the columns $\mathbf{E}_0, \ldots, \mathbf{E}_3$ of F_4^{H} are eigenvectors for S, where F_4 is the Fourier matrix. By Theorem 2.1 the inverse of F_4 is $(1/4)F_4^{\mathrm{H}}$. Hence `(1/4)*F4*S*F4'` should be a diagonal matrix (remember that the conjugate-transpose matrix A^{H} is given in MATLAB by `A'`). Check this by MATLAB. From the MATLAB calculation, what are the eigenvalues of S? Explain how your answer is predicted by Theorem 2.3.

(b) Circulant matrices

Use your random vector v from *(a)* and powers of the matrix S to construct a 4×4 *circulant matrix* C whose first column is v (see Theorem 2.2).

(i) Since the eigenvectors of S are also eigenvectors for C, the matrix `(1/4)*F4*C*F4'` should be in diagonal form. Check this by MATLAB.
(ii) From the MATLAB calculation, what are the eigenvalues of C?
(iii) How are the eigenvalues of C obtained from the eigenvalues of S? Use Theorem 2.4 and a hand calculation to obtain the same values as in (ii).

(c) Circular convolution

Consider the vector space of real periodic signals of length 32. The value of the signal $\mathbf{y}$ at discrete time j is denoted by $\mathbf{y}[j]$. The periodicity means that $\mathbf{y}[j+32] = \mathbf{y}[j]$ for all integers j. For purposes of calculation identify $\mathbf{y}$ with the column vector

$$\begin{bmatrix} \mathbf{y}[0] \\ \mathbf{y}[1] \\ \vdots \\ \mathbf{y}[31] \end{bmatrix} \in \mathbb{R}^{32}$$

(see Section 2.4.1). Let C be the *three-step moving average* operator defined by

$$C\mathbf{y}[j] = (1/3)\big(\mathbf{y}[j-1] + \mathbf{y}[j] + \mathbf{y}[j+1]\big) \quad \text{for } j = 0, 1, \ldots, 31 . \tag{2.36}$$

Notice that when $j = 0$ or $j - 31$ the periodicity of $\mathbf{y}$ is used in this definition. You will study the effect of C and C^2 on a signal.

(i) Find the periodic function $\mathbf{f}[j]$ of period 32 such that $C\mathbf{y}[j] = (\mathbf{f} \star \mathbf{y})[j]$, where $\star$ means circular convolution (see Definition 2.4). The function $\mathbf{f}$ is called the *filter*.

(ii) Write out by hand an explicit symbolic formula for $C^2\,\mathbf{y}[0]$ in terms of the values $\mathbf{y}[0], \ldots, \mathbf{y}[31]$. How many sample values of $\mathbf{y}$ are used to calculate $C^2\,\mathbf{y}[0]$?

(iii) Use your formula from (ii) and the shift-invariance property to obtain an explicit formula for $C^2\mathbf{y}[j]$ from the formula for $C^2\mathbf{y}[0]$.

Use MATLAB to create a 32×32 circulant matrix C which acts on column vectors y according to (2.36), as follows: Generate the shift operator S = shift(32) and the identity matrix I = eye(32); then write C as a linear combination of I and powers of S as in equation (2.24); in this case the inverse of S is S^31. Be sure to put a semicolon after each MATLAB command so that these matrices are not displayed on screen.

Generate a time vector t, a random signal y, and plot the signal signal as a solid red line:

```
t = [0:31]; y = rand(32,1);
figure, plot(t,y, 'r-'); hold on
```

Compare the signal with its three-step moving average (plotted in blue with a dotted line):

```
plot(t, C*y, 'b:')
```

Now compare with a double three-step averaging (plotted in green with a dashed line):

```
plot(t, C*C*y, 'g--')
```

Notice that the averaging process acts as a *lowpass filter* to reduce the variability in the signal: the graph of $C\,\mathbf{y}$ has less oscillation than the graph of $\mathbf{y}$, and the graph of $C^2\,\mathbf{y}$ has even less oscillation. Put the title Random Signal with 3-Step Moving Averages on the graph. Insert arrows and labels to identify the signal $\mathbf{y}$ and the moving averages $C\mathbf{y}$ and $C^2\mathbf{y}$.

2.7.4 *Fast Fourier transform*

This section uses MATLAB to explore the results in Section 2.6.

(a) Downsampling

The first step in the FFT (and a step in wavelet analysis) is to split a signal vector $\mathbf{y}$ (of length $2n$) into two vectors $\mathbf{y}_{\text{even}}$ and $\mathbf{y}_{\text{odd}}$, each of length n. This is called *downsampling*. The following function m-file generates the permutation matrix P_{2n} that does this (see Section 2.6 for the case $n = 2$):

```
function P = downsamp(n)    % Downsampling matrix size 2n x 2n
P = zeros(2*n, 2*n);        % Create 2n x 2n matrix of zeros
for j = 1:n
    P(j, 2*j - 1) = 1;      % Sort entries 1, 3, ..., 2n-1
                            % into rows 1,..., n
    P(n + j, 2*j) = 1;      % Sort entries 2, 4, ..., 2n
                            % into rows n+1, ..., 2n
end
```

Create and save this function file under the name downsamp.m. Test it by creating a random vector y = fix(20*rand(8,1)) and the 8×8 downsampling matrix P = downsamp(4). Calculate P*y and check that the entries in positions 1, 3, 5, and 7 of y are the first four entries in P*y, and the entries in positions 2, 4, 6, and 8 of y are the last four entries in P*y.

(b) Block decomposition of the Fourier matrix

The block form in the FFT uses a diagonal matrix D_n (see equation (2.33)) that is generated by the following function m-file:

```
function D = fftdiag(n)
D = zeros(n,n);
w = exp(pi*i/n);
for j = 1:n
D(j, j) = w^(-j+1) ;
end
```

Create and save this function file under the name fftdiag.m.

With the two function files you have made you can now create the Fourier matrix F_4 from the Fourier matrix F_2, and then the Fourier matrix F_8 from F_4. Type

```
P4 = downsamp(2); D2 = fftdiag(2); F2 = fft(eye(2));
F4 = [F2  D2*F2; F2  -D2*F2]*P4
```

Compare F4 with the Fourier matrix F_4 (see Example 2.11). Now repeat the process to generate the 8×8 Fourier matrix by first generating P8 = downsamp(4) and D4 = fftdiag(4). Use these and the matrix F4 you just generated to obtain the 8×8 Fourier matrix F8 (see equation (2.34)). Compare with the MATLAB-generated Fourier matrix by calculating the distance norm(F8 - fft(eye(8))) between the two matrices. This should be of size 10^{-15}, thus zero to the limits of machine computation.

(c) Speed comparison

Generate the Fourier matrix F_N for $N = 2^{12} = 4096$ and a random vector $\mathbf{x} \in \mathbb{R}^{4096}$ by the following commands:

```
F = fft(eye(4096)); x = rand(4096,1);
```

Calculate the computation time to obtain Fourier transform of $\mathbf{x}$ in two ways: first by direct matrix multiplication and then by the fast Fourier transform (be sure to type the semicolons as indicated, so the vectors will not appear on the screen), as follows.

```
tic; F*x; matrixtime = toc
tic; fft(x); ffttime = toc
speedup = floor(matrixtime/ffttime)
```

Now compare your observed speedup using the FFT with the theoretical prediction. If $N = 2^k$, then computing the matrix $\times$ vector F*x requires $N^2 = 2^{2k}$ scalar multiplications. Computing fft(x) needs at most $k2^{k-1}$ scalar multiplications by Equation (2.35). The theoretical speed advantage of the FFT over the plain (matrix $\times$ vector) multiplication, in terms of the scalar multiplications required, is thus

$$2^{2k}/[k2^{k-1}] = 2^{k+1}/k . \qquad (\star)$$

A similar speed advantage holds for the scalar additions required (see the end of Section 2.6). Compute the theoretical speed advantage $(\star)$ for the value of k that you have used, and compare it with the observed value speedup. Scalar multiplications require much more computing time than scalar additions, so it is the reduction in the number of multiplications that is crucial for the speedup given by the FFT.

2.8   Exercises

(1) Let $\mathbf{C}_k$ be the vector in $\mathbb{R}^N$ obtained by sampling the function $\cos(2\pi kt)$ at the points $t = 0, 1/N, 2/N, \ldots, (N-1)/N$. Let $\mathbf{S}_k \in \mathbb{R}^N$ be similarly defined by sampling the function $\sin(2\pi kt)$. Let $\mathbf{E}_k \in \mathbb{C}^N$ be defined by (2.3).

(a) Prove the following vector analogs of equations (2.2) and (1.34):

$$\mathbf{E}_k = \mathbf{C}_k + i\mathbf{S}_k , \quad \mathbf{C}_k = \tfrac{1}{2}(\mathbf{E}_k + \overline{\mathbf{E}}_k) , \quad \mathbf{S}_k = \tfrac{1}{2i}(\mathbf{E}_k - \overline{\mathbf{E}}_k) .$$

In particular, $\mathbf{C}_0 = \mathbf{E}_0$ and $\mathbf{S}_0 = \mathbf{0}$.

(b) Prove the aliasing relation $\mathbf{E}_{N-k} = \overline{\mathbf{E}}_k$ and use it and **(a)** to obtain the aliasing relations $\mathbf{C}_{N-k} = \mathbf{C}_k$ and $\mathbf{S}_{N-k} = -\mathbf{S}_k$. In particular, when $N = 2m$ is even, then $\mathbf{S}_m = -\mathbf{S}_m$, and so $\mathbf{S}_m = \mathbf{0}$.

(c) Calculate the vectors $\mathbf{C}_k$ and $\mathbf{S}_k$ for $k = 0, 1, 2, 3$ when $N = 4$ and check that they satisfy the relations in (b).

(2) Suppose that $f(t) = \sum_{-m \le k \le m} c_k f_k(t)$ is a trigonometric polynomial with frequencies at most m, where $f_k(t) = e^{2\pi i k t}$. Let $\mathbf{y} \in \mathbb{C}^N$ be the vector obtained by sampling $f(t)$ at $t = 0, 1/N, 2/N, \ldots, (N-1)/N$.

(a) Show that $\mathbf{y} = \sum_{-m \le k \le m} c_k \mathbf{E}_k$ with $\mathbf{E}_k$ defined by (2.3).

(b) Suppose $N > 2m$. Show that $c_k = (1/N)\langle \mathbf{E}_k , \mathbf{y} \rangle$. Conclude that the analog signal $f(t)$ is uniquely determined by the sampled digital signal $\mathbf{y}$ in this case and there is no aliasing. This proves the *Nyquist sampling criterion*.

(c) Suppose $N \le 2m$. Give an example of a nonzero trigonometric polynomial $f(t)$ with frequencies at most m such that $\mathbf{y} = \mathbf{0}$.

(3) Let F_N be the $N \times N$ Fourier matrix and set $T = (F_N)^2$.

(a) Calculate T and T^2 when $N = 4$.

(b) Enumerate the standard basis for $\mathbb{C}^N$ as $\mathbf{e}_0, \ldots, \mathbf{e}_{N-1}$. Show that $(1/N)T\mathbf{e}_0 = \mathbf{e}_0$ and $(1/N)T\mathbf{e}_k = \mathbf{e}_{N-k}$ for $k = 1, \ldots, N-1$.

(c) Consider the discrete Fourier transform on N-periodic functions (see Definition 2.27). If $\mathbf{y} \in \ell_{\mathbb{C}}^2[\mathbb{Z}/N\mathbb{Z}]$ and $\mathbf{g} = \widehat{\mathbf{y}}$, show that $\widehat{\mathbf{g}}[k] = N\mathbf{y}[-k]$ for all $k \in \mathbb{Z}$.

(d) Use (b) to show that $(F_N)^4 = N^2 I_N$.

(e) When $N \ge 4$ then the eigenvalues λ and their multiplicities for the matrix $(1/\sqrt{N})F_N$ are given by in Table 2.1 (the multiplicities add up to N since F_N is diagonalizable).[13] Use (d) to explain why the only possible eigenvalues are ± 1 and $\pm i$. Use the MATLAB command `fft(eye(4))` to generate F_4 and the command `eig` to verify the table entries for the case $N = 4$.

Table 2.1   Eigenvalues of unitary Fourier matrix

N	$\lambda = +1$	$\lambda = i$	$\lambda = -1$	$\lambda = -i$
$4m$	$m + 1$	$m - 1$	m	m
$4m + 1$	$m + 1$	m	m	m
$4m + 2$	$m + 1$	m	$m + 1$	m
$4m + 3$	$m + 1$	m	$m + 1$	$m + 1$

(4) Let T be the linear transformation $T\mathbf{y}[j] = \mathbf{y}[j-1] - 2\mathbf{y}[j] + \mathbf{y}[j+1]$ where $\mathbf{y}$ is a periodic function of period N.

---

[13]See the Wikipedia article on the *Discrete Fourier Transform* for more information.

(a) Find the N-periodic function $\mathbf{f}$ such that $T\mathbf{y} = \mathbf{f} \star \mathbf{y}$. (HINT: Write T in terms of the shift operator.)

(b) Find the Fourier transforms $\widehat{\mathbf{f}}$ and $\widehat{\mathbf{f} \star \mathbf{y}}$ of the functions in (a). (HINT: Use Theorem 2.5.)

(c) Suppose $N = 4$. Give the 4×4 circulant matrix C corresponding to T. What are the eigenvalues of C? How are they related to the values of $\widehat{\mathbf{f}}$?

(d) Suppose $N = 4$ and $\mathbf{y}$ corresponds to the vector $\begin{bmatrix} 2\ 3\ 1\ 5 \end{bmatrix}^{\mathrm{T}} \in \mathbb{C}^4$. Use the matrix C from part (c) to calculate the vectors corresponding to $T\mathbf{y}$, $\widehat{\mathbf{y}}$, and $\widehat{T\mathbf{y}}$. Compare with (b).

(5) Let $S = \begin{bmatrix} 0\ 0\ 1 \\ 1\ 0\ 0 \\ 0\ 1\ 0 \end{bmatrix}$ be the matrix for the shift operator relative to the standard basis for $\mathbb{C}^3$. Suppose the matrix $C = \begin{bmatrix} 4\ *\ * \\ 7\ *\ * \\ 5\ *\ * \end{bmatrix}$ satisfies $CS = SC$.

(a) Write C as a polynomial in S and fill in the entries $*$ in C.

(b) View a vector $\mathbf{y} = \begin{bmatrix} \mathbf{y}[0]\ \mathbf{y}[1]\ \mathbf{y}[2] \end{bmatrix}^{\mathrm{T}} \in \mathbb{C}^3$ as a 3-periodic function on the integers by setting $\mathbf{y}[j] = \mathbf{y}[j + 3]$. Let T be the linear transformation on 3-periodic functions corresponding to the matrix C above. Give explicit formulas (in terms of $\mathbf{y}[0]$, $\mathbf{y}[1]$, and $\mathbf{y}[2]$) for $T\mathbf{y}[j]$ for $j = 0, 1, 2$.

(c) Find the eigenvalues of C in terms of ω and ω^2.

(6) Let $\mathbf{f}$ be the 3-periodic function on the integers defined by $\mathbf{f}[0] = 4$, $\mathbf{f}[1] = 5$, $\mathbf{f}[2] = 6$, and $\mathbf{f}[k + 3] = f[k]$. Let $\mathbf{g}$ be another 3-periodic function on the integers and let $\mathbf{h} = \mathbf{f} \star \mathbf{g}$ (circular convolution).

(a) Express $\mathbf{h}[0]$, $\mathbf{h}[1]$, and $\mathbf{h}[2]$ in terms of $\mathbf{g}[0]$, $\mathbf{g}[1]$, and $\mathbf{g}[2]$.

(b) Let $\mathbf{y} = \begin{bmatrix} \mathbf{g}[0] \\ \mathbf{g}[1] \\ \mathbf{g}[2] \end{bmatrix}$ and $\mathbf{z} = \begin{bmatrix} \mathbf{f} \star \mathbf{g}[0] \\ \mathbf{f} \star \mathbf{g}[1] \\ \mathbf{f} \star \mathbf{g}[2] \end{bmatrix}$. Find the matrix C such that $C\mathbf{y} = \mathbf{z}$.

(7) Let C be an $N \times N$ circulant matrix.

(a) Show that C^{H} is also a circulant matrix. (HINT: Show that C^{H} commutes with the shift matrix.)

(b) If the first column of C is $\begin{bmatrix} c_0 & c_1 & c_2 & \cdots & c_{N-1} \end{bmatrix}^{\mathrm{T}}$, show that the first column of C^{H} is $\begin{bmatrix} \overline{c_0} & \overline{c_{N-1}} & \cdots & \overline{c_2} & \overline{c_1} \end{bmatrix}^{\mathrm{T}}$.

(8) Let $\mathbf{f}$, $\mathbf{g}$, and $\mathbf{h}$ be N-periodic functions on the integers.

(a) Let $C_{\mathbf{f}}$ and $C_{\mathbf{g}}$ be the $N \times N$ circulant matrices corresponding to $\mathbf{f}$ and $\mathbf{g}$, respectively. Show that $C_{\mathbf{f}} C_{\mathbf{g}} = C_{\mathbf{f} \star \mathbf{g}}$ (matrix product on the left and circular convolution on the right).

(b) Show that circular convolution is associative: $(\mathbf{f} \star \mathbf{g}) \star \mathbf{h} = \mathbf{f} \star (\mathbf{g} \star \mathbf{h})$.

(c) Show that circular convolution multiplication is commutative: $\mathbf{f} \star \mathbf{g} = \mathbf{g} \star \mathbf{f}$.

Chapter 3

Discrete Wavelet Transforms

3.1 Overview

The discrete Fourier transform from Chapter 2 is the tool of choice for analyzing the *frequency* content of a digital signal, as we illustrated by the train whistle (Example 2.10). We now introduce *one-scale wavelet transforms* that separate a signal into two parts (subsignals), each with half as many components as the original signal: a long-term *trend*, such as a steadily oscillating wave, and a short-term *detail* or *fluctuation*, such as a spike or random noise. Thus for wavelet transforms we distinguish between long-term and short-term events in the time domain, rather than low frequencies and high frequencies.

To obtain a *multiple-scale wavelet transform* we repeat the one-scale transformation on the trend subsignal to obtain shorter-term trend and detail subsignals. This process, called the *pyramid algorithm*, gives the *multiresolution representation* of a signal. We illustrate an important application of wavelet transforms that is not possible with the discrete Fourier transform, namely to signal compression and noise removal.

Although there is only one discrete Fourier transform, there is a vast collection of discrete wavelet transforms that have been created to achieve better signal processing results. In this chapter we begin with the Haar transform, and then present two recent examples (the CDF(2, 2) and Daub4 transforms) in detail. The transforms are obtained via the *lifting method* by first splitting the digital signal into even and odd subsignals, and then applying successive *prediction* and *update* elementary transformations. This matrix factorization approach furnishes a computationally fast implementation of discrete wavelet transforms. The origin of the formulas used in these transforms is deferred to Chapter 4 for the sake of giving many examples and signal-processing applications in this chapter.

The matrix for a one-scale discrete wavelet transform for analysis (or synthesis) has some resemblance to a circulant matrix. It has two blocks (one for trend and one for detail), and each block is determined from its first row (or column) by successive shifts by *two* positions. These double shifts come from the downsampling process already encountered in the fast Fourier transform algorithm in Chapter 2.

73

For two-dimensional digital images we obtain a discrete wavelet transform by operating simultaneously on the rows and columns of the matrix that encodes the image. Left multiplication by a wavelet matrix used for a one-dimensional signal transforms the columns of the image matrix, while right multiplication by the transpose of the same wavelet matrix transforms the rows. The resulting transformed matrix is thus partitioned into four blocks. One block is the trend (low-resolution version of the image), as for one-dimensional signals. The other three blocks emphasize the horizontal, vertical, and diagonal details (the one-dimensional edges in the image). To obtain a multiscale 2D wavelet transform we repeat this left and right multiplication procedure on the trend block. Multiscale wavelet transforms give an effective tool for compression and edge detection for image files.

3.2   Haar Wavelet Transform for Digital Signals

We use an example from [Mulcahy (1996)] to introduce the multiresolution decomposition for finite digital signals by the *Haar wavelet transform*.[1] Then we show how the calculations can be viewed as *prediction* and *update* linear transformations.

3.2.1   *Basic example*

Consider the digital signal $\mathbf{s}_3[n]$ with 2^3 nonzero values plotted in Figure 3.1 from the data in Table 3.1. We use a piecewise-linear plot, rather than a stem plot as in Chapters 1 and 2, to show the fluctuations in a signal more clearly and as a reminder that digital signals are used to model analog signals. We will analyze analog signals directly by wavelet methods in Chapter 5.

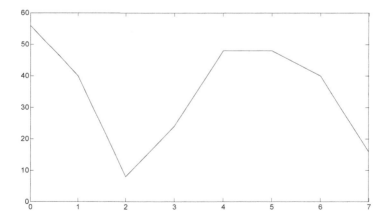

Fig. 3.1   Level-three signal

---

[1]The Hungarian mathematician A. Haar introduced this type of multiresolution analysis for analog signals in 1910; we will study it in detail in Chapter 5.

Table 3.1   Signal

n :	0	1	2	3	4	5	6	7
$\mathbf{s}_3[n]$:	56	40	8	24	48	48	40	16

There are two features of the signal that we want to analyze:

(1) The overall *trend* of the signal. For example, there is no change in the signal when n goes from 4 to 5. Thus the value at $n = 4$ is a good predictor of the value at $n = 5$.
(2) The *detail* (also called the *fluctuation*) in the signal. For example, there is a large change in the signal between $n = 0$ and $n = 1$. The value of the signal at $n = 0$ is not an accurate predictor of the value at $n = 1$.

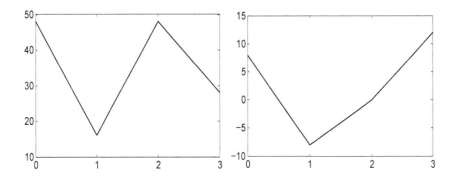

Fig. 3.2   Level-two trend (left) and detail (right) of level-three signal

We will adopt the strategy of forming a simplified version of the signal using the average value at adjacent points $2n$, $2n + 1$. This gives the *trend* of the signal. The fluctuations in the signal can then be measured by how much the actual signal value at $2n$ differs from the average value. This difference gives the *detail* in the signal.

Table 3.2   Level-two trend/detail

n :	0	1	2	3	n :	0	1	2	3
$(\mathbf{s}_3)_{\text{even}}[n]$:	56	8	48	40	$(\mathbf{s}_3)_{\text{odd}}[n]$:	40	24	48	16
$\mathbf{s}_2[n]$:	48	16	48	28	$\mathbf{d}_2[n]$:	8	-8	0	12

To carry out the computations we *split* the signal into two half-length signals

$$(\mathbf{s}_3)_{\text{even}}[n] = \mathbf{s}_3[2n] \quad \text{and} \quad (\mathbf{s}_3)_{\text{odd}}[n] = \mathbf{s}_3[2n + 1]$$

by taking every other value of $\mathbf{s}_3$ (this is also called *downsampling*).
We then calculate the *level-two trend*

$$\mathbf{s}_2[n] = \frac{1}{2}\Big\{(\mathbf{s}_3)_{\text{even}}[n] + (\mathbf{s}_3)_{\text{odd}}[n]\Big\}$$

(the average of the values of $\mathbf{s}_3$ at $2n$ and $2n+1$) and the *level-two detail*

$$\mathbf{d}_2[n] = (\mathbf{s}_3)_{\text{even}}[n] - \mathbf{s}_2[n]$$

(the difference between the value of $\mathbf{s}_3$ at $2n$ and the average of the values of $\mathbf{s}_3$ at $2n$ and $2n+1$). These signals of length 2^2 are given in Table 3.2. Notice the value $\mathbf{d}_2[2] = 0$ corresponding to $\mathbf{s}_3[4] = \mathbf{s}_3[5]$ and the large value $\mathbf{d}_2[3] = 12$ corresponding to the big difference between $\mathbf{s}_3[6]$ and $\mathbf{s}_3[7]$. The level-two trend and the level-two detail are plotted in Figure 3.2 as piecewise linear curves.

We can repeat this analysis on the trend vector $\mathbf{s}_2$. Define the *level-one trend*

$$\mathbf{s}_1[n] = \frac{1}{2}\left\{(\mathbf{s}_2)_{\text{even}}[n] + (\mathbf{s}_2)_{\text{odd}}[n]\right\}$$

and the *level-one detail*

$$\mathbf{d}_1[n] = (\mathbf{s}_2)_{\text{even}}[n] - \mathbf{s}_1[n].$$

These signals of length 2^1 are given in Table 3.3.

Table 3.3   Level-one trend/detail

n :	0	1	n :	0	1
$(\mathbf{s}_2)_{\text{even}}[n]$:	48	48	$(\mathbf{s}_2)_{\text{odd}}[n]$:	16	28
$\mathbf{s}_1[n]$:	32	38	$\mathbf{d}_1[n]$:	16	10

Finally, we repeat this analysis on the trend vector $\mathbf{s}_1$. Define the *level-zero trend*

$$\mathbf{s}_0[n] = \frac{1}{2}\left\{(\mathbf{s}_1)_{\text{even}}[n] + (\mathbf{s}_1)_{\text{odd}}[n]\right\}$$

and the *level-zero detail*

$$\mathbf{d}_0[n] = (\mathbf{s}_1)_{\text{even}}[n] - \mathbf{s}_0[n].$$

These signals of length 2^0 are given in Table 3.4. We can assemble this three-level decomposition into the *multiresolution analysis* of the original signal in Table 3.5.

Table 3.4   Level-zero trend/detail

n :	0	n :	0
$(\mathbf{s}_1)_{\text{even}}[n]$:	32	$(\mathbf{s}_1)_{\text{odd}}[n]$:	38
$\mathbf{s}_0[n]$:	35	$\mathbf{d}_0[n]$:	-3

The flow-chart to calculate the multiresolution analysis is shown in Figure 3.3. Note that details get saved at each level and only the trends go to the top. We shall return to this *pyramid algorithm* in Section 3.3.

Table 3.5   Multiresolution analysis

$\mathbf{s}_0$	$\mathbf{d}_0$	$\mathbf{d}_1$		$\mathbf{d}_2$			
35	-3	16	10	8	-8	0	12

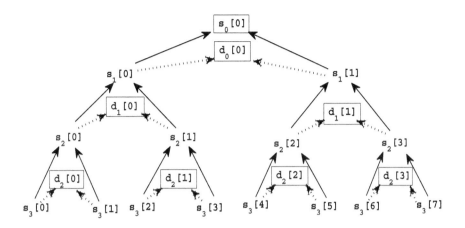

Fig. 3.3   Three-scale Haar analysis pyramid for signal of length 8

In the multiresolution analysis the first entry $s_0[0]$ is the *average* of the original signal: $35 = (56 + 40 + 8 + 24 + 48 + 48 + 40 + 16)/8$. This property is clear from the chain of equations

$$s_0[0] = \frac{1}{2}\sum_{n=0}^{1} s_1[n] = \frac{1}{4}\sum_{n=0}^{3} s_2[n] = \frac{1}{8}\sum_{n=0}^{7} s_3[n].$$

3.2.2   *Prediction and update transformations*

We now introduce some linear algebra to put the calculations of Section 3.2.1 into the general framework of the Haar wavelet transform. Suppose $\mathbf{x}$ is a real-valued signal of even length $N = 2k$ with values $\mathbf{x}[0], \mathbf{x}[1], \ldots, \mathbf{x}[N-1]$. We identify $\mathbf{x}$ with the $N \times 1$ column vector with these components (notice that we are indexing the components from 0 to $N-1$, whereas in MATLAB the indexing would go from 1 to N). Following the same pattern as for the fast Fourier transform, we split (downsample) $\mathbf{x}$ into even and odd subsignals of length $N/2$:

$$\mathbf{x}_{\text{even}}[n] = \mathbf{x}[2n] \quad \text{and} \quad \mathbf{x}_{\text{odd}}[n] = \mathbf{x}[2n+1] \quad \text{for } n = 0, 1, \ldots, N/2 - 1.$$

These formulas define a linear transformation that we denote by $\boxed{\text{split}}$:

$$\boxed{\text{split}}\,\mathbf{x} = \begin{bmatrix} \mathbf{x}_{\text{even}} \\ \mathbf{x}_{\text{odd}} \end{bmatrix}.$$

When we make $\mathbf{x}$ into a column vector the transformation $\boxed{\text{split}}$ is given by the permutation matrix P_N in Section 2.6. We define $\boxed{\text{merge}}$ to be the inverse permutation matrix (the transpose of $\boxed{\text{split}}$):

$$\boxed{\text{merge}} \begin{bmatrix} \mathbf{x}_{\text{even}} \\ \mathbf{x}_{\text{odd}} \end{bmatrix} = \mathbf{x}.$$

Define the *trend* vector $\mathbf{s}$ and *detail* vector $\mathbf{d}$ of $\mathbf{x}$ just as in Section 3.2.1 by

$$\mathbf{s} = \frac{1}{2}\left(\mathbf{x}_{\text{even}} + \mathbf{x}_{\text{odd}}\right) \quad \text{and} \quad \mathbf{d} = \mathbf{x}_{\text{even}} - \mathbf{s}$$

(notice that $\mathbf{s}$ and $\mathbf{d}$ are vectors with $N/2$ components). Let $\mathbf{T_a}$ be the linear transformation on $\mathbb{R}^N$ given by

$$\mathbf{T_a x} = \begin{bmatrix} \mathbf{s} \\ \mathbf{d} \end{bmatrix}$$

(the subscript $\mathbf{a}$ is for *analysis*). We can rewrite the formulas for $\mathbf{s}$ and $\mathbf{d}$ as

$$2\mathbf{d} = \mathbf{x}_{\text{even}} - \mathbf{x}_{\text{odd}} \quad \text{and} \quad \mathbf{s} = \mathbf{x}_{\text{even}} - \mathbf{d}.$$

We use these formulas to factor $\mathbf{T_a}$ as a product of $\boxed{\text{split}}$ and the following upper and lower triangular matrices (where I denotes the $N/2 \times N/2$ identity matrix and $\mathbf{0}$ the $N/2 \times N/2$ zero matrix):

$$\begin{bmatrix} \mathbf{x}_{\text{even}} \\ -2\mathbf{d} \end{bmatrix} = \begin{bmatrix} \mathbf{x}_{\text{even}} \\ \mathbf{x}_{\text{odd}} - \mathbf{x}_{\text{even}} \end{bmatrix} = \begin{bmatrix} I & 0 \\ -I & I \end{bmatrix} \begin{bmatrix} \mathbf{x}_{\text{even}} \\ \mathbf{x}_{\text{odd}} \end{bmatrix}, \qquad \text{(Prediction)}$$

$$\begin{bmatrix} \mathbf{s} \\ -2\mathbf{d} \end{bmatrix} = \begin{bmatrix} \mathbf{x}_{\text{even}} - \mathbf{d} \\ -2\mathbf{d} \end{bmatrix} = \begin{bmatrix} I & \frac{1}{2}I \\ 0 & I \end{bmatrix} \begin{bmatrix} \mathbf{x}_{\text{even}} \\ -2\mathbf{d} \end{bmatrix}, \qquad \text{(Update)}$$

$$\begin{bmatrix} \mathbf{s} \\ \mathbf{d} \end{bmatrix} = \begin{bmatrix} \mathbf{s} \\ \frac{1}{2}(2\mathbf{d}) \end{bmatrix} = \begin{bmatrix} I & 0 \\ 0 & \frac{1}{2}I \end{bmatrix} \begin{bmatrix} \mathbf{s} \\ 2\mathbf{d} \end{bmatrix}. \qquad \text{(Normalization)}$$

We define $N \times N$ matrices P, U, and D (in 2×2 block form) by

$$P = \begin{bmatrix} I & 0 \\ -I & I \end{bmatrix}, \quad U = \begin{bmatrix} I & \frac{1}{2}I \\ 0 & I \end{bmatrix}, \quad \text{and} \quad D = \begin{bmatrix} I & 0 \\ 0 & -\frac{1}{2}I \end{bmatrix}.$$

Then

$$\mathbf{T_a x} = DUP\,\boxed{\text{split}}\,\mathbf{x}$$

(notice the order of multiplication of the factors). Thus we have factored $\mathbf{T}_a$ as the product of matrices that just do elementary row operations. These elementary matrices are invertible and the inverses have the same triangular form:

$$P^{-1} = \begin{bmatrix} I & 0 \\ I & I \end{bmatrix}, \quad U^{-1} = \begin{bmatrix} I & -\frac{1}{2}I \\ 0 & I \end{bmatrix}, \quad \text{and} \quad D^{-1} = \begin{bmatrix} I & 0 \\ 0 & -2I \end{bmatrix}.$$

Hence the factorization shows that $\mathbf{T_a}$ is invertible with inverse

$$\mathbf{T_a}^{-1} = \boxed{\text{merge}}\,P^{-1}U^{-1}D^{-1}.$$

We call the inverse matrix the *synthesis matrix* and write $\mathbf{T_a}^{-1} = \mathbf{T_s}$. For any signal $\mathbf{x}$ of length N, we set $\mathbf{y} = \mathbf{T_a x}$. Then we can reconstruct $\mathbf{x}$ from $\mathbf{y}$ by $\mathbf{x} = \mathbf{T_s y}$.

Using block multiplication of matrices, we calculate

$$\mathbf{T_a} = DUP\,\boxed{\text{split}} = \frac{1}{2} \begin{bmatrix} I & I \\ I & -I \end{bmatrix} \boxed{\text{split}} \qquad (3.1)$$

and

$$\mathbf{T_s} = \boxed{\text{merge}} \, P^{-1} U^{-1} D^{-1} = \boxed{\text{merge}} \begin{bmatrix} I & I \\ I & -I \end{bmatrix}. \tag{3.2}$$

We call $\mathbf{T_a}$ and $\mathbf{T_s}$ the *one-scale Haar analysis* and *one-scale Haar synthesis* matrices. Notice that $2\mathbf{T_a^T} = \mathbf{T_s}$, so that

$$(\sqrt{2}\mathbf{T_a})(\sqrt{2}\mathbf{T_a})^T = \mathbf{T_s}\mathbf{T_a} = I.$$

Thus the *normalized Haar matrix* $\sqrt{2}\mathbf{T_a}$ is orthogonal.

Example 3.1. If $N = 4$, then

$$\mathbf{T_a} = \frac{1}{2} \begin{bmatrix} 1 & 0 & 1 & 0 \\ 0 & 1 & 0 & 1 \\ 1 & 0 & -1 & 0 \\ 0 & 1 & 0 & -1 \end{bmatrix} \begin{bmatrix} 1 & 0 & 0 & 0 \\ 0 & 0 & 1 & 0 \\ 0 & 1 & 0 & 0 \\ 0 & 0 & 0 & 1 \end{bmatrix} = \frac{1}{2} \begin{bmatrix} 1 & 1 & 0 & 0 \\ 0 & 0 & 1 & 1 \\ 1 & -1 & 0 & 0 \\ 0 & 0 & 1 & -1 \end{bmatrix}$$

and

$$\mathbf{T_s} = \begin{bmatrix} 1 & 0 & 0 & 0 \\ 0 & 0 & 1 & 0 \\ 0 & 1 & 0 & 0 \\ 0 & 0 & 0 & 1 \end{bmatrix} \begin{bmatrix} 1 & 0 & 1 & 0 \\ 0 & 1 & 0 & 1 \\ 1 & 0 & -1 & 0 \\ 0 & 1 & 0 & -1 \end{bmatrix} = \begin{bmatrix} 1 & 0 & 1 & 0 \\ 1 & 0 & -1 & 0 \\ 0 & 1 & 0 & 1 \\ 0 & 1 & 0 & -1 \end{bmatrix}.$$

The calculation of $\mathbf{s}_1$ and $\mathbf{d}_1$ from $\mathbf{s}_2$ in Section 3.2.1 can then be obtained as the matrix-vector product

$$\begin{bmatrix} \mathbf{s}_1 \\ \mathbf{d}_1 \end{bmatrix} = \mathbf{T_a s_2} = \frac{1}{2} \begin{bmatrix} 1 & 1 & 0 & 0 \\ 0 & 0 & 1 & 1 \\ 1 & -1 & 0 & 0 \\ 0 & 0 & 1 & -1 \end{bmatrix} \begin{bmatrix} 48 \\ 16 \\ 48 \\ 28 \end{bmatrix} = \begin{bmatrix} 32 \\ 38 \\ 16 \\ 10 \end{bmatrix}.$$

In the opposite direction, we can obtain $\mathbf{s}_2$ from $\mathbf{s}_1$ and $\mathbf{d}_1$ by

$$\mathbf{s_2} = \mathbf{T_s} \begin{bmatrix} \mathbf{s}_1 \\ \mathbf{d}_1 \end{bmatrix} = \begin{bmatrix} 1 & 0 & 1 & 0 \\ 1 & 0 & -1 & 0 \\ 0 & 1 & 0 & 1 \\ 0 & 1 & 0 & -1 \end{bmatrix} \begin{bmatrix} 32 \\ 38 \\ 16 \\ 10 \end{bmatrix} = \begin{bmatrix} 48 \\ 16 \\ 48 \\ 28 \end{bmatrix}.$$

There is an important pattern to observe in these matrices: rows 2 and 4 of $\mathbf{T_a}$ are obtained from rows 1 and 3 by shifting to the right by two positions. Likewise, columns 2 and 4 of $\mathbf{T_s}$ are obtained from columns 1 and 3 by shifting down by two positions. We will study this pattern in more detail in Section 3.5. ■

Example 3.2. A signal $\mathbf{x}$ of length $2^{10} = 1024$ is shown on the left in Figure 3.4. It consists of four cycles of a sine wave followed by a step function. Random noise has been added to both parts of the signal.[2] Applying the Haar analysis matrix $\mathbf{T_a}$ to $\mathbf{x}$ (using MATLAB), we obtain the one-scale trend $\mathbf{s}_9$ and detail $\mathbf{d}_9$ of length $2^9 = 512$ shown on the right in Figure 3.4. It is evident that the trend shows the essential features of the signal, while the detail consists mostly of random fluctuations. ■

---

[2] The finite Fourier transform is not an effective tool for analyzing this type of signal.

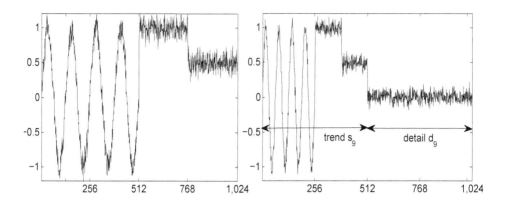

Fig. 3.4   Signal with random noise (left) and one-scale Haar transform of signal (right)

3.3   Multiple Scale Haar Wavelet Transform

In Section 3.2.1 we took a signal $\mathbf{s}_3$ of length 2^3 and transformed it into a *trend* $\mathbf{s}_2$ and a *detail* $\mathbf{d}_2$, each of half the length. We then repeated this operation on the trend portion. We now describe this so-called *pyramid algorithm* in terms of the analysis and synthesis matrices introduced in Section 3.2.

3.3.1   *Matrix description of multiresolution representation*

Write $\mathbf{T}_{\mathbf{a}}^{(k)}$ and $\mathbf{T}_{\mathbf{s}}^{(k)}$ for the $2^k \times 2^k$ Haar analysis and synthesis matrices:

$$\mathbf{T}_{\mathbf{a}}^{(k)} = \frac{1}{2} \left[\begin{array}{cc} I^{(k-1)} & I^{(k-1)} \\ I^{(k-1)} & -I^{(k-1)} \end{array} \right] \boxed{\text{split}} , \qquad \mathbf{T}_{\mathbf{s}}^{(k)} = \left[\begin{array}{cc} I^{(k-1)} & I^{(k-1)} \\ I^{(k-1)} & -I^{(k-1)} \end{array} \right] \boxed{\text{split}} . \qquad (3.3)$$

Here we use the notation $I^{(p)}$ for the identity matrix of size $2^p \times 2^p$. Taking $k = 3$, we can write calculations in Section 3.2.1 in block-matrix form as

$$\left[\begin{array}{c} \mathbf{s}_2 \\ \mathbf{d}_2 \end{array} \right] = \mathbf{T}_{\mathbf{a}}^{(3)} \mathbf{s}_3 , \qquad \left[\begin{array}{c} \mathbf{s}_1 \\ \mathbf{d}_1 \\ \mathbf{d}_2 \end{array} \right] = \left[\begin{array}{cc} \mathbf{T}_{\mathbf{a}}^{(2)} & 0 \\ 0 & I^{(2)} \end{array} \right] \left[\begin{array}{c} \mathbf{s}_2 \\ \mathbf{d}_2 \end{array} \right] ,$$

$$\left[\begin{array}{c} \mathbf{s}_0 \\ \mathbf{d}_0 \\ \mathbf{d}_1 \\ \mathbf{d}_2 \end{array} \right] = \left[\begin{array}{ccc} \mathbf{T}_{\mathbf{a}}^{(1)} & 0 & 0 \\ 0 & I^{(1)} & 0 \\ 0 & 0 & I^{(2)} \end{array} \right] \left[\begin{array}{c} \mathbf{s}_1 \\ \mathbf{d}_1 \\ \mathbf{d}_2 \end{array} \right] .$$

In these formulas $\mathbf{0}$ denotes matrices of zeros of the appropriate sizes.

Let $\mathbf{s}_k$ be a signal of length 2^k (level k). We describe the general three-scale Haar multiresolution analysis pyramid algorithm by the diagram

$$\mathbf{s}_k \longrightarrow \boxed{\mathbf{T}_{\mathbf{a}}^{(k)}} \overset{\mathbf{s}_{k-1}}{\longrightarrow} \boxed{\mathbf{T}_{\mathbf{a}}^{(k-1)}} \overset{\mathbf{s}_{k-2}}{\longrightarrow} \boxed{\mathbf{T}_{\mathbf{a}}^{(k-2)}} \longrightarrow \mathbf{s}_{k-3} \qquad (3.4)$$
$$ \searrow \searrow \searrow$$
$$ \mathbf{d}_{k-1} \mathbf{d}_{k-2} \mathbf{d}_{k-3}$$

where the boxes denote the analysis transformations and the labeled arrows show inputs and outputs from each box. Note that the outputs are the *trend* $\mathbf{s}_{k-3}$ at level $k-3$ together with the three *details* $\mathbf{d}_{k-3}$, $\mathbf{d}_{k-2}$, $\mathbf{d}_{k-1}$ at levels $k-3$, $k-2$, and $k-1$. We can express this pyramid algorithm (3.4) in matrix form as

$$
\begin{bmatrix} \mathbf{s}_{k-3} \\ \mathbf{d}_{k-3} \\ \mathbf{d}_{k-2} \\ \mathbf{d}_{k-1} \end{bmatrix} = \begin{bmatrix} \mathbf{T}_\mathbf{a}^{(k-2)} & \mathbf{0} & \mathbf{0} \\ \mathbf{0} & I^{(k-2)} & \mathbf{0} \\ \mathbf{0} & \mathbf{0} & I^{(k-1)} \end{bmatrix} \begin{bmatrix} \mathbf{T}_\mathbf{a}^{(k-1)} & \mathbf{0} \\ \mathbf{0} & I^{(k-1)} \end{bmatrix} \mathbf{T}_\mathbf{a}^{(k)} \mathbf{s}_k = \mathbf{W}_\mathbf{a}^{(3)} \mathbf{s}_k, \quad (3.5)
$$

where we use the notation $\mathbf{W}_\mathbf{a}^{(3)}$ for the $2^k \times 2^k$ *three-scale Haar analysis matrix.*

The inverse to $\mathbf{W}_\mathbf{a}^{(3)}$ is the *three-scale Haar synthesis matrix* $\mathbf{W}_\mathbf{s}^{(3)}$. Since the one-scale synthesis matrices are the inverses of the one-scale analysis matrices, the factorization (3.5) gives

$$
\mathbf{s}_k = \mathbf{W}_\mathbf{s}^{(3)} \begin{bmatrix} \mathbf{s}_{k-3} \\ \mathbf{d}_{k-3} \\ \mathbf{d}_{k-2} \\ \mathbf{d}_{k-1} \end{bmatrix}, \quad \text{with}
$$

$$
\mathbf{W}_\mathbf{s}^{(3)} = \mathbf{T}_\mathbf{s}^{(k)} \begin{bmatrix} \mathbf{T}_\mathbf{s}^{(k-1)} & \mathbf{0} \\ \mathbf{0} & I^{(k-1)} \end{bmatrix} \begin{bmatrix} \mathbf{T}_\mathbf{s}^{(k-2)} & \mathbf{0} & \mathbf{0} \\ \mathbf{0} & I^{(k-2)} & \mathbf{0} \\ \mathbf{0} & \mathbf{0} & I^{(k-1)} \end{bmatrix}. \quad (3.6)
$$

We can describe the reconstruction of $\mathbf{s}_k$ using the multiresolution synthesis pyramid algorithm by the diagram

$$
\mathbf{s}_{k-3} \longrightarrow \boxed{\mathbf{T}_\mathbf{s}^{(k-2)}} \overset{\mathbf{s}_{k-2}}{\longrightarrow} \boxed{\mathbf{T}_\mathbf{s}^{(k-1)}} \overset{\mathbf{s}_{k-1}}{\longrightarrow} \boxed{\mathbf{T}_\mathbf{s}^{(k)}} \longrightarrow \mathbf{s}_k \qquad (3.7)
$$

$$
\mathbf{d}_{k-3} \qquad\qquad \mathbf{d}_{k-2} \qquad\quad \mathbf{d}_{k-1}
$$

Note that the inputs are the trend $\mathbf{s}_{k-3}$ at level $k-3$ together with the three details $\mathbf{d}_{k-3}$, $\mathbf{d}_{k-2}$, $\mathbf{d}_{k-1}$ at levels $k-3$, $k-2$, and $k-1$.

In equations (3.5) and (3.6) the matrix blocks are of different sizes, so the matrix multiplication cannot be simplified while still in this block form.[3] It is instructive, however, to examine the case $k=3$ more closely by carrying out the multiplication in (3.5) entry-by-entry to obtain the 8×8 three-scale Haar wavelet analysis matrix

$$
\mathbf{W}_\mathbf{a}^{(3)} = \begin{bmatrix} \frac{1}{8} & \frac{1}{8} & \frac{1}{8} & \frac{1}{8} & \frac{1}{8} & \frac{1}{8} & \frac{1}{8} & \frac{1}{8} \\ \frac{1}{8} & \frac{1}{8} & \frac{1}{8} & \frac{1}{8} & -\frac{1}{8} & -\frac{1}{8} & -\frac{1}{8} & -\frac{1}{8} \\ \frac{1}{4} & \frac{1}{4} & -\frac{1}{4} & -\frac{1}{4} & 0 & 0 & 0 & 0 \\ 0 & 0 & 0 & 0 & \frac{1}{4} & \frac{1}{4} & -\frac{1}{4} & -\frac{1}{4} \\ \frac{1}{2} & -\frac{1}{2} & 0 & 0 & 0 & 0 & 0 & 0 \\ 0 & 0 & \frac{1}{2} & -\frac{1}{2} & 0 & 0 & 0 & 0 \\ 0 & 0 & 0 & 0 & \frac{1}{2} & -\frac{1}{2} & 0 & 0 \\ 0 & 0 & 0 & 0 & 0 & 0 & \frac{1}{2} & -\frac{1}{2} \end{bmatrix} \begin{bmatrix} u_0 \\ v_1 \\ v_2 \\ v_3 \\ v_4 \\ v_5 \\ v_6 \\ v_7 \end{bmatrix}, \quad (3.8)
$$

---

[3]In Section 3.5 we will separate each one-scale analysis and synthesis matrix into a *trend block* and a *detail block* of the correct size for block multiplication.

where we have labeled the rows as $\mathbf{u}_0, \mathbf{v}_1, \ldots, \mathbf{v}_7$.

For signals of length 2^3 the three-scale synthesis matrix $\mathbf{W}_s^{(3)} = \begin{bmatrix} \mathbf{h}_0 & \mathbf{h}_1 & \cdots & \mathbf{h}_7 \end{bmatrix}$ has columns

$$
\mathbf{h}_0 = \begin{bmatrix} 1 \\ 1 \\ 1 \\ 1 \\ 1 \\ 1 \\ 1 \\ 1 \end{bmatrix}, \quad
\mathbf{h}_1 = \begin{bmatrix} 1 \\ 1 \\ 1 \\ 1 \\ -1 \\ -1 \\ -1 \\ -1 \end{bmatrix}, \quad
\mathbf{h}_2 = \begin{bmatrix} 1 \\ 1 \\ -1 \\ -1 \\ 0 \\ 0 \\ 0 \\ 0 \end{bmatrix}, \quad
\mathbf{h}_3 = \begin{bmatrix} 0 \\ 0 \\ 0 \\ 0 \\ 1 \\ 1 \\ -1 \\ -1 \end{bmatrix},
$$

and

$$
\mathbf{h}_4 = \begin{bmatrix} 1 \\ -1 \\ 0 \\ 0 \\ 0 \\ 0 \\ 0 \\ 0 \end{bmatrix}, \quad
\mathbf{h}_5 = \begin{bmatrix} 0 \\ 0 \\ 1 \\ -1 \\ 0 \\ 0 \\ 0 \\ 0 \end{bmatrix}, \quad
\mathbf{h}_6 = \begin{bmatrix} 0 \\ 0 \\ 0 \\ 0 \\ 1 \\ -1 \\ 0 \\ 0 \end{bmatrix}, \quad
\mathbf{h}_7 = \begin{bmatrix} 0 \\ 0 \\ 0 \\ 0 \\ 0 \\ 0 \\ 1 \\ -1 \end{bmatrix}.
$$

Note that $\mathbf{h}_3 = S^4 \mathbf{h}_2$ while $\mathbf{h}_5 = S^2 \mathbf{h}_4$, $\mathbf{h}_6 = S^4 \mathbf{h}_4$, and $\mathbf{h}_7 = S^6 \mathbf{h}_4$, where S is the 8×8 shift matrix (we are enumerating the columns by 0 to 7, just as we did for the Fourier matrix).

Remark 3.1. The vector $\mathbf{h}_0$ (we call this the *scaling vector*) describes a signal that is constantly 1; this is the *direct current* (DC) component. The vector $\mathbf{h}_1$ (we will call it the *coarse* wavelet vector) describes a signal that is 1 for four time units, then switches sign and is -1 for four time units; this is the slow *alternating current* (AC) component. The vector $\mathbf{h}_2$ (we will call it the *medium* wavelet vector) describes a signal that is 1 for two time units, switches sign and is -1 for two time units, and then is zero (faster AC component). The vector $\mathbf{h}_4$ (we will call it the *fine* wavelet vector) describes a signal that is 1 for one time unit, switches sign and is -1 for one time unit, and then is zero (fastest AC component). The other vectors are shifts of these vectors by an even number of positions. In Section 5.3 we will extend this construction to obtain the Haar wavelet transform of analog signals.

The rows of the analysis matrix $\mathbf{W}_a^{(3)}$ give the dual basis to the columns of the synthesis matrix $\mathbf{W}_s^{(3)}$:

$$\mathbf{u}_0 \mathbf{h}_0 = 1, \qquad\qquad\qquad \mathbf{u}_0 \mathbf{h}_k = 0 \quad \text{for } k = 1, \ldots, 7,$$

$$\mathbf{v}_j \mathbf{h}_0 = 0 \quad \text{for } j = 1, \ldots, 7, \qquad \mathbf{v}_j \mathbf{h}_k = \begin{cases} 1 & \text{for } j = k, \\ 0 & \text{for } j \neq k. \end{cases}$$

These relations are evident from the explicit formulas for these vectors.

To obtain the *multiresolution representation* of the original signal $\mathbf{s}_3$, recall that the product of a matrix $\mathbf{A}$ and a vector $\mathbf{b}$ is the linear combination of the columns of $\mathbf{A}$ that has the entries of $\mathbf{b}$ as coefficients. Thus

$$\mathbf{s}_3 = \mathbf{W}_s^{(3)} \begin{bmatrix} \mathbf{s}_0 \\ \mathbf{d}_0 \\ \mathbf{d}_1 \\ \mathbf{d}_2 \end{bmatrix} = \mathbf{s}_0[0]\mathbf{h}_0 + \mathbf{d}_0[0]\mathbf{h}_1 + \big(\mathbf{d}_1[0]\mathbf{h}_2 + \mathbf{d}_1[1]\mathbf{h}_3\big) \\ + \big(\mathbf{d}_2[0]\mathbf{h}_4 + \mathbf{d}_2[1]\mathbf{h}_5 + \mathbf{d}_2[2]\mathbf{h}_6 + \mathbf{d}_2[3]\mathbf{h}_7\big).$$

In this formula we build up the signal by taking the overall average $\mathbf{s}_0[0]\mathbf{h}_0$, then add the slow fluctuation term $\mathbf{d}_0[0]\mathbf{h}_1$, followed by a sum of two faster and shorter fluctuations, and finally a sum of the four fastest and shortest fluctuations. For the signal in Section 3.2.1 the multiresolution representation is thus

$$\mathbf{s}_3 = 35\mathbf{h}_0 - 3\mathbf{h}_1 + \big(16\mathbf{h}_2 + 10\mathbf{h}_3\big) + \big(8\mathbf{h}_4 - 8\mathbf{h}_5 + 0\mathbf{h}_6 + 12\mathbf{h}_7\big). \qquad (3.9)$$

Remark 3.2. Diagrams (3.4) and (3.7) fit together to describe the three-scale pyramid algorithm. This algorithm is the basic tool for multiresolution wavelet analysis. For a signal of length 2^k and any integer m between 1 and k we can construct an m-scale pyramid algorithm in the same way. The key idea is to split the information in the signal into $m+1$ subsignals: one *trend* signal that gives a low-resolution approximation to the original signal, together with m *detail* signals that contain the short-term changes in the signal at successively shorter intervals of measurement. The sum of the lengths of the subsignals will be still be 2^k, and the algorithm can be executed in place by replacing the trend at each scale by the trend and detail at the next coarser scale.

3.3.2 *Signal processing using the multiresolution representation*

The multiresolution representation of a signal reveals features that are hidden in the original (time-domain) representation. We can use the new information about the signal to modify it in various desirable ways. This is called *signal processing*.

One of the main applications of the multiresolution representation is *compression*. We choose a *threshold* ϵ and set to zero all coefficients in the multiresolution representation of the signal whose absolute value is less than ϵ.

Example 3.3. Take the signal $\mathbf{s}_3$ from Section 3.2.1 and apply the compression threshold $\epsilon = 4$. This gives the modified signal

$$\mathbf{y} = 35\mathbf{h}_0 + \big(16\mathbf{h}_2 + 10\mathbf{h}_3\big) + \big(8\mathbf{h}_4 - 8\mathbf{h}_5 + 12\mathbf{h}_7\big)$$

(see Figure 3.5). Notice that the graphs of the modified signal and the original signal are almost the same. To measure the relative difference between the two graphs, we use the ratios of the energies (the square of the norms):

$$\text{relative compression error} = \frac{||\mathbf{s}_3 - \mathbf{y}||^2}{||\mathbf{s}_3||^2}.$$

For compression with $\epsilon = 4$ every entry in $\mathbf{y}$ happens to differ from the corresponding entry in $\mathbf{s}_3$ by ± 3, so the relative compression error is

$$\frac{3^2 + 3^2 + 3^2 + 3^2 + 3^2 + 3^2 + 3^2 + 3^2}{56^2 + 40^2 + 8^2 + 24^2 + 48^2 + 48^2 + 40^2 + 16^2} = 0.6\% \,.$$

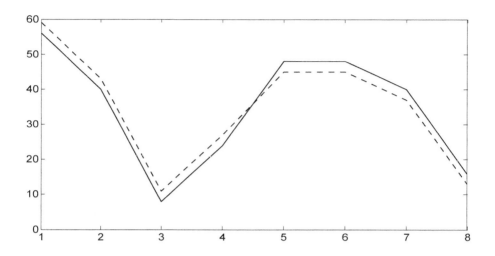

Fig. 3.5   Signal (solid line) and compressed signal (dashed line) with threshold 4

If we apply the same compression algorithm but with $\epsilon = 9$ to the multiresolution representation (3.9), we get the modified signal

$$\mathbf{z} = 35\mathbf{h}_0 + 16\mathbf{h}_2 + 10\mathbf{h}_3 + 12\mathbf{h}_7 \,.$$

Now the graphs of the modified signal and the original signal have the same general shape, but differ in details. The relative compression error is

$$\frac{5^2 + 11^2 + 11^2 + 5^2 + 3^2 + 3^2 + 3^2 + 3^2}{56^2 + 40^2 + 8^2 + 24^2 + 48^2 + 48^2 + 40^2 + 16^2} = 2.8\% \,.$$

This is about five times larger than the compression error with $\epsilon = 4$. On the other hand, the signal $\mathbf{z}$ only has four nonzero coefficients in its multiresolution representation, so we have compressed the original signal by a ratio of 2:1 while still retaining most of its essential features. ∎

The multiresolution process has been described as a *mathematical microscope* for examining signals (and images). To understand why this terminology is appropriate, we look again at the signal from Example 3.2 (sine wave and step function with random noise).

Example 3.4. We apply the $2^9 \times 2^9$ Haar analysis matrix $\mathbf{T}_{\mathrm{a}}^{(9)}$ to the trend $\mathbf{s}_9$ in Figure 3.4 to obtain the trend $\mathbf{s}_8$ and detail $\mathbf{d}_8$ shown on the left side of Figure 3.6. These two vectors in $\mathbb{R}^{256}$ together with the vector $\mathbf{d}_9 \in \mathbb{R}^{512}$ from Figure 3.4 furnish the two-scale Haar multiresolution analysis of this signal.

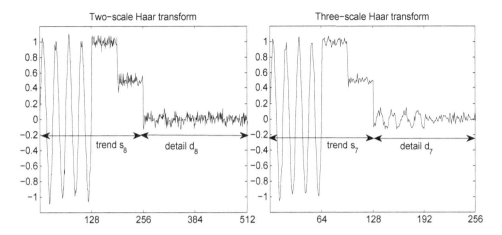

Fig. 3.6   Multiresolution analysis of sine wave and step function signal with random noise

To continue, we apply the $2^8 \times 2^8$ Haar analysis matrix $\mathbf{T}_a^{(8)}$ to the trend $\mathbf{s}_8$ to obtain the trend $\mathbf{s}_7$ and detail $\mathbf{d}_7$ shown on the right side of Figure 3.6. These two vectors in $\mathbb{R}^{128}$ together with $\mathbf{d}_8 \in \mathbb{R}^{256}$ from the left side of this figure and the vector $\mathbf{d}_9 \in \mathbb{R}^{512}$ from Figure 3.4 furnish the three-scale Haar multiresolution analysis.

We observe that the trend $\mathbf{s}_7$ is approximately a sine wave followed by a step function, with much less noise than the trend $\mathbf{s}_8$. In the detail $\mathbf{d}_7$ we can see a difference between the wide slow oscillation associated with the sine wave (entries 129 to 192) and the small purely random noise associated with the step function (entries 193 to 256). This illustrates why the multiresolution process is a much more powerful tool than simply zooming in on the graph of the signal. ∎

3.4   Wavelet Transforms for Periodic Signals by Lifting

The approach to discrete wavelet transforms that we have introduced in Sections 3.2 and 3.3 is called the *lifting method*. We split a signal of even length into *even* and *odd* parts, apply successive *prediction* and *update* linear transformations to the two parts of the signal, and finally apply a normalization transformation. All of these transformations are given by invertible upper-triangular or lower-triangular matrices, so the computations are very efficient and no information in the signal is lost. We illustrate this procedure with two new discrete wavelet transforms that we will study in considerable detail.

3.4.1   *CDF*(2, 2) *transform*

Suppose $\mathbf{x}$ is a signal of even length N with values $\mathbf{x}[0]$, $\mathbf{x}[1]$, ..., $\mathbf{x}[N-1]$; we identify $\mathbf{x}$ with the vector in $\mathbb{R}^N$ having these components. Let $\mathbf{x}_{\text{even}}$ and $\mathbf{x}_{\text{odd}}$ be the corresponding signals of length $N/2$, which we likewise identify with vectors in $\mathbb{R}^{N/2}$ as in Section 3.2. For the Haar transform we predicted the value $\mathbf{x}[2n+1]$ for $n = 0, 1, \ldots, N/2 - 1$ by using the previous value $\mathbf{x}[2n]$. We can modify this scheme by predicting $\mathbf{x}[2n+1]$ to be the *average* of the neighboring values $\mathbf{x}[2n]$ and $\mathbf{x}[2n+2]$. The deviation $\mathbf{d}$ from the predicted value is thus

$$\mathbf{d}[n] = \mathbf{x}[2n+1] - \frac{1}{2}\big(\mathbf{x}[2n] + \mathbf{x}[2n+2]\big) = \mathbf{x}_{\text{odd}}[n] - \frac{1}{2}\big(\mathbf{x}_{\text{even}}[n] + \mathbf{x}_{\text{even}}[n+1]\big). \quad (3.10)$$

Notice that if $\mathbf{x}[k] = ak + b$ for $k = 2n,\ 2n+1,\ 2n+2$, where a and b are constants, then

$$\mathbf{d}[n] = a(2n+1) + b - \frac{1}{2}\big(a(2n + 2n + 2) + 2b\big) = 0.$$

Hence when $\mathbf{x}$ is a linear function of n then the deviation $\mathbf{d}$ should be zero.

There is a problem with the formula for $\mathbf{d}$, however. To calculate the last deviation $\mathbf{d}[N/2 - 1]$ we need the value $\mathbf{x}[N]$, but we only have the values $\mathbf{x}[n]$ for $n = 0, \ldots, N-1$. We shall avoid this difficulty by extending the signal $\mathbf{x}$ to be *periodic* of period N:

$$\mathbf{x}[k + N] = \mathbf{x}[k] \quad \text{for all integers } k.$$

Likewise, we extend $\mathbf{x}_{\text{even}}$ and $\mathbf{x}_{\text{odd}}$ to periodic functions of period $N/2$:

$$\mathbf{x}_{\text{even}}[k + N/2] = \mathbf{x}_{\text{even}}[k] \quad \text{and} \quad \mathbf{x}_{\text{odd}}[k + N/2] = \mathbf{x}_{\text{odd}}[k] \quad \text{for all integers } k.$$

Let S be the shift operator. Then $\mathbf{x}_{\text{even}}[n+1] = \big(S^{-1}\mathbf{x}_{\text{even}}\big)[n]$, so the formula for $\mathbf{d}$ can be written in vector form as

$$\mathbf{d} = \mathbf{x}_{\text{odd}} - \frac{1}{2}\big(\mathbf{x}_{\text{even}} + S^{-1}\mathbf{x}_{\text{even}}\big). \quad (3.11)$$

Since $\mathbf{x}_{\text{even}}$ and $\mathbf{x}_{\text{odd}}$ are periodic of period $N/2$, so is $\mathbf{d}$:

$$\mathbf{d}[k + N/2] = \mathbf{d}[k] \quad \text{for all integers } k.$$

Define the *prediction* transformation by $P \begin{bmatrix} \mathbf{x}_{\text{even}} \\ \mathbf{x}_{\text{odd}} \end{bmatrix} = \begin{bmatrix} \mathbf{x}_{\text{even}} \\ \mathbf{d} \end{bmatrix}$. From formula (3.11) we can write the matrix for P in block form as

$$P = \begin{bmatrix} I & \mathbf{0} \\ -\frac{1}{2}(I + S^{-1}) & I \end{bmatrix}, \quad (3.12)$$

where I is the $N/2 \times N/2$ identity matrix.

As the next step in the lifting procedure, we use the detail vector $\mathbf{d}$ to update $\mathbf{x}_{\text{even}}$ and obtain the *trend vector* $\mathbf{s}$:

$$\mathbf{s}[n] = \mathbf{x}_{\text{even}}[n] + \frac{1}{4}\big(\mathbf{d}[n] + \mathbf{d}[n-1]\big). \quad (3.13)$$

Thus $\mathbf{s}[n] = \mathbf{x}_{\text{even}}[n]$ whenever $\mathbf{d}[n] + \mathbf{d}[n-1] = 0$. The constant $1/4$ appears in the formula so that $\mathbf{s}$ and $\mathbf{x}$ will have the same average value:

$$\frac{1}{(N/2)} \sum_{n=0}^{(N/2)-1} \mathbf{s}[n] = \frac{1}{N} \sum_{n=0}^{N-1} \mathbf{x}[n] \quad (3.14)$$

(see Exercises 3.8 #4). The formula for **s** can be written in vector form as

$$\mathbf{s} = \mathbf{x}_{\text{even}} + \frac{1}{4}(\mathbf{d} + S\mathbf{d}). \tag{3.15}$$

Since $\mathbf{x}_{\text{even}}$ and $\mathbf{d}$ are periodic of period $N/2$, so is **s**:

$$\mathbf{s}[k + N/2] = \mathbf{s}[k] \quad \text{for all integers } k.$$

Define the *update* transformation by $U\begin{bmatrix} \mathbf{x}_{\text{even}} \\ \mathbf{d} \end{bmatrix} = \begin{bmatrix} \mathbf{s} \\ \mathbf{d} \end{bmatrix}$. From formula (3.15) we can write the matrix for U in block form as

$$U = \begin{bmatrix} I & \frac{1}{4}(I + S) \\ \mathbf{0} & I \end{bmatrix}. \tag{3.16}$$

The final step in the lifting process is a *normalization* $D\begin{bmatrix} \mathbf{s} \\ \mathbf{d} \end{bmatrix} = \begin{bmatrix} \sqrt{2}\mathbf{s} \\ (1/\sqrt{2})\mathbf{d} \end{bmatrix}$. Thus D is given by the diagonal matrix

$$D = \begin{bmatrix} \sqrt{2}I & \mathbf{0} \\ \mathbf{0} & (1/\sqrt{2})I \end{bmatrix}. \tag{3.17}$$

The *one-scale* CDF$(2,2)$ *analysis transform* is the product of these transformations:

$$\mathbf{T_a} = DUP\boxed{\text{split}}. \tag{3.18}$$

Example 3.5. Suppose $N = 4$. In this case $S = S^{-1}$ on 2-periodic functions, and $I + S^{-1}$ has the matrix $\begin{bmatrix} 1 & 1 \\ 1 & 1 \end{bmatrix}$. Hence

$$P = \begin{bmatrix} 1 & 0 & 0 & 0 \\ 0 & 1 & 0 & 0 \\ -\frac{1}{2} & -\frac{1}{2} & 1 & 0 \\ -\frac{1}{2} & -\frac{1}{2} & 0 & 1 \end{bmatrix} \quad \text{and} \quad U = \begin{bmatrix} 1 & 0 & \frac{1}{4} & \frac{1}{4} \\ 0 & 1 & \frac{1}{4} & \frac{1}{4} \\ 0 & 0 & 1 & 0 \\ 0 & 0 & 0 & 1 \end{bmatrix}.$$

Carrying out the matrix multiplication, we find that

$$\mathbf{T_a} = \begin{bmatrix} \sqrt{2} & 0 & 0 & 0 \\ 0 & \sqrt{2} & 0 & 0 \\ 0 & 0 & \frac{1}{\sqrt{2}} & 0 \\ 0 & 0 & 0 & \frac{1}{\sqrt{2}} \end{bmatrix} \begin{bmatrix} 1 & 0 & \frac{1}{4} & \frac{1}{4} \\ 0 & 1 & \frac{1}{4} & \frac{1}{4} \\ 0 & 0 & 1 & 0 \\ 0 & 0 & 0 & 1 \end{bmatrix} \begin{bmatrix} 1 & 0 & 0 & 0 \\ 0 & 1 & 0 & 0 \\ -\frac{1}{2} & -\frac{1}{2} & 1 & 0 \\ -\frac{1}{2} & -\frac{1}{2} & 0 & 1 \end{bmatrix} \boxed{\text{split}}$$

$$= \frac{1}{2\sqrt{2}} \begin{bmatrix} 3 & 1 & -1 & 1 \\ -1 & 1 & 3 & 1 \\ -1 & 2 & -1 & 0 \\ -1 & 0 & -1 & 2 \end{bmatrix} = \begin{bmatrix} \mathbf{u}_0 \\ \mathbf{u}_1 \\ \mathbf{v}_0 \\ \mathbf{v}_1 \end{bmatrix},$$

where at the last step we have written $\mathbf{T_a}$ in terms of its rows. We observe that these row vectors are related as follows:

(1) The second row is obtained by shifting the first row to the right two positions (using periodic wraparound).

(2) The fourth is obtained by shifting the third row to the right two positions (using periodic wraparound).

We already saw this pattern in the Haar transform, and we will show in Section 3.5 that it holds for all one-scale wavelet analysis matrices. ∎

As in the case of the Haar transform, it is easy to construct the inverse (synthesis) transform $\mathbf{T_s}$ by inverting the prediction, update, and normalization transforms:

$$P^{-1} = \begin{bmatrix} I & 0 \\ \frac{1}{2}(I + S^{-1}) & I \end{bmatrix}, \quad U^{-1} = \begin{bmatrix} I & -\frac{1}{4}(I + S) \\ 0 & I \end{bmatrix}, \quad D^{-1} = \begin{bmatrix} (1/\sqrt{2})I & 0 \\ 0 & \sqrt{2}I \end{bmatrix}.$$

Hence we obtain the *one-scale* CDF$(2,2)$ *synthesis transform* as

$$\mathbf{T_s} = \boxed{\text{merge}}\, P^{-1}U^{-1}D^{-1}. \tag{3.19}$$

Example 3.6. Suppose $N = 4$. Then we calculate that

$$\mathbf{T_s} = \boxed{\text{merge}} \begin{bmatrix} 1 & 0 & 0 & 0 \\ 0 & 1 & 0 & 0 \\ \frac{1}{2} & \frac{1}{2} & 1 & 0 \\ \frac{1}{2} & \frac{1}{2} & 0 & 1 \end{bmatrix} \begin{bmatrix} 1 & 0 & -\frac{1}{4} & -\frac{1}{4} \\ 0 & 1 & -\frac{1}{4} & -\frac{1}{4} \\ 0 & 0 & 1 & 0 \\ 0 & 0 & 0 & 1 \end{bmatrix} \begin{bmatrix} \frac{1}{\sqrt{2}} & 0 & 0 & 0 \\ 0 & \frac{1}{\sqrt{2}} & 0 & 0 \\ 0 & 0 & \sqrt{2} & 0 \\ 0 & 0 & 0 & \sqrt{2} \end{bmatrix}$$

$$= \frac{\sqrt{2}}{4} \begin{bmatrix} 2 & 0 & -1 & -1 \\ 1 & 1 & 3 & -1 \\ 0 & 2 & -1 & -1 \\ 1 & 1 & -1 & 3 \end{bmatrix} = \begin{bmatrix} \tilde{\mathbf{u}}_0 & \tilde{\mathbf{u}}_1 & \tilde{\mathbf{v}}_0 & \tilde{\mathbf{v}}_1 \end{bmatrix},$$

where at the last step we have written $\mathbf{T_s}$ in terms of its columns. We observe that the columns are related as follows:

(1) The second column is obtained by shifting the first column down two positions (using periodic wraparound).
(2) The fourth column is obtained by shifting the third column down two positions (using periodic wraparound).

We already saw this pattern in the Haar transform; we will show in Section 3.5 that it holds for all one-scale wavelet synthesis matrices. ∎

Remark 3.3. The CDF$(2,2)$ discrete wavelet transform that we have just constructed is part of the CDF (Cohen–Daubechies–Feauveau) family of transforms. The analysis matrices for the CDF transforms are not orthogonal, unlike the case of the (normalized) Haar transform, so the columns of the synthesis matrix are quite different from the rows of the analysis matrix. However, this family of transforms gives fast computations, since moving the $\sqrt{2}$ factors onto the synthesis matrix makes all matrix coefficients binary rational numbers (denominators that are powers of 2). These transforms also have many desirable properties relating to smoothness and feature detection; we shall study them in detail.

3.4.2 *Daub4 transform*

Our next example of a wavelet transform belongs to a family of orthogonal wavelet transforms that we shall study in detail in Section 4.10. The creation of this family of transforms by Ingrid Daubechies [Daubechies (1988)] was a decisive advance in discrete wavelet theory and practice.

Let V be the N-dimensional vector space of all vectors of the form $\begin{bmatrix} \mathbf{u} \\ \mathbf{v} \end{bmatrix}$, where $\mathbf{u}$ and $\mathbf{v}$ are real-valued $N/2$-periodic functions on $\mathbb{Z}$. Every linear transformation T on V is given in block form as $T = \begin{bmatrix} A & B \\ C & D \end{bmatrix}$, where A, B, C, and D are linear transformations on the vector space of real-valued $N/2$-periodic functions on $\mathbb{Z}$ (see Section 1.5.3). If we identify an $N/2$-periodic function on $\mathbb{Z}$ with a column vector in $\mathbb{R}^{N/2}$ as in Section 3.2.2, then A, B, C, and D become $N/2 \times N/2$ matrices.

Following the same pattern as for the Haar and CDF$(2,2)$ transform, we split a digital signal $\mathbf{x}$ of even length $N \geq 4$ into $\mathbf{x}_{\text{even}}$ and $\mathbf{x}_{\text{odd}}$. We extend $\mathbf{x}$ to be a periodic of period N, so that $\mathbf{x}_{\text{even}}$ and $\mathbf{x}_{\text{odd}}$ are periodic of period $N/2$. Thus the vector $\begin{bmatrix} \mathbf{x}_{\text{even}} \\ \mathbf{x}_{\text{odd}} \end{bmatrix}$ is in V. We will perform successive linear transformations on this vector to obtain the Daub4 transform of $\mathbf{x}$.

We begin with an *update* operation to define the *first trend*

$$\mathbf{s}^{(1)}[n] = \mathbf{x}[2n] + \sqrt{3}\,\mathbf{x}[2n+1] = \mathbf{x}_{\text{even}}[n] + \sqrt{3}\mathbf{x}_{\text{odd}}[n]. \tag{3.20}$$

Define the *first update* transformation U_1 on the vector space V by

$$U_1 \begin{bmatrix} \mathbf{x}_{\text{even}} \\ \mathbf{x}_{\text{odd}} \end{bmatrix} = \begin{bmatrix} \mathbf{s}^{(1)} \\ \mathbf{x}_{\text{odd}} \end{bmatrix}.$$

From formula (3.20) we can write U_1 in block form as

$$U_1 = \begin{bmatrix} I & \sqrt{3}I \\ \mathbf{0} & I \end{bmatrix}, \tag{3.21}$$

where I is the identity transformation. Next we predict $\mathbf{x}_{\text{odd}}[n]$ using two adjacent values of $\mathbf{s}^{(1)}$:

$$\text{prediction of } \mathbf{x}_{\text{odd}}[n] = \frac{1}{4} \left\{ \sqrt{3}\mathbf{s}^{(1)}[n] + (\sqrt{3} - 2)\mathbf{s}^{(1)}[n-1] \right\}.$$

The difference between the prediction of $\mathbf{x}_{\text{odd}}$ and the actual value is the (unnormalized) detail

$$\mathbf{d}^{(1)}[n] = \mathbf{x}_{\text{odd}}[n] - \frac{1}{4} \left\{ \sqrt{3}\mathbf{s}^{(1)}[n] + (\sqrt{3} - 2)\mathbf{s}^{(1)}[n-1] \right\}. \tag{3.22}$$

Since $\mathbf{s}^{(1)}[n-1] = S\mathbf{s}^{(1)}[n]$, where S is the shift operator, we can write equation (3.22) in vector form as

$$\mathbf{d}^{(1)} = \mathbf{x}_{\text{odd}} - \frac{1}{4}\left(\sqrt{3}I + (\sqrt{3} - 2)S \right)\mathbf{s}^{(1)}. \tag{3.23}$$

Define the *prediction* transformation by $P \begin{bmatrix} \mathbf{s}^{(1)} \\ \mathbf{x}_{\text{odd}} \end{bmatrix} = \begin{bmatrix} \mathbf{s}^{(1)} \\ \mathbf{d}^{(1)} \end{bmatrix}$. From formula (3.23) we can write P in block form as

$$P = \begin{bmatrix} I & 0 \\ -(1/4)(\sqrt{3}I - (\sqrt{3}-2)S) & I \end{bmatrix}. \tag{3.24}$$

As the next step in the lifting procedure, we use $\mathbf{d}^{(1)}$ to update the first trend $\mathbf{s}^{(1)}$ and obtain the *second trend* $\mathbf{s}^{(2)}$:

$$\mathbf{s}^{(2)}[n] = \mathbf{s}^{(1)}[n] - \mathbf{d}^{(1)}[n+1].$$

The formula for $\mathbf{s}^{(2)}$ can be written in vector form as

$$\mathbf{s}^{(2)} = \mathbf{s}^{(1)} - S^{-1}\mathbf{d}^{(1)}. \tag{3.25}$$

Define the *second update* transformation by $U_2 \begin{bmatrix} \mathbf{s}^{(1)} \\ \mathbf{d}^{(1)} \end{bmatrix} = \begin{bmatrix} \mathbf{s}^{(2)} \\ \mathbf{d}^{(1)} \end{bmatrix}$. From formula (3.25) we can write U_2 in block form as

$$U_2 = \begin{bmatrix} I & -S^{-1} \\ 0 & I \end{bmatrix}. \tag{3.26}$$

The final step in the lifting process is a *normalization* $\begin{bmatrix} \mathbf{s} \\ \mathbf{d} \end{bmatrix} = D \begin{bmatrix} \mathbf{s}^{(2)} \\ \mathbf{d}^{(1)} \end{bmatrix}$, where

$$D = \frac{1}{\sqrt{2}} \begin{bmatrix} (\sqrt{3}-1)I & 0 \\ 0 & (\sqrt{3}+1)I \end{bmatrix} \tag{3.27}$$

is diagonal. Note that $(\sqrt{3}-1)(\sqrt{3}+1) = 2$, so $\det D = 1$. The one-scale Daub4 analysis transform is the product of these transformations:

$$\mathbf{T_a} = DU_2PU_1 \boxed{\text{split}}. \tag{Daub4}$$

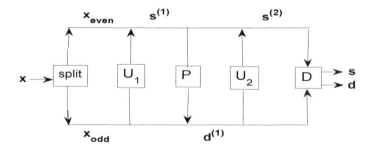

Fig. 3.7   Daub4 analysis lifting steps (split, update, predict, update, normalize)

Figure 3.7 is a flow chart that indicates the lifting steps. Notice that the update transformations only change the signals in the top branch, while the prediction transformation only changes the signal in the bottom branch, as indicated by the arrows.

We calculate (using block multiplication of the matrices) that

$$DU_2 PU_1 = \frac{1}{4\sqrt{2}} \begin{bmatrix} (aI + cS^{-1}) & (bI + dS^{-1}) \\ -(bI + dS) & (aI + cS) \end{bmatrix}, \tag{3.28}$$

where $a = 1+\sqrt{3}$, $b = 3+\sqrt{3}$, $c = 3-\sqrt{3}$, and $d = 1-\sqrt{3}$. We call the transformation (3.28) the Daub4 *polyphase analysis matrix in the time-domain*. Notice that each block is a polynomial in the shift operator. When we identify an $N/2$-periodic function on $\mathbb{Z}$ with a column vector in $\mathbb{R}^{N/2}$, then each block in the time-domain polyphase matrix becomes an $N/2 \times N/2$ circulant matrix.

Theorem 3.1. *The* Daub4 *wavelet analysis matrix* $\mathbf{T_a}$ *is orthogonal.*

Proof. From (3.28) we calculate

$$\mathbf{T_a T_a^T} = (DU_2 PU_1) \boxed{\text{split}} \boxed{\text{split}}^{\mathrm{T}} (DU_2 PU_1)^{\mathrm{T}}$$

$$= \frac{1}{32} \begin{bmatrix} (aI + cS^{-1}) & (bI + dS^{-1}) \\ -(bI + dS) & (aI + cS) \end{bmatrix} \begin{bmatrix} (aI + cS^{-1}) & -(bI + dS) \\ (bI + dS^{-1}) & (aI + cS) \end{bmatrix},$$

since the permutation matrix $\boxed{\text{split}}$ is orthogonal. Carrying out the block multiplication, we find that

$$\mathbf{T_a T_a^T} = \frac{1}{32} \begin{bmatrix} \alpha I + \beta(S + S^{-1}) & 0 \\ 0 & \alpha I + \beta(S + S^{-1}) \end{bmatrix},$$

where $\alpha = a^2 + b^2 + c^2 + d^2$ and $\beta = ac + bd$. An easy calculation shows that $\alpha = 32$ and $\beta = 0$.[4] Hence $\mathbf{T_a T_a^T}$ is the identity matrix. □

Example 3.7. When $N = 4$, then from (3.28) we find that the Daub4 analysis matrix is

$$\mathbf{T_a} = \frac{1}{4\sqrt{2}} \begin{bmatrix} a & c & b & d \\ c & a & d & b \\ -b & -d & a & c \\ -d & -b & c & a \end{bmatrix} \boxed{\text{split}} = \frac{1}{4\sqrt{2}} \begin{bmatrix} a & b & c & d \\ c & d & a & b \\ -b & a & -d & c \\ -d & c & -b & a \end{bmatrix},$$

since right multiplication by $\boxed{\text{split}}$ interchanges columns 2 and 3. Note that the second row is obtained from the first row by shifting right two places (with wraparound). The same relation holds between the third and fourth rows. Also, the third row is obtained from the second row by reversing the entries and inserting alternating signs (likewise for the fourth and the first rows); this procedure on vectors of even length automatically produces pairs of mutually orthogonal vectors. Since the analysis matrix is orthogonal, the synthesis matrix is the transpose:

$$\mathbf{T_s} = \frac{1}{4\sqrt{2}} \begin{bmatrix} a & c & -b & -d \\ b & d & a & c \\ c & a & -d & -b \\ d & b & c & a \end{bmatrix}.$$

The columns of $\mathbf{T_s}$ have the same shift pattern as the rows of $\mathbf{T_a}$. ∎

---

[4] In Section 4.10 we will solve these quadratic equations to obtain a, b, c, d.

We now have introduced three different wavelet transforms: Haar, CDF$(2,2)$, and Daub4. It is instructive to compare the one-scale trend and detail that each of these transforms creates for the same test signal.

Example 3.8. We create a digital signal by taking 64 samples of the function $f(t)$ $(0 \leq t \leq 1)$ that is a parabolic arc (for $0 \leq t \leq 1/4$), followed by a step function (for $1/4 < t \leq 3/4$), and terminating with a linear function (for $3/4 < t \leq 1$); see Figure 3.8. The function $f(t)$ is chosen to exhibit several types of behavior: it has a discontinuity at $t = 1/2$ (corresponding to $n = 32$ for the digital signal) and its derivative has discontinuities at $t = 1/4$ (corresponding to $n = 16$) and $t = 3/4$ (corresponding to $n = 48$).

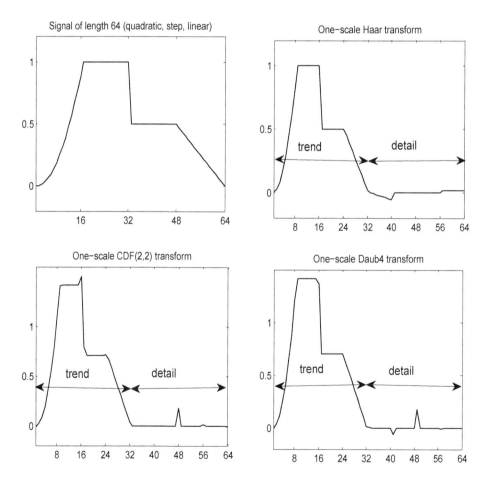

Fig. 3.8   Signal of length 64 with Haar, CDF$(2,2)$, and Daub4 transforms

The graphs of three different one-scale wavelet transforms of the signal are shown in Figure 3.8. The trend of the Haar transform is very close to the original signal, while the trends in the CDF$(2, 2)$ and Daub4 transforms are distorted around $n = 16$ by the jump discontinuity in the analogue test signal at $t = 1/2$. This jump discontinuity appears in the Daub4 detail (around $n = 40$) but not in the CDF$(2, 2)$ detail. The slope discontinuity in the analogue test signal at $t = 3/4$ appears in both the CDF$(2, 2)$ and Daub4 details around $n = 48$. Notice that the detail portions of all three transforms contain information about the location (in time) of the singularities in the sampled analogue signal, with the most precise information coming from the Daub4 transform. This is a feature that the finite Fourier transform lacks. We shall return to this example in Section 5.5.4. ∎

Remark 3.4. Many types of discrete wavelet transforms have been created starting with the pioneering work of Ingrid Daubechies, Stéphane Mallat, Yves Meyer, and many other mathematicians in the late 1980's; we shall construct some of these transforms in Chapter 4 and their analog versions in Chapter 5. The choice of the transform to use in a particular situation depends on the characteristics of the signal or image to be analyzed. This adaptability is one of chief advantages of wavelet methods over the finite Fourier transform for signal and image processing.

3.5 Wavelet Bases for Periodic Signals

We now study the decomposition of a signal into trend and details given by multi-scale wavelet transforms via the pyramid algorithm. The building blocks in this algorithm are the one-scale wavelet analysis and synthesis matrices. In Section 3.2 we examined some examples of these transforms (Haar, CDF$(2, 2)$, Daub4) and found that their matrices had some features in common. We now consider the general case.

3.5.1 *Lifting steps and polyphase matrices*

A one-scale periodic wavelet analysis transform $\mathbf{T_a}$ is obtained by the lifting procedure as

$$\mathbf{T_a} = D \cdot (\text{product of updates and predictions}) \cdot \boxed{\text{split}} \,. \qquad (3.29)$$

Here $\boxed{\text{split}}$ separates the input signal $\mathbf{x}$ of even length N into subsignals $\mathbf{x}_{\text{even}}$ and $\mathbf{x}_{\text{odd}}$ of length $N/2$, which we identify with real-valued $N/2$-periodic functions on $\mathbb{Z}$. The *update* transformations U and the *prediction* transformations P that occur in the lifting steps have 2×2 block form

$$U = \begin{bmatrix} I & p(S) \\ 0 & I \end{bmatrix} \quad \text{and} \quad P = \begin{bmatrix} I & 0 \\ q(S) & I \end{bmatrix}.$$

Here S is the shift operator and $p(z)$ and $q(z)$ are Laurent polynomials in the variable z. The normalization matrix is $D = \begin{bmatrix} \alpha I & 0 \\ 0 & \beta I \end{bmatrix}$ with $\alpha\beta \neq 0$ (usually

$\alpha\beta = 1$ so that $\det D = 1$). When we carry out the matrix multiplication in (3.29) we get a matrix of the form

$$\mathbf{T_a} = \begin{bmatrix} p_{00}(S) & p_{01}(S) \\ p_{10}(S) & p_{11}(S) \end{bmatrix} \boxed{\text{split}}\,, \tag{3.30}$$

with $p_{ij}(z)$ Laurent polynomials with real coefficients. These polynomials don't depend on the signal length N, so the transformation $\mathbf{T_a}$ is defined by the same lifting-step algorithm on signals of any even length.[5]

Definition 3.1. The 2×2 matrix $\mathbf{H}_p(z) = \begin{bmatrix} p_{00}(z^{-1}) & p_{01}(z^{-1}) \\ p_{10}(z^{-1}) & p_{11}(z^{-1}) \end{bmatrix}$ obtained by replacing[6] the shift operator S by the variable z^{-1} in (3.30) is called the *polyphase analysis matrix in the frequency domain* for the wavelet transform.

The polyphase analysis matrix is determined by the four Laurent polynomials $p_{ij}(z)$ obtained by multiplying out the lifting step matrices. Going in the opposite direction, we will construct a lifting step factorization starting from the polyphase matrix in Section 4.7.

Each of the factors in (3.29) is an invertible linear transformation, so the product $\mathbf{T_a}$ is invertible. The inverse wavelet synthesis transform $\mathbf{T_s} = \mathbf{T_a}^{-1}$ also can be expressed as a product of prediction, update, and normalization matrices with a final merge transformation. Thus it has the form

$$\mathbf{T_s} = \boxed{\text{merge}} \begin{bmatrix} q_{00}(S) & q_{01}(S) \\ q_{10}(S) & q_{11}(S) \end{bmatrix} \quad \text{with } q_{ij}(z) \text{ Laurent polynomials.} \tag{3.31}$$

Definition 3.2. The 2×2 matrix $\mathbf{G}_p(z) = \begin{bmatrix} q_{00}(z^{-1}) & q_{01}(z^{-1}) \\ q_{10}(z^{-1}) & q_{11}(z^{-1}) \end{bmatrix}$ is called the *polyphase synthesis matrix in the frequency domain* for the wavelet transform.

For all wavelet transforms the polyphase analysis and synthesis matrices (either in the time domain or in the frequency domain) are mutual inverses: $\mathbf{H}_p(z)^{-1} = \mathbf{G}_p(z)$ for all $z \neq 0$. For an orthogonal wavelet transform, such as Daub4, we have $\mathbf{G}_p(z) = \mathbf{H}_p(z^{-1})^{\mathrm{T}}$ since $S^{\mathrm{T}} = S^{-1}$. Hence $q_{ij}(z) = p_{ji}(z^{-1})$ in this case.

Example 3.9. For the Daub4 transform, we see from equation (3.28) that the polyphase analysis matrix is

$$\mathbf{H}_p(z) = \frac{1}{4\sqrt{2}} \begin{bmatrix} (a + cz) & (b + dz) \\ -(b + dz) & (a + cz) \end{bmatrix},$$

where $a = 1 + \sqrt{3}$, $b = 3 + \sqrt{3}$, $c = 3 - \sqrt{3}$, $d = 1 - \sqrt{3}$. ∎

---

[5]For fixed N, however, we can express $p(S)$ and $q(S)$ as polynomials with only positive powers of S using the relation $S^{-1} = S^{(N/2)-1}$ on $N/2$-periodic signals.

[6]We substitute z^{-1} (rather than z) for S in the definition of the polyphase analysis matrix to be consistent with Theorem 2.3 and the z-transform in Chapter 4.

Example 3.10. For the CDF$(2, 2)$ analysis transform we have

$$\mathbf{T_a} = \begin{bmatrix} \sqrt{2}I & 0 \\ 0 & (1/\sqrt{2})I \end{bmatrix} \begin{bmatrix} I & \frac{1}{4}(I+S) \\ 0 & I \end{bmatrix} \begin{bmatrix} I & 0 \\ -\frac{1}{2}(I+S^{-1}) & I \end{bmatrix} \boxed{\text{split}}$$

$$= \frac{1}{4\sqrt{2}} \begin{bmatrix} (-S^{-1}+6I-S) & (2I+2S) \\ -(2S^{-1}+2I) & 4I \end{bmatrix} \boxed{\text{split}} . \qquad (3.32)$$

Thus the polyphase analysis matrix in the frequency domain is

$$\mathbf{H}_p(z) = \frac{1}{4\sqrt{2}} \begin{bmatrix} (-z+6-z^{-1}) & (2+2z^{-1}) \\ -(2z+2) & 4 \end{bmatrix} .$$

Note that the polyphase analysis matrix has determinant 1, since the prediction, update, and normalization matrices all have determinant 1. We can obtain the CDF$(2, 2)$ synthesis transform from the analysis transform by inverting the prediction, update, and normalization matrices:

$$\mathbf{T_s} = \boxed{\text{merge}} P^{-1}U^{-1}D^{-1} = \boxed{\text{merge}} \frac{1}{4\sqrt{2}} \begin{bmatrix} 4I & -(2I+2S) \\ (2S^{-1}+2I) & (-S^{-1}+6I-S) \end{bmatrix} .$$

Thus the polyphase synthesis matrix in the frequency domain is

$$\mathbf{G}_p(z) = \frac{1}{4\sqrt{2}} \begin{bmatrix} 4 & -(2+2z^{-1}) \\ (2z+2) & (-z+6-z^{-1}) \end{bmatrix} .$$

Note that $\mathbf{G}_p(z) \neq \mathbf{H}_p(z^{-1})^{\mathrm{T}}$, so the CDF$(2, 2)$ transform is not orthogonal. ∎

Example 3.11. As another illustration of the lifting procedure consider the CDF$(3, 1)$ transform. This transform starts with an update step followed by a prediction step. Define

$$\mathbf{s}^{(1)}[n] = \mathbf{x}_{\text{even}}[n] - (1/3)\mathbf{x}_{\text{odd}}[n-1] ,$$

$$\mathbf{d}^{(1)}[n] = \mathbf{x}_{\text{odd}}[n] - (9/8)\mathbf{s}^{(1)}[n] - (3/8)\mathbf{s}^{(1)}[n+1] .$$

Here $\mathbf{s}^{(1)}$ and $\mathbf{d}^{(1)}$ are the first versions of the trend and detail. We write these equations in vector form using the shift operator as

$$\mathbf{s}^{(1)} = \mathbf{x}_{\text{even}} - (1/3)S\,\mathbf{x}_{\text{odd}} ,$$

$$\mathbf{d}^{(1)} = \mathbf{x}_{\text{odd}} - (9/8)I + (3/8)S^{-1})\,\mathbf{s}^{(1)} .$$

In 2×2 block-matrix form these equations are

$$\begin{bmatrix} \mathbf{s}^{(1)} \\ \mathbf{x}_{\text{odd}} \end{bmatrix} = \begin{bmatrix} I & -(1/3)S \\ 0 & I \end{bmatrix} \begin{bmatrix} \mathbf{x}_{\text{even}} \\ \mathbf{x}_{\text{odd}} \end{bmatrix} = U_1 \begin{bmatrix} \mathbf{x}_{\text{even}} \\ \mathbf{x}_{\text{odd}} \end{bmatrix} ,$$

$$\begin{bmatrix} \mathbf{s}^{(1)} \\ \mathbf{d}^{(1)} \end{bmatrix} = \begin{bmatrix} I & 0 \\ -((9/8)I + (3/8)S^{-1}) & I \end{bmatrix} \begin{bmatrix} \mathbf{s}^{(1)} \\ \mathbf{d}^{(1)} \end{bmatrix} = P \begin{bmatrix} \mathbf{s}^{(1)} \\ \mathbf{x}_{\text{odd}} \end{bmatrix} ,$$

where U_1 is the first update matrix and P is the prediction matrix.

For this transform there is a second update step to obtain the second version $\mathbf{s}^{(2)}$ of the trend:

$$\mathbf{s}^{(2)} = \mathbf{s}^{(1)} + (4/9)\mathbf{d}^{(1)} .$$

We write this step in block-matrix form as

$$\begin{bmatrix} \mathbf{s}^{(2)} \\ \mathbf{d}^{(1)} \end{bmatrix} = \begin{bmatrix} I & (4/9)I \\ 0 & I \end{bmatrix} \begin{bmatrix} \mathbf{s}^{(1)} \\ \mathbf{d}^{(1)} \end{bmatrix} = U_2 \begin{bmatrix} \mathbf{s}^{(1)} \\ \mathbf{d}^{(1)} \end{bmatrix},$$

where U_2 is the second update matrix.

We finish with a normalization step to obtain the final trend $\mathbf{s}$ and detail $\mathbf{d}$:

$$\mathbf{s} = (3/\sqrt{2})\,\mathbf{s}^{(2)}, \quad \mathbf{d} = (\sqrt{2}/3)\,\mathbf{d}^{(1)}.$$

In block-matrix form

$$\begin{bmatrix} \mathbf{s} \\ \mathbf{d} \end{bmatrix} = \begin{bmatrix} (3/\sqrt{2})I & \mathbf{0} \\ \mathbf{0} & (\sqrt{2}/3)I \end{bmatrix} \begin{bmatrix} \mathbf{s}^{(2)} \\ \mathbf{d}^{(1)} \end{bmatrix} = D \begin{bmatrix} \mathbf{s}^{(2)} \\ \mathbf{d}^{(1)} \end{bmatrix},$$

where D is a diagonal matrix with determinant one.

The CDF$(3,1)$ transform is the product of these lifting steps:

$$\begin{bmatrix} \mathbf{s} \\ \mathbf{d} \end{bmatrix} = D U_2 P U_1 \boxed{\text{split}}\, \mathbf{x}.$$

Thus the CDF$(3,1)$ analysis matrix is $\mathbf{T_a} = D U_2 P U_1 \boxed{\text{split}}$. Carrying out the matrix multiplications, we find that

$$\mathbf{T_a} = \frac{1}{4\sqrt{2}} \begin{bmatrix} (6I - 2S^{-1}) & (-2S + 6I) \\ -(3I + S^{-1}) & (S + 3I) \end{bmatrix} \boxed{\text{split}}. \tag{3.33}$$

Hence the polyphase analysis matrix in the frequency domain is

$$\mathbf{H}_p(z) = \frac{1}{4\sqrt{2}} \begin{bmatrix} (6 - 2z) & (-2z^{-1} + 6) \\ -(3 + z) & (z^{-1} + 3) \end{bmatrix}.$$

The lifting steps occur in the same order as for the Daub4 transform (see Figure 3.7), and the polyphase matrix has determinant 1.

To obtain the synthesis transform, we replace each lifting step by its inverse. Thus synthesis matrix is $\mathbf{T_s} = \boxed{\text{merge}}\, U_1^{-1} P^{-1} U_2^{-1} D^{-1}$, where

$$U_1^{-1} = \begin{bmatrix} I & (1/3)S \\ 0 & I \end{bmatrix}, \qquad P^{-1} = \begin{bmatrix} I & 0 \\ ((9/8)I + (3/8)S^{-1}) & I \end{bmatrix},$$

$$U_2^{-1} = \begin{bmatrix} I & -(4/9)I \\ 0 & I \end{bmatrix}, \qquad D^{-1} = \begin{bmatrix} (\sqrt{2}/3)I & 0 \\ 0 & (3/\sqrt{2})I \end{bmatrix}.$$

Notice that the inverse of a prediction matrix is still a prediction matrix, and likewise for an update matrix or normalization matrix. ∎

There is a CDF(p,q) wavelet transform for each pair (p,q) of positive integers with $p+q$ an even integer. This family of transforms was first constructed in the article [Cohen, Daubechies, and Feauveau (1992)]; we will study them systematically in Section 4.5.3.[7]

---

[7]See Exercises 3.8 #7 and #8 for more examples of the CDF family of transforms constructed by lifting.

3.5.2   One-scale wavelet matrices

We return to a general wavelet transform of real N-periodic signals, where $N = 2m$ is even. Let $\mathbf{T_a}$ be a one-scale wavelet analysis matrix given by a lifting-step formula of the type (3.29). Write $\mathbf{T_a}$ in terms of its rows as

$$
\mathbf{T_a} =
\begin{bmatrix}
\mathbf{u}_0 \\
\vdots \\
\mathbf{u}_{m-1} \\
\mathbf{v}_0 \\
\vdots \\
\mathbf{v}_{m-1}
\end{bmatrix}
=
\begin{bmatrix}
\mathbf{U} \\
\mathbf{V}
\end{bmatrix} ,
\tag{3.34}
$$

where $\mathbf{u}_j$, $\mathbf{v}_j$ are $1 \times 2m$ row vectors and $\mathbf{U}$, $\mathbf{V}$ are $m \times 2m$ matrices. Let $\mathbf{T_s}$ be the inverse synthesis matrix. We write $\mathbf{T_s}$ in terms of its columns as

$$
\mathbf{T_s} = \left[\, \widetilde{\mathbf{u}}_0 , \, \dots , \, \widetilde{\mathbf{u}}_{m-1} , \widetilde{\mathbf{v}}_0 , \, \dots , \, \widetilde{\mathbf{v}}_{m-1} \right] = \left[\, \widetilde{\mathbf{U}} \quad \widetilde{\mathbf{V}} \,\right] ,
\tag{3.35}
$$

where $\widetilde{\mathbf{u}}_j$, $\widetilde{\mathbf{v}}_j$ are $2m \times 1$ column vectors and $\widetilde{\mathbf{U}}$, $\widetilde{\mathbf{V}}$ are $2m \times m$ matrices. Recall from Section 1.4 that the matrix inverse property $\mathbf{T_a T_s} = I$ can be expressed as the *biorthogonality relations*

$$
\begin{aligned}
\mathbf{u}_j \widetilde{\mathbf{u}}_k &= \delta[j - k], & \mathbf{u}_j \widetilde{\mathbf{v}}_k &= 0, \\
\mathbf{v}_j \widetilde{\mathbf{u}}_k &= 0, & \mathbf{v}_j \widetilde{\mathbf{v}}_k &= \delta[j - k],
\end{aligned}
\tag{3.36}
$$

for $j, k = 0, 1, \dots, m - 1$, where $\delta[n] = 1$ if $n = 0$ and $\delta[n] = 0$ if $n \neq 0$. These relations are the same as the matrix equations

$$
\begin{aligned}
\mathbf{U} \widetilde{\mathbf{U}} &= I, & \mathbf{U} \widetilde{\mathbf{V}} &= 0, \\
\mathbf{V} \widetilde{\mathbf{U}} &= 0, & \mathbf{V} \widetilde{\mathbf{V}} &= I.
\end{aligned}
$$

When $\mathbf{T_a}$ is an orthogonal matrix (as for the Haar or Daub4 transforms) we have $\widetilde{\mathbf{U}} = \mathbf{U}^{\mathrm{T}}$ and $\widetilde{\mathbf{V}} = \mathbf{V}^{\mathrm{T}}$. In general, these matrices are all different (see Example 3.12).

We found in Section 2.4 that a circulant matrix is completely determined by its first column (or first row). We now show that the wavelet analysis matrix $\mathbf{T_a}$ is completely determined by the two row vectors $\mathbf{u}_0$ and $\mathbf{v}_0$. Likewise, the wavelet synthesis matrix $\mathbf{T_s}$ is completely determined by the two column vectors $\widetilde{\mathbf{u}}_0$ and $\widetilde{\mathbf{v}}_0$. We have already seen this pattern in the Haar, CDF(2, 2) and Daub4 transforms.

Theorem 3.2. *Let* $\mathbf{T_a}$ *(analysis) and* $\mathbf{T_s}$ *(synthesis) be one-scale wavelet matrices with rows and columns as in* (3.34) *and* (3.35). *Let* S_N *be the* $N \times N$ *shift matrix* ($N = 2m$).

(1) Call $\mathbf{u}_0$ *the scaling vector and* $\mathbf{v}_0$ *the wavelet vector. For* $k = 1, \dots, m - 1$

$$
\begin{aligned}
\mathbf{u}_k &= \mathbf{u}_0(S_N)^{-2k} && \text{(shift components of } \mathbf{u}_0 \text{ right } 2k \text{ positions with wraparound)}, \\
\mathbf{v}_k &= \mathbf{v}_0(S_N)^{-2k} && \text{(shift components of } \mathbf{v}_0 \text{ right } 2k \text{ positions with wraparound)}.
\end{aligned}
$$

(2) Call $\widetilde{\mathbf{u}}_0$ the trend vector *and $\widetilde{\mathbf{v}}_0$ the* detail vector. *For $k = 1, \ldots, m-1$*

$$\widetilde{\mathbf{u}}_k = (S_N)^{2k}\widetilde{\mathbf{u}}_0 \quad (\text{shift components of } \widetilde{\mathbf{u}}_0 \text{ down } 2k \text{ positions with wraparound}),$$

$$\widetilde{\mathbf{v}}_k = (S_N)^{2k}\widetilde{\mathbf{v}}_0 \quad (\text{shift components of } \widetilde{\mathbf{v}}_0 \text{ down } 2k \text{ positions with wraparound}).$$

Proof. The key point is the relation between shifting and splitting (or merging). Let $\mathbf{x} \in \mathbb{R}^N$. Shifting the components of $\mathbf{x}$ down by *two positions* (with wraparound) is the same as shifting the even and odd parts of $\mathbf{x}$ down by *one position* (with wraparound). Thus with S_m denoting the $m \times m$ shift matrix,

$$\boxed{\text{split}}\,(S_N)^2\mathbf{x} = \begin{bmatrix} S_m\mathbf{x}_{\text{even}} \\ S_m\mathbf{x}_{\text{odd}} \end{bmatrix} = \begin{bmatrix} S_m & 0 \\ 0 & S_m \end{bmatrix}\boxed{\text{split}}\,\mathbf{x}.$$

Since this relation holds for all vectors $\mathbf{x}$, it implies the matrix equation

$$\boxed{\text{split}}\,(S_N)^2 = \begin{bmatrix} S_m & \mathbf{0} \\ \mathbf{0} & S_m \end{bmatrix}\boxed{\text{split}}. \tag{3.37}$$

Take $\mathbf{T_a}$ as in (3.30), write $C_{ij} = p_{ij}(S_m)$, and use (3.37) to calculate

$$\mathbf{T_a}(S_N)^2 = \begin{bmatrix} C_{00} & C_{01} \\ C_{10} & C_{11} \end{bmatrix}\boxed{\text{split}}\,(S_N)^2 = \begin{bmatrix} C_{00} & C_{01} \\ C_{10} & C_{11} \end{bmatrix}\begin{bmatrix} S_m & \mathbf{0} \\ \mathbf{0} & S_m \end{bmatrix}\boxed{\text{split}}$$

$$= \begin{bmatrix} C_{00}S_m & C_{01}S_m \\ C_{10}S_m & C_{11}S_m \end{bmatrix}\boxed{\text{split}} = \begin{bmatrix} S_m & \mathbf{0} \\ \mathbf{0} & S_m \end{bmatrix}\begin{bmatrix} C_{00} & C_{01} \\ C_{10} & C_{11} \end{bmatrix}\boxed{\text{split}}.$$

Here we have used the fundamental property $C_{ij}S_m = S_mC_{ij}$ of a circulant matrix. Thus

$$\mathbf{T_a}(S_N)^2 = \begin{bmatrix} S_m & \mathbf{0} \\ \mathbf{0} & S_m \end{bmatrix}\mathbf{T_a}. \tag{3.38}$$

Now write $\mathbf{T_a}$ in terms of the matrices $\mathbf{U}$, $\mathbf{V}$ and apply the last equation:

$$\begin{bmatrix} \mathbf{U} \\ \mathbf{V} \end{bmatrix}(S_N)^2 = \begin{bmatrix} S_m & \mathbf{0} \\ \mathbf{0} & S_m \end{bmatrix}\begin{bmatrix} \mathbf{U} \\ \mathbf{V} \end{bmatrix}.$$

Carrying out the block multiplication on each side of this equation, we see that

$$\begin{bmatrix} \mathbf{U}(S_N)^2 \\ \mathbf{V}(S_N)^2 \end{bmatrix} = \begin{bmatrix} S_m\mathbf{U} \\ S_m\mathbf{V} \end{bmatrix}.$$

Hence

$$S_m\mathbf{U} = \mathbf{U}(S_N)^2 \quad \text{and} \quad S_m\mathbf{V} = \mathbf{V}(S_N)^2 \tag{3.39}$$

(all the matrix products in these equations are defined since $\mathbf{U}$ and $\mathbf{V}$ are of size $m \times N$). Now left multiplication by S_m shifts the rows of $\mathbf{U}$ and $\mathbf{V}$ down one position, so matching up rows on both sides of equation (3.39) gives

$$\begin{bmatrix} \mathbf{u}_{m-1} \\ \mathbf{u}_0 \\ \vdots \\ \mathbf{u}_{m-2} \end{bmatrix} = \begin{bmatrix} \mathbf{u}_0(S_N)^2 \\ \mathbf{u}_1(S_N)^2 \\ \vdots \\ \mathbf{u}_{m-1}(S_N)^2 \end{bmatrix} \quad \text{and} \quad \begin{bmatrix} \mathbf{v}_{m-1} \\ \mathbf{v}_0 \\ \vdots \\ \mathbf{v}_{m-2} \end{bmatrix} = \begin{bmatrix} \mathbf{v}_0(S_N)^2 \\ \mathbf{v}_1(S_N)^2 \\ \vdots \\ \mathbf{v}_{m-1}(S_N)^2 \end{bmatrix}. \tag{3.40}$$

Comparing the rows in equations (3.40) and multiplying each row on the right by $(S_N)^{-2}$, we see that

$$\mathbf{u}_1 = \mathbf{u}_0(S_N)^{-2}, \quad \mathbf{u}_2 = \mathbf{u}_1(S_N)^{-2} = \mathbf{u}_0(S_N)^{-4}, \ldots,$$
$$\mathbf{v}_1 = \mathbf{v}_0(S_N)^{-2}, \quad \mathbf{v}_2 = \mathbf{v}_1(S_N)^{-2} = \mathbf{v}_0(S_N)^{-4}, \ldots.$$

Multiplying a row vector $\mathbf{u}$ on the right by $(S_N)^{-2}$ shifts the components of $\mathbf{u}$ to the right two positions (with periodic wraparound). This proves that the rows of $\mathbf{T_a}$ follow the pattern asserted by Theorem 3.2.

The proof for the inverse matrix $\mathbf{T_s}$ follows the same pattern. Taking inverses in (3.38), we obtain the relation

$$(S_N)^{-2}\mathbf{T_s} = \mathbf{T_s} \begin{bmatrix} (S_m)^{-1} & \mathbf{0} \\ \mathbf{0} & (S_m)^{-1} \end{bmatrix}.$$

Writing $\mathbf{T_s}$ in block form and multiplying by $(S_N)^2$, we obtain

$$\begin{bmatrix} \widetilde{\mathbf{U}} & \widetilde{\mathbf{V}} \end{bmatrix} = (S_N)^2 \begin{bmatrix} \widetilde{\mathbf{U}} & \widetilde{\mathbf{V}} \end{bmatrix} \begin{bmatrix} (S_m)^{-1} & \mathbf{0} \\ \mathbf{0} & (S_m)^{-1} \end{bmatrix}.$$

Carrying out the block multiplication on each side of this equation gives

$$\begin{bmatrix} \widetilde{\mathbf{U}} & \widetilde{\mathbf{V}} \end{bmatrix} = \begin{bmatrix} (S_N)^2\widetilde{\mathbf{U}}(S_m)^{-1} & (S_N)^2\widetilde{\mathbf{V}}(S_m)^{-1} \end{bmatrix}.$$

Hence

$$\widetilde{\mathbf{U}} = (S_N)^2\widetilde{\mathbf{U}}(S_m)^{-1} \quad \text{and} \quad \widetilde{\mathbf{V}} = (S_N)^2\widetilde{\mathbf{V}}(S_m)^{-1}.$$

Since right multiplication by $(S_m)^{-1}$ shifts the columns of $\widetilde{\mathbf{U}}$ and $\widetilde{\mathbf{V}}$ to the right one position, we find that

$$\begin{bmatrix} \widetilde{\mathbf{u}}_0, \widetilde{\mathbf{u}}_1, \ldots, \widetilde{\mathbf{u}}_{m-1} \end{bmatrix} = \begin{bmatrix} (S_N)^2\widetilde{\mathbf{u}}_{m-1}, (S_N)^2\widetilde{\mathbf{u}}_0, \ldots, (S_N)^2\widetilde{\mathbf{u}}_{m-2} \end{bmatrix},$$
$$\begin{bmatrix} \widetilde{\mathbf{v}}_0, \widetilde{\mathbf{v}}_1, \ldots, \widetilde{\mathbf{v}}_{m-1} \end{bmatrix} = \begin{bmatrix} (S_N)^2\widetilde{\mathbf{v}}_{m-1}, (S_N)^2\widetilde{\mathbf{v}}_0, \ldots, (S_N)^2\widetilde{\mathbf{v}}_{m-2} \end{bmatrix}.$$

Matching up columns in these equations, we obtain

$$\widetilde{\mathbf{u}}_1 = (S_N)^2\,\widetilde{\mathbf{u}}_0, \quad \widetilde{\mathbf{u}}_2 = (S_N)^2\,\widetilde{\mathbf{u}}_1 = (S_N)^4\,\widetilde{\mathbf{u}}_0, \ldots,$$
$$\widetilde{\mathbf{v}}_1 = (S_N)^2\,\widetilde{\mathbf{v}}_0, \quad \widetilde{\mathbf{v}}_2 = (S_N)^2\,\widetilde{\mathbf{v}}_1 = (S_N)^4\,\widetilde{\mathbf{v}}_0, \ldots.$$

This proves that the columns of $\mathbf{T_s}$ follow the pattern asserted by Theorem 3.2. □

Remark 3.5. If $\mathbf{T_a}$ is an orthogonal matrix (e.g. for the Daub4 or for the normalized Haar transforms), then the trend vector is the transpose of the scaling vector, and the detail vector is the transpose of the wavelet vector.

We showed in Section 2.4 that the product of a circulant matrix with a column vector can be expressed in terms of circular convolution with the first column of the matrix. A similar result holds for wavelet matrices, but now we must also use the downsampling operation $\mathbf{y} \longrightarrow \mathbf{y}_{\text{even}}$ to obtain the separation into trend and detail.

Because the matrices $\mathbf{U}$ and $\mathbf{V}$ have a special row structure (rather than column structure), we also need to use the *flip* $\mathbf{u} \longrightarrow \overset{\vee}{\mathbf{u}}$ of $1 \times N$ row vectors defined by

$$\overset{\vee}{\mathbf{u}} = [\mathbf{u}[N-1], \ldots, \mathbf{u}[1], \mathbf{u}[0]] \quad \text{if} \quad \mathbf{u} = [\mathbf{u}[0], \mathbf{u}[1], \ldots, \mathbf{u}[N-1]] \; .$$

Corollary 3.1. *Let* $\mathbf{x} \in \mathbb{R}^N$. *View* $\mathbf{x}$, *the scaling vector* $\mathbf{u}_0$, *and the wavelet vector* $\mathbf{v}_0$ *as* N-*periodic functions on the integers* $\mathbb{Z}$. *Then*

$$\mathbf{T_a}\,\mathbf{x} = \begin{bmatrix} (\overset{\vee}{\mathbf{u}}_0 \star \mathbf{x})_{\text{even}} \\ (\overset{\vee}{\mathbf{v}}_0 \star \mathbf{x})_{\text{even}} \end{bmatrix}, \tag{3.41}$$

where $\star$ *denotes circular convolution.*

Proof. Since $\left(\mathbf{u}_0 S_N^{-2j}\right)[k] = \mathbf{u}_0[k-2j]$ when we view $\mathbf{u}_0$ as an N-periodic function on $\mathbb{Z}$, Theorem 3.2 gives

$$(\mathbf{Ux})[j] = \left(\mathbf{u}_0 S_N^{-2j}\right)\mathbf{x} = \sum_{k=0}^{N-1} \mathbf{u}_0[k-2j]\,\mathbf{x}[k] = \sum_{k=0}^{N-1} \overset{\vee}{\mathbf{u}}_0\,[2j-k]\,\mathbf{x}[k]$$

$$= (\overset{\vee}{\mathbf{u}}_0 \star \mathbf{x}[k])[2j] = (\overset{\vee}{\mathbf{u}}_0 \star \mathbf{x}[k])_{\text{even}}[j]\,.$$

The same calculation applies to $\mathbf{Vx}$. □

Example 3.12. Consider the CDF$(2,2)$ wavelet transform on $\mathbb{R}^8$. Multiply the polyphase matrix (3.32) by the permutation matrix $\boxed{\text{split}}$ and put the normalization step into the synthesis matrix. This gives (with the help of MATLAB) the one-step analysis and synthesis matrices

$$\mathbf{T_a} = \begin{bmatrix} 6 & 2 & -1 & 0 & 0 & 0 & -1 & 2 \\ -1 & 2 & 6 & 2 & -1 & 0 & 0 & 0 \\ 0 & 0 & -1 & 2 & 6 & 2 & -1 & 0 \\ -1 & 0 & 0 & 0 & -1 & 2 & 6 & 2 \\ -2 & 4 & -2 & 0 & 0 & 0 & 0 & 0 \\ 0 & 0 & -2 & 4 & -2 & 0 & 0 & 0 \\ 0 & 0 & 0 & 0 & -2 & 4 & -2 & 0 \\ -2 & 0 & 0 & 0 & 0 & 0 & -2 & 4 \end{bmatrix}, \quad \mathbf{T_s} = \frac{1}{32}\begin{bmatrix} 4 & 0 & 0 & 0 & -2 & 0 & 0 & -2 \\ 2 & 2 & 0 & 0 & 6 & -1 & 0 & -1 \\ 0 & 4 & 0 & 0 & -2 & -2 & 0 & 0 \\ 0 & 2 & 2 & 0 & -1 & 6 & -1 & 0 \\ 0 & 0 & 4 & 0 & 0 & -2 & -2 & 0 \\ 0 & 0 & 2 & 2 & 0 & -1 & 6 & -1 \\ 0 & 0 & 0 & 4 & 0 & 0 & -2 & -2 \\ 2 & 0 & 0 & 2 & -1 & 0 & -1 & 6 \end{bmatrix}.$$

In this case the scaling and wavelet vectors are

$$\mathbf{u}_0 = \begin{bmatrix} 6 & 2 & -1 & 0 & 0 & 0 & -1 & 2 \end{bmatrix} \quad \text{and} \quad \mathbf{v}_0 = \begin{bmatrix} -2 & 4 & -2 & 0 & 0 & 0 & 0 & 0 \end{bmatrix} \;.$$

Notice how the other rows of $\mathbf{T_a}$ are obtained from $\mathbf{u}_0$ and $\mathbf{v}_0$ by shifting as described in Theorem 3.2. The trend and detail vectors are

$$\widetilde{\mathbf{u}}_0 = \frac{1}{32}\begin{bmatrix} 4 & 2 & 0 & 0 & 0 & 0 & 0 & 2 \end{bmatrix}^{\mathrm{T}} \quad \text{and} \quad \widetilde{\mathbf{v}}_0 = \frac{1}{32}\begin{bmatrix} -2 & 6 & -2 & -1 & 0 & 0 & 0 & -1 \end{bmatrix}^{\mathrm{T}}\;.$$

The other columns of $\mathbf{T_s}$ are obtained from $\widetilde{\mathbf{u}}_0$ and $\widetilde{\mathbf{v}}_0$ by shifting as described in Theorem 3.2. The biorthogonality relations (3.36) imply that for $k = 0, \ldots, 7$,

$$\mathbf{u}_0\,(S_8)^{2k}\widetilde{\mathbf{u}}_0 = \delta[k]\,, \qquad \mathbf{u}_0\,(S_8)^{2k}\widetilde{\mathbf{v}}_0 = 0\,,$$
$$\mathbf{v}_0\,(S_8)^{2k}\widetilde{\mathbf{u}}_0 = 0\,, \qquad \mathbf{v}_0\,(S_8)^{2k}\widetilde{\mathbf{v}}_0 = \delta[k]\,. \tag{3.42}$$

Equations (3.42) can be checked directly from the formulas above, of course. ■

Remark 3.6. Wavelet matrices of any even size can be constructed using scaling and wavelet vectors with a fixed number of nonzero entries, as we will see in Chapter 4. For example, in the CDF$(2, 2)$ analysis matrix of size $N \geq 6$ the scaling vector has 5 nonzero entries and the wavelet vector 3 nonzero entries. For such wavelet matrices equation (3.41) shows that the computation of $\mathbf{T_a}\mathbf{x}$ only requires cN arithmetic operations, where c is a constant (independent of N) that depends on the number of nonzero entries in the scaling and wavelet vector. Thus the computation of wavelet transforms using such matrices is N times faster than ordinary multiplication of an $N \times N$ matrix with a vector in $\mathbb{R}^N$, which requires on the order of N^2 arithmetic operations.

Remark 3.7. In Example 3.12 we see an important computational feature[8] of the CDF$(2, 2)$ wavelet matrices: the entries in the analysis matrix can be chosen as integers, and the entries in the synthesis matrix as rational numbers with denominator 2^k for a fixed integer k (here $k = 5$). This means that when the signal has integer entries (which occurs in image processing, as we will see in Section 3.6), then all calculations done in binary (base 2) arithmetic are *exact* and *fast*, since division by 2 is just a bit shift.

3.5.3 *Trend and detail subspaces*

Given $\mathbf{x} \in \mathbb{R}^N$, we define the one-scale *trend transform* $\mathbf{s}$ and *detail transform* $\mathbf{d}$ of $\mathbf{x}$ using the analysis matrix $\mathbf{T_a}$:

$$\begin{bmatrix} \mathbf{s} \\ \mathbf{d} \end{bmatrix} = \mathbf{T_a}\mathbf{x} .$$

We will view vectors with m components as m-periodic functions, and use the symbol S (instead of S_m) for the shift operator. With this convention, Corollary 3.1 allows us to calculate $\mathbf{s}$ and $\mathbf{d}$ by circular convolution:

$$\mathbf{s} = (\overset{\vee}{\mathbf{u}}_0 \star \mathbf{x})_{\text{even}} \quad \text{and} \quad \mathbf{d} = (\overset{\vee}{\mathbf{v}}_0 \star \mathbf{x})_{\text{even}} ,$$

where $\mathbf{u}_0$ is the scaling vector and $\mathbf{v}_0$ is the wavelet vector for $\mathbf{T_a}$.

Since $\mathbf{x} = \mathbf{T_s}(\mathbf{T_a}\mathbf{x}) = \mathbf{T_s}\begin{bmatrix} \mathbf{s} \\ \mathbf{d} \end{bmatrix}$, we can reconstruct $\mathbf{x}$ from $\mathbf{s}$ and $\mathbf{d}$ by the synthesis matrix:

$$\mathbf{x} = \begin{bmatrix} \widetilde{\mathbf{U}} & \widetilde{\mathbf{V}} \end{bmatrix} \begin{bmatrix} \mathbf{s} \\ \mathbf{d} \end{bmatrix} = \widetilde{\mathbf{U}}\mathbf{s} + \widetilde{\mathbf{V}}\mathbf{d} . \tag{3.43}$$

We define

$$\mathbf{x}_s = \widetilde{\mathbf{U}}\mathbf{s} = \mathbf{T_s}\begin{bmatrix} \mathbf{s} \\ \mathbf{0} \end{bmatrix} \quad \text{and} \quad \mathbf{x}_d = \widetilde{\mathbf{V}}\mathbf{d} = \mathbf{T_s}\begin{bmatrix} \mathbf{0} \\ \mathbf{d} \end{bmatrix} . \tag{3.44}$$

From (3.43) we have $\mathbf{x} = \mathbf{x}_s + \mathbf{x}_d$.

---

[8]This property holds for the entire CDF family of wavelet transforms that we will construct in Section 4.5.

We call the subspace of $\mathbb{R}^N$ with basis $\{\widetilde{\mathbf{u}}_0, S^2\widetilde{\mathbf{u}}_0, \ldots, S^{N-2}\widetilde{\mathbf{u}}_0\}$ (the even shifts of the trend vector) the *trend subspace* for the wavelet transform $\mathbf{T_a}$. This is the column space of $\widetilde{\mathbf{U}}$. It contains $\mathbf{x}_s$ for all $\mathbf{x} \in \mathbb{R}^N$, since

$$\mathbf{x}_s = \widetilde{\mathbf{U}}\mathbf{s} = (\mathbf{u}_0\mathbf{x})\,\widetilde{\mathbf{u}}_0 + (\mathbf{u}_0 S^2\mathbf{x})\,S^2\widetilde{\mathbf{u}}_0 + \cdots + (\mathbf{u}_0 S^{N-2}\mathbf{x})\,S^{N-2}\widetilde{\mathbf{u}}_0. \qquad (3.45)$$

We call the subspace of $\mathbb{R}^N$ with basis $\{\widetilde{\mathbf{v}}_0, S^2\widetilde{\mathbf{v}}_0, \ldots, S^{N-2}\widetilde{\mathbf{v}}_0\}$ (the even shifts of the detail vector) the *detail subspace* for the wavelet transform $\mathbf{T_a}$. This is the column space of $\widetilde{\mathbf{V}}$. It contains $\mathbf{x}_d$ for all $\mathbf{x} \in \mathbb{R}^N$, since

$$\mathbf{x}_d = \widetilde{\mathbf{V}}\mathbf{d} = (\mathbf{v}_0\mathbf{x})\widetilde{\mathbf{v}}_0 + (\mathbf{v}_0 S^2\mathbf{x})\,S^2\widetilde{\mathbf{v}}_0 + \cdots + (\mathbf{v}_0 S^{N-2}\mathbf{x})\,S^{N-2}\widetilde{\mathbf{v}}_0. \qquad (3.46)$$

The decomposition $\mathbf{x} = \mathbf{x}_s + \mathbf{x}_d$ is valid for every vector $\mathbf{x}$ and is unique since the combined set of N vectors

$$\{\widetilde{\mathbf{u}}_0, S^2\widetilde{\mathbf{u}}_0, \ldots, S^{N-2}\widetilde{\mathbf{u}}_0\} \cup \{\widetilde{\mathbf{v}}_0, S^2\widetilde{\mathbf{v}}_0, \ldots, S^{N-2}\widetilde{\mathbf{v}}_0\}$$

is a basis for $\mathbb{R}^N$. This means that $\mathbb{R}^N$ is the *direct sum* of the trend and detail subspaces (see Section 1.5.2). When $\mathbf{T_a}$ is an orthogonal matrix, for example the Haar or Daub4 matrices, then the trend and detail subspaces are mutually orthogonal.

We call $\mathbf{x}_s$ the *projection of* $\mathbf{x}$ *onto the trend subspace*, and $\mathbf{x}_d$ the *projection of* $\mathbf{x}$ *onto the detail subspace*. The vectors $\mathbf{x}_s$ and $\mathbf{x}_s$ in $\mathbb{R}^N$ are the *inverse transforms* via equations (3.44) of the trend transform $\mathbf{s}$ and detail transform $\mathbf{d}$ of $\mathbf{x}$, which are vectors in $\mathbb{R}^{N/2}$.

Example 3.13. Consider the CDF(2, 2) transform with $N = 8$. We determined the matrices $\mathbf{T_a}$ and $\mathbf{T_s}$ in Example 3.12. The trend and detail vectors are

$$\widetilde{\mathbf{u}}_0 = \frac{1}{32}\begin{bmatrix} 4 & 2 & 0 & 0 & 0 & 0 & 0 & 2 \end{bmatrix}^{\mathrm{T}} \quad \text{and} \quad \widetilde{\mathbf{v}}_0 = \frac{1}{32}\begin{bmatrix} -2 & 6 & -2 & -1 & 0 & 0 & 0 & -1 \end{bmatrix}^{\mathrm{T}}.$$

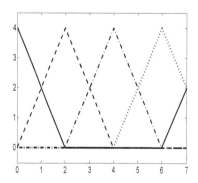
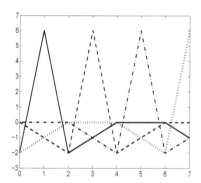

Fig. 3.9   One-scale CDF(2, 2) trend subspace basis (left) and detail subspace basis (right)

In Figure 3.9 we have plotted the piecewise linear functions on $0 \le t \le 7$ corresponding to these vectors and their shifts by 2 (rescaled by a factor of 32) as solid lines, dashed, dashed-dotted, and dotted lines. The vectors are obtained by sampling these functions at $t = 0, 1, \ldots, 7$.

Take the signal $\mathbf{x} = \begin{bmatrix} 0\ 1\ 2\ 3\ 4\ 5\ 6\ 7 \end{bmatrix}^{\mathrm{T}}$ obtained from sampling the linear function $f(t) = t$ at $t = 0, 1, \ldots, 7$. Calculating $\mathbf{T_a x}$ using MATLAB, we find that the trend and detail are $\mathbf{s} = \begin{bmatrix} 8\ 16\ 32\ 56 \end{bmatrix}^{\mathrm{T}}$ and $\mathbf{d} = \begin{bmatrix} 0\ 0\ 0\ 16 \end{bmatrix}^{\mathrm{T}}$. The projections of $\mathbf{x}$ onto the trend and the detail subspaces are

$$\mathbf{x}_s = \mathbf{T_s} \begin{bmatrix} 8 \\ 16 \\ 32 \\ 56 \\ 0 \\ 0 \\ 0 \end{bmatrix} = \begin{bmatrix} 1 \\ 1.5 \\ 2 \\ 3 \\ 4 \\ 5 \\ 7 \\ 4 \end{bmatrix} \quad \text{and} \quad \mathbf{x}_d = \mathbf{T_s} \begin{bmatrix} 0 \\ 0 \\ 0 \\ 0 \\ 0 \\ 0 \\ 16 \end{bmatrix} = \begin{bmatrix} -1 \\ -0.5 \\ 0 \\ 0 \\ 0 \\ -0.5 \\ -1 \\ 3 \end{bmatrix}.$$

It is evident that $\mathbf{x} = \mathbf{x}_s + \mathbf{x}_d$ (sum of trend projection and detail projection).

Recall that we constructed the CDF$(2, 2)$ transform in Section 3.4 so that the detail projection would be zero for linear signals. In this example the signal $\mathbf{x}$ has linearly increasing entries, but the detail projection $\mathbf{x}_d$ is not zero. This is due to the periodic wraparound: the 8-periodic extension of the function $f(t)$ is a sawtooth wave; sampling this wave gives the periodic extension of $\mathbf{x}$ which has a jump of 7 between $n = 7$ and $n = 8$. This jump influences the trend projection, which is only piecewise linear. Note that the 8-periodic extension of the trend projection has a smaller jump of 3 between $n = 7$ and $n = 8$. If we use the trend projection to approximate the signal, then the relative compression error is

$$\frac{\|\mathbf{x} - \mathbf{x}_s\|^2}{\|\mathbf{x}\|^2} = \frac{\|\mathbf{x}_d\|^2}{\|\mathbf{x}\|^2} \approx 8\%.$$

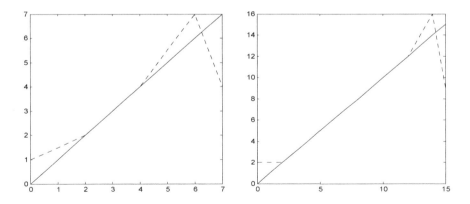

Fig. 3.10   Linear signal (solid line) and CDF$(2, 2)$ trend projections (dotted lines)

The left-hand plot in Figure 3.10 shows the piecewise linear functions on $0 \le t \le 7$ corresponding to $\mathbf{x}$ and $\mathbf{x}_s$ by sampling at integer points. The right-hand plot in this figure shows piecewise linear functions on $0 \leq t \leq 15$ corresponding to $f(t)$ and its trend projection based on sampling at integer points using the CDF$(2,2)$ analysis and synthesis matrices of size 16×16. As in the plot on the left, the boundary effects give nonzero entries in the detail vector for two values at the beginning and three values at the end of the signal. Thus the trend fits the signal exactly at the remaining 11 values. The relative compression error now is approximately 4%.   ■

3.5.4 *Multiscale wavelet matrices*

Now that we know how a one-scale wavelet matrix is built from a scaling vector and a wavelet vector, we can obtain a similar description of a general multiscale wavelet matrix that includes the case of the Haar three-scale wavelet matrix (3.8) from Section 3.3. The *pyramid algorithm* that is used for multiresolution analysis is shown in Figure 3.11 for three scales and a signal $\mathbf{x}$ of length $N = 2^k$ as an illustration (the diagram extends in an obvious way to more scales to obtain successively coarser compressions and details). Here we have chosen one-scale wavelet analysis matrices $\mathbf{T}_{\mathrm{a}}^{(j)}$ of size $2^j \times 2^j$ for $j = k, k-1, k-2$ and the signal length 2^k is fixed.[9]

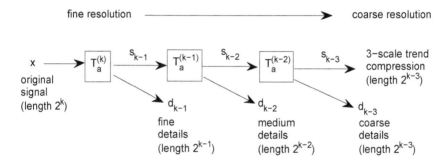

Fig. 3.11   Three-scale multiresolution analysis pyramid algorithm

Extending what we did for one-scale wavelet matrices in (3.34), we can write the three-scale wavelet analysis matrix $\mathbf{W}_{\mathrm{a}}^{(3)}$ of size $N \times N$ in terms of the following submatrices:

(1) three-scale trend analysis matrix $\mathbf{U}_3$ of size $(N/8) \times N$,
(2) coarse detail analysis matrix $\mathbf{V}_3$ of size $(N/8) \times N$,
(3) medium detail analysis matrix $\mathbf{V}_2$ of size $(N/4) \times N$,
(4) fine detail analysis matrix $\mathbf{V}_1$ of size $(N/2) \times N$.

---

[9]In Figure 3.11 the level of *fine resolution* depends on the original digital signal $\mathbf{x}$. If this signal is obtained by sampling an analog signal, for example, then the choice of sampling rate will affect the amount of detail.

For example, the three-scale Haar transform with $N = 8$ in (3.8) is built from the matrices

$$\mathbf{U}_3 = \begin{bmatrix} \frac{1}{2} & \frac{1}{2} & \frac{1}{2} & \frac{1}{2} & \frac{1}{2} & \frac{1}{2} & \frac{1}{2} & \frac{1}{2} \end{bmatrix},$$

$$\mathbf{V}_3 = \begin{bmatrix} \frac{1}{2} & \frac{1}{2} & \frac{1}{2} & \frac{1}{2} & -\frac{1}{2} & -\frac{1}{2} & -\frac{1}{2} & -\frac{1}{2} \end{bmatrix},$$

$$\mathbf{V}_2 = \begin{bmatrix} \frac{1}{2} & \frac{1}{2} & -\frac{1}{2} & -\frac{1}{2} & 0 & 0 & 0 & 0 \\ 0 & 0 & 0 & 0 & \frac{1}{2} & \frac{1}{2} & -\frac{1}{2} & -\frac{1}{2} \end{bmatrix}, \tag{3.47}$$

$$\mathbf{V}_1 = \begin{bmatrix} \frac{1}{2} & -\frac{1}{2} & 0 & 0 & 0 & 0 & 0 & 0 \\ 0 & 0 & \frac{1}{2} & -\frac{1}{2} & 0 & 0 & 0 & 0 \\ 0 & 0 & 0 & 0 & \frac{1}{2} & -\frac{1}{2} & 0 & 0 \\ 0 & 0 & 0 & 0 & 0 & 0 & \frac{1}{2} & -\frac{1}{2} \end{bmatrix}.$$

The three-scale multiresolution decomposition of $\mathbf{x}$ shown in Figure 3.11 is then

$$\mathbf{W}_a^{(3)}\mathbf{x} = \begin{bmatrix} \mathbf{U}_3 \\ \mathbf{V}_3 \\ \mathbf{V}_2 \\ \mathbf{V}_1 \end{bmatrix} \mathbf{x} = \begin{bmatrix} \mathbf{s}_{k-3} \\ \mathbf{d}_{k-3} \\ \mathbf{d}_{k-2} \\ \mathbf{d}_{k-1} \end{bmatrix}. \tag{3.48}$$

Here the three-scale trend $\mathbf{s}_{k-3} \in \mathbb{R}^{N/8}$, the coarse detail $\mathbf{d}_{k-3} \in \mathbb{R}^{N/8}$, the medium detail $\mathbf{d}_{k-2} \in \mathbb{R}^{N/4}$, and the fine detail $\mathbf{d}_{k-1} \in \mathbb{R}^{N/2}$.

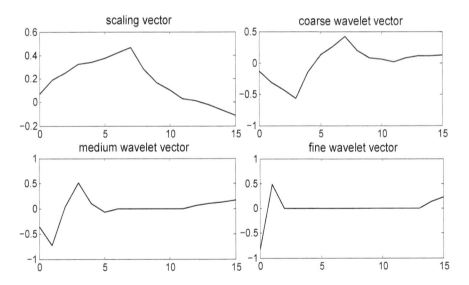

Fig. 3.12 Scaling and wavelet vectors for three-scale Daub4 transform

Let $\mathbf{u}^{(3)}$ be the first row of $\mathbf{U}_3$. We call this the *scaling vector* for the matrix $\mathbf{W}_a^{(3)}$. Let $\mathbf{v}^{(j)}$ be the first row of $\mathbf{V}_j$ for $j = 1, 2, 3$. We call $\mathbf{v}^{(1)}$ the *fine wavelet vector*, $\mathbf{v}^{(2)}$ the *medium wavelet vector*, and $\mathbf{v}^{(3)}$ the *coarse wavelet vector*. These vectors are plotted in Figure 3.12 for the three-scale Daub4 transform with $N = 2^4$.

For the graphs we treat vectors as digital signals, index their entries by $0, 1, \ldots, 15$, and use linear interpolation between entries.

The coarse scaling and wavelet vectors for the three-scale Daub4 transform have all entries nonzero, whereas the medium wavelet vector has 6 consecutive zero entries and the fine wavelet vector has 12 consecutive zero entries.[10] Also the wavelet vectors have entries that sum to zero, so they measure changes in $\mathbf{x}$. The scalar $\mathbf{v}^{(1)}\mathbf{x}$ detects the changes in $\mathbf{x}$ over 3 time steps, whereas the scalar $\mathbf{v}^{(2)}\mathbf{x}$ detects the changes in $\mathbf{x}$ over 11 time steps and the scalar $\mathbf{v}^{(3)}\mathbf{x}$ detects overall changes in $\mathbf{x}$, especially in the range $0 \leq n \leq 8$.

Theorem 3.3. *The rows of a three-scale wavelet analysis matrix $\mathbf{W}_{\mathrm{a}}^{(3)}$ are obtained from the scaling and wavelet vectors by shifting (with N-periodic wraparound) as follows:*

(1) Successive right shifts of the scaling vector by 2^3 positions for each shift generate the rows of the matrix $\mathbf{U}_3$.
(2) Successive right shifts of the coarse wavelet vector by 2^3 positions for each shift generate the rows of the matrix $\mathbf{V}_3$.
(3) Successive right shifts of the medium wavelet vector by 2^2 positions for each shift generate the rows of the matrix $\mathbf{V}_2$.
(4) Successive right shifts of the fine wavelet vector by 2 positions for each shift generate the rows of the matrix $\mathbf{V}_1$.

Proof. Write the one-scale analysis matrices in block form $\mathbf{T}_{\mathrm{a}}^{(j)} = \begin{bmatrix} \mathbf{U}^{(j)} \\ \mathbf{V}^{(j)} \end{bmatrix}$. Following the pyramid algorithm in Figure 3.11 we calculate

$$\mathbf{s}_{k-1} = \mathbf{U}^{(k)}\mathbf{x}, \qquad\qquad \mathbf{d}_{k-1} = \mathbf{V}^{(k)}\mathbf{x},$$
$$\mathbf{s}_{k-2} = \mathbf{U}^{(k-1)}\mathbf{U}^{(k)}\mathbf{x}, \qquad\qquad \mathbf{d}_{k-2} = \mathbf{V}^{(k-1)}\mathbf{U}^{(k)}\mathbf{x},$$
$$\mathbf{s}_{k-3} = \mathbf{U}^{(k-2)}\mathbf{U}^{(k-1)}\mathbf{U}^{(k)}\mathbf{x}, \qquad\qquad \mathbf{d}_{k-3} = \mathbf{V}^{(k-2)}\mathbf{U}^{(k-1)}\mathbf{U}^{(k)}\mathbf{x}.$$

Comparing this with (3.48), we see that the blocks in the three-scale analysis matrix are built from the blocks in the one-scale analysis matrices in the following way:

$$\mathbf{U}_3 = \mathbf{U}^{(k-2)}\mathbf{U}^{(k-1)}\mathbf{U}^{(k)}, \qquad \mathbf{V}_3 = \mathbf{V}^{(k-2)}\mathbf{U}^{(k-1)}\mathbf{U}^{(k)},$$
$$\mathbf{V}_2 = \mathbf{V}^{(k-1)}\mathbf{U}^{(k)}, \qquad \mathbf{V}_1 = \mathbf{V}^{(k)}. \tag{3.49}$$

Note that in these matrix products all trend blocks are on the right with one detail block on the left.

The proof of Theorem 3.2 is based on equation (3.38), which we write as

$$\mathbf{U}^{(j)}(S_{(j)})^2 = S_{(j-1)}\mathbf{U}^{(j)}, \qquad \mathbf{V}^{(j)}(S_{(j)})^2 = S_{(k-1)}\mathbf{V}^{(j)},$$

where $S_{(j)}$ is the shift matrix of size $2^j \times 2^j$. Using this relation repeatedly in (3.49), we get

$$\mathbf{U}_3(S_{(k)})^8 = S_{(k-3)}\mathbf{U}_3, \qquad \mathbf{V}_3(S_{(k)})^8 = S_{(k-3)}\mathbf{V}_3,$$
$$\mathbf{V}_2(S_{(k)})^4 = S_{(k-2)}\mathbf{V}_2, \qquad \mathbf{V}_1(S_{(k)})^2 = S_{(k-1)}\mathbf{V}_1. \tag{3.50}$$

---

[10]Compare this pattern with that of the three-scale Haar matrix (3.47).

Now we use the argument in the proof of Theorem 3.2, but with right shifts by 8, 4, or 2 positions as determined by (3.50). □

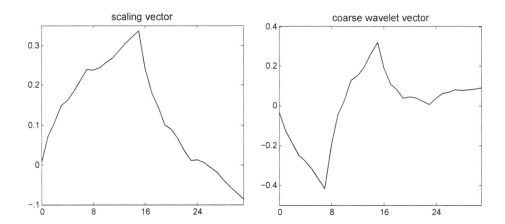

Fig. 3.13   Scaling and coarse wavelet vector for four-scale Daub4 transform

Remark 3.8. Figure 3.12 reveals a significant difference between the multiscale Haar and multiscale Daub4 scaling and wavelet vectors. For the Haar transform the nonzero entries in these vectors follow a simple pattern of ± 1. For the Daub4 transform the piecewise-linear functions that interpolate the scaling vector and detail vectors change in a more subtle way at each level of resolution. For a four-scale multiresolution analysis of a signal of length 2^5, the scaling and coarse wavelet vectors (plotted as piecewise linear functions) are shown in Figure 3.13. The graphs have the same general outline as the corresponding graphs in Figure 3.12, but are more jagged. In Section 5.4 we will examine the behavior of these graphs as the number of scales keeps increasing.

We reconstruct the signal $\mathbf{x}$ from its multiscale analysis transform using the one-scale synthesis matrices $\mathbf{T_s}^{(j)}$ that reverse the steps in the analysis pyramid. We write the three-scale synthesis matrix in block form as

$$\mathbf{W_s}^{(3)} = \begin{bmatrix} \widetilde{\mathbf{U}}_3 & \widetilde{\mathbf{V}}_3 & \widetilde{\mathbf{V}}_2 & \widetilde{\mathbf{V}}_1 \end{bmatrix}, \tag{3.51}$$

where the blocks have $N/8$, $N/8$, $N/4$, and $N/2$ columns, respectively. We call the first column of $\widetilde{\mathbf{U}}_3$ the *coarse trend* vector, and the first columns of $\widetilde{\mathbf{V}}_3$, $\widetilde{\mathbf{V}}_2$, and $\widetilde{\mathbf{V}}_1$ the *coarse detail, medium detail,* and *fine detail* vectors.

The three-scale multiresolution synthesis of $\mathbf{x}$ shown in Figure 3.14 is built up by adding together the pieces at each resolution:

$$\mathbf{x} = \underbrace{\widetilde{\mathbf{U}}_3\,\mathbf{s}_{k-3}}_{\substack{\text{coarse} \\ \text{trend}}} + \underbrace{\widetilde{\mathbf{V}}_3\,\mathbf{d}_{k-3}}_{\substack{\text{coarse} \\ \text{detail}}} + \underbrace{\widetilde{\mathbf{V}}_2\,\mathbf{d}_{k-2}}_{\substack{\text{medium} \\ \text{detail}}} + \underbrace{\widetilde{\mathbf{V}}_1\,\mathbf{d}_{k-1}}_{\substack{\text{fine} \\ \text{detail}}}. \tag{3.52}$$

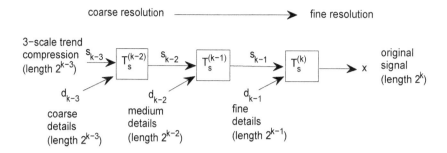

Fig. 3.14 Three-scale multiresolution synthesis pyramid algorithm

For an orthogonal wavelet transform, such as (normalized) Haar or Daub4, the block matrices in the synthesis matrix $\mathbf{W}_s^{(3)}$ are the transposes of the corresponding blocks in the analysis matrix $\mathbf{W}_a^{(3)}$. The trend vector is the transpose of the scaling vector, and the detail vectors are the transposes of the corresponding wavelet vectors.

Theorem 3.4. *The columns of a three-scale wavelet synthesis matrix* $\mathbf{W}_s^{(3)}$ *are obtained from the coarse trend vector and the three detail vectors by shifting (with N-periodic wraparound) as follows:*

(1) Successive down shifts of the coarse trend vector by 2^3 *positions for each shift generate the columns of the matrix* $\widetilde{\mathbf{U}}_3$. *These columns give a basis for the* $N/8$*-dimensional coarse trend subspace.*

(2) Successive down shifts of the coarse detail vector by 2^3 *positions for each shift generate the columns of the matrix* $\widetilde{\mathbf{V}}_3$. *These columns give a basis for the* $N/8$*-dimensional coarse details subspace.*

(3) Successive down shifts of the medium detail vector by 2^2 *positions for each shift generate the columns of the matrix* $\widetilde{\mathbf{V}}_2$. *These columns give a basis for the* $N/4$*-dimensional medium details subspace.*

(4) Successive down shifts of the fine detail vector by 2 *positions for each shift generate the columns of the matrix* $\widetilde{\mathbf{V}}_1$. *These columns give a basis for the* $N/2$*-dimensional fine details subspace.*

Proof. Write $\mathbf{T}_s^{(j)} = \left[\widetilde{\mathbf{U}}^{(j)} \ \ \widetilde{\mathbf{V}}^{(j)} \right]$ as in Theorem 3.2. Then from (3.52) we find that

$$
\begin{aligned}
\widetilde{\mathbf{U}}_3 &= \widetilde{\mathbf{U}}^{(k)} \widetilde{\mathbf{U}}^{(k-1)} \widetilde{\mathbf{U}}^{(k-2)}, & \widetilde{\mathbf{V}}_3 &= \widetilde{\mathbf{U}}^{(k)} \widetilde{\mathbf{U}}^{(k-1)} \widetilde{\mathbf{V}}^{(k-2)}, \\
\widetilde{\mathbf{V}}_2 &= \widetilde{\mathbf{U}}^{(k)} \widetilde{\mathbf{V}}^{(k-1)}, & \widetilde{\mathbf{V}}_1 &= \widetilde{\mathbf{V}}^{(k)}.
\end{aligned}
\tag{3.53}
$$

The rest of the proof follows from (3.53) by the same argument as in Theorem 3.3.   □

Example 3.14. Consider the test signal $\mathbf{x}$ of length 64 from Example 3.8. The coefficients of the three-scale Daub4 transform of this signal are shown in Figure

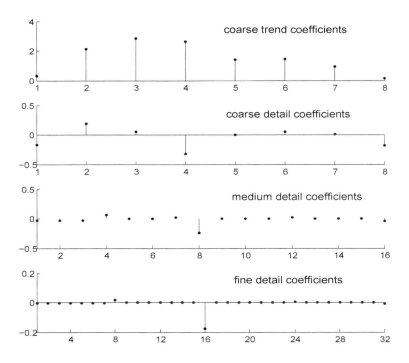

Fig. 3.15   Three-scale Daub4 transform of piecewise polynomial signal

3.15. Of the 8 coarse trend coefficients, 2 are close to 0, whereas 4 of the 8 coarse detail coefficients, 15 of the 16 medium detail coefficients, and 31 of the 32 fine detail coefficients are nearly zero. Thus most of the information in the signal is contained in 11 of the 64 coefficients. The largest detail coefficients are all at the midpoint where the original analog signal has a jump discontinuity.

The projections of this signal onto the trend and detail subspaces are plotted in Figure 3.16. These four projections are mutually perpendicular since Daub4 is an orthogonal transformation, and the square of the length of $\mathbf{x}$ is the sum of the squares of their lengths:

$$\|\mathbf{x}\|^2 = 24.9000 = 24.5714 + 0.2287 + 0.0679 + 0.0320$$

$$= \|\mathbf{x}_{\text{coarse trend}}\|^2 + \|\mathbf{x}_{\text{coarse detail}}\|^2 + \|\mathbf{x}_{\text{medium detail}}\|^2 + \|\mathbf{x}_{\text{fine detail}}\|^2.$$

These four projections of $\mathbf{x}$ are linear combinations of the shifted basis vectors from Figure 3.12 with coefficients from Figure 3.15. The coarse trend projection is a rough version of the original signal and has 98.7% of the energy. The coarse detail projection has 0.9% of the energy and is essentially a linear combination of four shifted coarse wavelet vectors in Figure 3.16, corresponding to the four significant coarse coefficients. The medium and fine detail projections resemble the medium and fine detail wavelet vectors in Figure 3.16 shifted to the midpoint $n = 32$, corresponding to the one significant medium detail coefficient and one significant fine detail coefficient. ∎

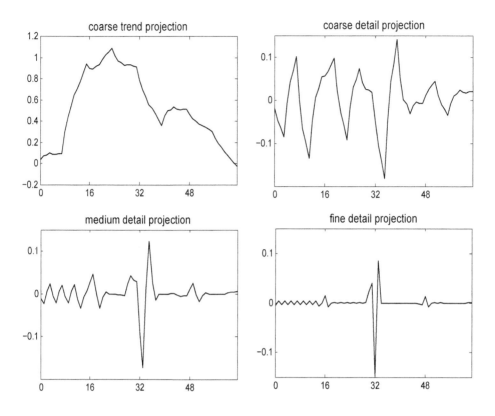

Fig. 3.16   Three-scale Daub4 projections of piecewise polynomial signal

Example 3.15.   We now perform some signal processing on the signal **x** in Example 3.14.

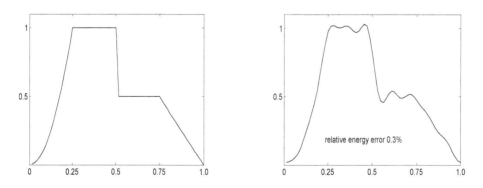

Fig. 3.17   Signal (left) and 4:1 Fourier compression (right)

The digital signal **x** of length 64 was obtained by sampling the analog signal $f(t)$ plotted on the left in Figure 3.17 for $0 \leq t \leq 1$. The right side shows a 4:1 Fourier-compressed version $\mathbf{x}_{\mathrm{comp}}$ of this signal (plotted as a line graph). To carry out the compression, we calculated the discrete Fourier transform $\mathbf{X}$ of $\mathbf{x}$, replaced the 48 smallest coefficients in $\mathbf{X}$ by zeros to obtain $\mathbf{X}_{\mathrm{comp}}$, and calculated $\mathbf{x}_{\mathrm{comp}}$ as the inverse Fourier transform of $\mathbf{X}_{\mathrm{comp}}$. The relative energy error $\|\mathbf{x} - \mathbf{x}_{\mathrm{comp}}\|^2/\|\mathbf{x}\|^2$ of the compressed signal is 0.3%, and comes mostly from the wiggles in the graph on both sides of the jump at $t = 0.5$. This is an unavoidable aspect of a Fourier expansion at a jump discontinuity, called the *Gibbs phenomenon*.

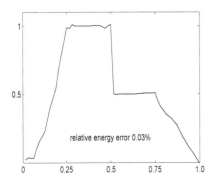 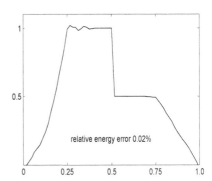

Fig. 3.18   4:1 compression by three-scale Daub4 (left) and Daub6 (right) wavelet transforms

Now we carry out compression of the same signal using orthogonal wavelet transforms. We have already introduced the Daub4 transform; in Section 4.10 we will construct the Daub2K transforms for all positive integers K. Figure 3.18 shows the 4:1 compressions of the signal (plotted as line graphs for purposes of comparison) using the 16 largest Daub4 or Daub6 three-scale wavelet transform coefficients. The wavelet compression is much superior to the Fourier compression—the Daub4 transform gives 10 times smaller relative energy error, whereas the Daub6 transform gives 15 times smaller error with no extraneous wiggles at $t = 0.5$. ∎

3.6   Two-Dimensional Wavelet Transforms

3.6.1   *Images as matrices*

A two-dimensional black and white image can be digitized as a matrix $\mathbf{X}$ of size $M \times N$ by imposing a rectangular grid with M horizontal strips and N vertical strips on the image. Each rectangle in the grid is called a *pixel* (picture element) and is given a numerical value (*gray scale*) corresponding to the average darkness or brightness of the image in the pixel.[11] This method of capturing an image is called *raster graphics*. Since $2^8 = 256$, eight-bit encoding of the pixels uses the integers 0 to 255 with 0 for black and 255 for white. The origin of coordinates is placed at

---

[11]Color images are digitized in a similar way, but using three matrices, one for each primary color—see [Van Fleet (2008), Chapter 3].

the upper left-hand corner of the image and the vertical axis points down. Thus the entry $\mathbf{X}[i,j]$ in $\mathbf{X}$ encodes the average gray scale level of the pixel that is i units *down* and j units to the *right* of the upper left-hand corner of the image (this system of coordinates agrees with the usual labeling of matrix entries). We shall only consider the case $M = N$ of square image matrices in the following.

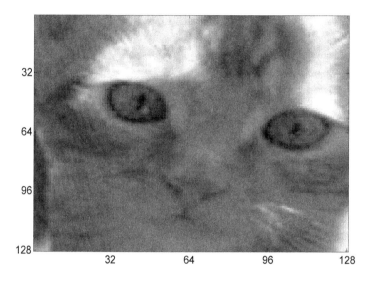

Fig. 3.19   Digital image

Figure 3.19 shows an example of a digital image, eight-bit encoded as a 128×128 gray scale matrix. The 4×4 submatrix A that encodes the dark upper left-hand corner of the image and the 4×4 submatrix B that encodes the bright lower right-hand corner are

$$A = \begin{bmatrix} 97 & 85 & 74 & 74 \\ 97 & 85 & 78 & 78 \\ 97 & 81 & 74 & 78 \\ 95 & 88 & 78 & 74 \end{bmatrix}, \quad B = \begin{bmatrix} 187 & 191 & 191 & 191 \\ 187 & 191 & 205 & 205 \\ 187 & 191 & 191 & 196 \\ 187 & 191 & 191 & 191 \end{bmatrix}. \tag{3.54}$$

The entries in B are more than twice as large as those of A because of the difference in brightness. Both A and B illustrate the redundancy in image encoding, with many entries that are the same number. This property makes it possible to compress image matrices very efficiently using wavelet transforms, as we will see at the end of Section 3.6.3.

3.6.2   *One-scale 2D wavelet transform*

Consider an image that is encoded by an $N \times N$ matrix $X = [x_{ij}]$ with N even. Let $\mathbf{W_a}$ be the $N \times N$ one-scale analysis matrix for a wavelet transform (of type Haar, CDF(2, 2), Daub4, ...). To perform a one-scale wavelet analysis transformation of

X, we first view X as N side-by-side column vectors and apply $\mathbf{W_a}$ to each column by left multiplication:

$$\mathbf{X} = \begin{bmatrix} \mathbf{x}_1 \cdots \mathbf{x}_N \end{bmatrix} \quad \text{and} \quad \mathbf{W_a X} = \begin{bmatrix} \mathbf{W_a x}_1 \cdots \mathbf{W_a x}_N \end{bmatrix}, \text{ where } \mathbf{x}_j = \begin{bmatrix} x_{1j} \\ \vdots \\ x_{Nj} \end{bmatrix}.$$

Treating the columns separately this way is not adequate for image processing, however, because it ignores the two-dimensional nature of an image; there are high correlations between the entries in adjacent rows and columns. One way to remedy this problem is to take another wavelet transform of the *rows* of the matrix $\mathbf{Z} = \mathbf{W_a X}$. Since the rows of $\mathbf{Z}$ are the columns of $\mathbf{Z}^T$, we can do this second transformation by multiplying on the left with $\mathbf{W_a}$ as before and using two transpositions:

$$\left(\mathbf{W_a Z}^T\right)^T = \mathbf{Z}\mathbf{W_a}^T = \mathbf{W_a X}\mathbf{W_a}^T.$$

With this motivation, we define the *one-scale 2D wavelet transform* of $\mathbf{X}$ (of type Haar, CDF$(2,2)$, Daub4, ...) to be the matrix

$$\mathbf{Y} = \mathbf{W_a X}\mathbf{W_a}^T. \tag{3.55}$$

Write $\mathbf{W_a} = \begin{bmatrix} \mathbf{U} \\ \mathbf{V} \end{bmatrix}$, where $\mathbf{U}$ consists of the $N/2$ trend rows and $\mathbf{V}$ consists of the $N/2$ detail rows (see Theorem 3.2). Then

$$\mathbf{Y} = \begin{bmatrix} \mathbf{U} \\ \mathbf{V} \end{bmatrix} \mathbf{X} \begin{bmatrix} \mathbf{U}^T & \mathbf{V}^T \end{bmatrix} = \begin{bmatrix} \mathbf{UXU}^T & \mathbf{UXV}^T \\ \mathbf{VXU}^T & \mathbf{VXV}^T \end{bmatrix}. \tag{3.56}$$

The four blocks in Y are denoted by

$$\mathbf{Y_{ss}} = \mathbf{UXU}^T, \quad \mathbf{Y_{sd}} = \mathbf{UXV}^T, \quad \mathbf{Y_{ds}} = \mathbf{VXU}^T, \quad \mathbf{Y_{dd}} = \mathbf{VXV}^T.$$

The first subscript refers to the wavelet block multiplying $\mathbf{X}$ on the *left* ($\mathbf{s}$ for the trend block $\mathbf{U}$ and $\mathbf{d}$ for the detail block $\mathbf{V}$); the second subscript refers to the wavelet block multiplying $\mathbf{X}$ on the *right* ($\mathbf{s}$ for the transposed trend block $\mathbf{U}^T$ and $\mathbf{d}$ for the transposed detail block $\mathbf{V}^T$).

Example 3.16. Let $\mathbf{W_a} = \frac{1}{2}\begin{bmatrix} 1 & 1 \\ 1 & -1 \end{bmatrix}$ be the (unnormalized) 2×2 Haar transform matrix and take $\mathbf{X} = \begin{bmatrix} a & b \\ c & d \end{bmatrix}$ an arbitrary 2×2 image matrix. Then

$$\mathbf{Y} = \frac{1}{4}\begin{bmatrix} 1 & 1 \\ 1 & -1 \end{bmatrix}\begin{bmatrix} a & b \\ c & d \end{bmatrix}\begin{bmatrix} 1 & 1 \\ 1 & -1 \end{bmatrix} = \frac{1}{4}\begin{bmatrix} (a+b+c+d) & (a-b+c-d) \\ (a-c+b-d) & (a-b-c+d) \end{bmatrix}.$$

We can describe the four blocks in $\mathbf{Y}$ (which are 1×1 in this simple example) as follows:

$$\mathbf{Y_{ss}} = \frac{1}{4}(a+b+c+d) \quad \text{(overall average)},$$

$$\mathbf{Y_{sd}} = \frac{1}{4}[(a-b)+(c-d)] \quad \text{(average of column-to-column differences)},$$

$$\mathbf{Y_{ds}} = \frac{1}{4}[(a-c)+(b-d)] \quad \text{(average of row-to-row differences)},$$

$$\mathbf{Y_{dd}} = \frac{1}{4}[(a-b)-(c-d)] \quad \text{(column difference of row differences)}.$$

Thus for the matrix $\mathbf{X} = \begin{bmatrix} 1 & 0 \\ 1 & 0 \end{bmatrix}$, the overall average $\mathbf{Y}_{\mathbf{ss}} = 1/2$ and the average column-to-column difference $\mathbf{Y}_{\mathbf{sd}} = 1/2$, whereas $\mathbf{Y}_{\mathbf{ds}} = \mathbf{Y}_{\mathbf{dd}} = 0$ since both rows are the same.   ∎

Remark 3.9. As illustrated by Example 3.16, each of the four submatrices of $\mathbf{Y}$ extracts different information from the image matrix:

- Multiplying on the left by the trend submatrix $\mathbf{U}$ and on the right by the transposed trend submatrix $\mathbf{U}^{\mathrm{T}}$ treats rows and columns of $\mathbf{X}$ in the same way and gives the overall *trend* in the image.
- Multiplying $\mathbf{X}$ on the right by the transposed detail submatrix $\mathbf{V}^{\mathrm{T}}$ detects changes along the *rows* of $\mathbf{X}$, corresponding to *vertical* details in the image; then left multiplication by $\mathbf{U}$ averages these details to obtain $\mathbf{X}_{\mathbf{sd}}$.
- Likewise, multiplying $\mathbf{X}$ on the left by the detail submatrix $\mathbf{V}$ detects changes along the *columns* of $\mathbf{X}$, corresponding to *horizontal* details in the image; then right multiplication by $\mathbf{U}$ averages these details to obtain $\mathbf{X}_{\mathbf{ds}}$.
- Using $\mathbf{V}$ on the left and $\mathbf{V}^{\mathrm{T}}$ on the right treats rows and columns of $\mathbf{X}$ in the same way and detects simultaneous horizontal and vertical changes, corresponding to *diagonal* details in the image.

Let $\mathbf{W_s} = \mathbf{W_a}^{-1}$ be the one-scale synthesis matrix. Then the original matrix $\mathbf{X}$ can be reconstructed from the transform $\mathbf{Y}$ by writing
$$\mathbf{X} = \mathbf{W_s}\mathbf{W_a}\mathbf{X}\mathbf{W_a}^{\mathrm{T}}\mathbf{W_s}^{\mathrm{T}} = \mathbf{W_s}\mathbf{Y}\mathbf{W_s}^{\mathrm{T}} .$$
Recall the block decomposition $\mathbf{W_s} = \begin{bmatrix} \widetilde{\mathbf{U}} & \widetilde{\mathbf{V}} \end{bmatrix}$ from Theorem 3.2 in terms of the trend and detail columns. Combining this with the block decomposition of $\mathbf{Y}$ in (3.56), we can write the reconstruction formula for $\mathbf{X}$ as
$$\mathbf{X} = \begin{bmatrix} \widetilde{\mathbf{U}} & \widetilde{\mathbf{V}} \end{bmatrix} \begin{bmatrix} \mathbf{Y}_{\mathbf{ss}} & \mathbf{Y}_{\mathbf{sd}} \\ \mathbf{Y}_{\mathbf{ds}} & \mathbf{Y}_{\mathbf{dd}} \end{bmatrix} \begin{bmatrix} \widetilde{\mathbf{U}}^{\mathrm{T}} \\ \widetilde{\mathbf{V}}^{\mathrm{T}} \end{bmatrix} = \begin{bmatrix} \widetilde{\mathbf{U}} & \widetilde{\mathbf{V}} \end{bmatrix} \begin{bmatrix} \mathbf{Y}_{\mathbf{ss}}\widetilde{\mathbf{U}}^{\mathrm{T}} + \mathbf{Y}_{\mathbf{sd}}\widetilde{\mathbf{V}}^{\mathrm{T}} \\ \mathbf{Y}_{\mathbf{ds}}\widetilde{\mathbf{U}}^{\mathrm{T}} + \mathbf{Y}_{\mathbf{dd}}\widetilde{\mathbf{V}}^{\mathrm{T}} \end{bmatrix}$$
$$= \widetilde{\mathbf{U}}\mathbf{Y}_{\mathbf{ss}}\widetilde{\mathbf{U}}^{\mathrm{T}} + \widetilde{\mathbf{U}}\mathbf{Y}_{\mathbf{sd}}\widetilde{\mathbf{V}}^{\mathrm{T}} + \widetilde{\mathbf{V}}\mathbf{Y}_{\mathbf{ds}}\widetilde{\mathbf{U}}^{\mathrm{T}} + \widetilde{\mathbf{V}}\mathbf{Y}_{\mathbf{dd}}\widetilde{\mathbf{V}}^{\mathrm{T}} . \qquad (3.57)$$
We call formula (3.57) the *multiresolution representation* of $\mathbf{X}$. From Example 3.16 and Remark 3.9 we see that each of the four matrices in the sum (3.57) emphasize a different aspect of the image encoded by $\mathbf{X}$:

$\mathbf{X}_{\mathbf{ss}} = \widetilde{\mathbf{U}}\mathbf{Y}_{\mathbf{ss}}\widetilde{\mathbf{U}}^{\mathrm{T}}$ column and row trend (blurred version of entire image),

$\mathbf{X}_{\mathbf{sd}} = \widetilde{\mathbf{U}}\mathbf{Y}_{\mathbf{sd}}\widetilde{\mathbf{V}}^{\mathrm{T}}$ row trend and column detail (vertical aspects of image),

$\mathbf{X}_{\mathbf{ds}} = \widetilde{\mathbf{V}}\mathbf{Y}_{\mathbf{ds}}\widetilde{\mathbf{U}}^{\mathrm{T}}$ row detail and column trend (horizontal aspects of image),

$\mathbf{X}_{\mathbf{dd}} = \widetilde{\mathbf{V}}\mathbf{Y}_{\mathbf{dd}}\widetilde{\mathbf{V}}^{\mathrm{T}}$ column and row details (diagonal aspects of image).

Example 3.17. Consider the 2×2 image matrix $\mathbf{X} = \begin{bmatrix} 14 & 2 \\ 4 & 0 \end{bmatrix}$. The one-scale Haar transform of $\mathbf{X}$ is
$$\mathbf{Y} = \frac{1}{4}\begin{bmatrix} 1 & 1 \\ 1 & -1 \end{bmatrix}\begin{bmatrix} 14 & 2 \\ 4 & 0 \end{bmatrix}\begin{bmatrix} 1 & 1 \\ 1 & -1 \end{bmatrix} = \begin{bmatrix} 5 & 4 \\ 3 & 2 \end{bmatrix} .$$

The 2×2 Haar synthesis matrix $\mathbf{W_s} = \begin{bmatrix} 1 & 1 \\ 1 & -1 \end{bmatrix}$ has trend block $\widetilde{\mathbf{U}} = \begin{bmatrix} 1 \\ 1 \end{bmatrix}$ and

detail block $\widetilde{\mathbf{V}} = \begin{bmatrix} 1 \\ -1 \end{bmatrix}$. Hence the one-scale multiresolution representation of $\mathbf{X}$ is

$$\begin{bmatrix} 14 & 2 \\ 4 & 0 \end{bmatrix} = 5\widetilde{\mathbf{U}}\widetilde{\mathbf{U}}^{\mathrm{T}} + 4\widetilde{\mathbf{U}}\widetilde{\mathbf{V}}^{\mathrm{T}} + 3\widetilde{\mathbf{V}}\widetilde{\mathbf{U}}^{\mathrm{T}} + 2\widetilde{\mathbf{V}}\widetilde{\mathbf{V}}^{\mathrm{T}}$$

$$= 5\begin{bmatrix} 1 & 1 \\ 1 & 1 \end{bmatrix} + 4\begin{bmatrix} 1 & -1 \\ 1 & -1 \end{bmatrix} + 3\begin{bmatrix} 1 & 1 \\ -1 & -1 \end{bmatrix} + 2\begin{bmatrix} 1 & -1 \\ -1 & 1 \end{bmatrix}$$

$$= \mathbf{X_{ss}} + \mathbf{X_{sd}} + \mathbf{X_{ds}} + \mathbf{X_{dd}} \ .$$

If we represent a matrix entry 1 by a white box and a matrix entry -1 by a black box, then this last equation can be displayed as

$$\mathbf{X} = 5\,\square\square \atop \square\square + 4\,\square\blacksquare \atop \square\blacksquare + 3\,\square\square \atop \blacksquare\blacksquare + 2\,\square\blacksquare \atop \blacksquare\square \ .$$

The coefficient 5 multiplying the first matrix is the average of the four entries in $\mathbf{X}$. The other three matrices display the pattern of vertical, horizontal, and diagonal detail in $\mathbf{X}$. ∎

Example 3.18. Consider the following 4×4 sample digital image:

is encoded (after rescaling gray levels) by $\quad \mathbf{X} = \begin{bmatrix} 0 & 0 & 0 & 0 \\ 0 & 0 & 2 & 2 \\ 0 & 2 & 0 & 2 \\ 0 & 2 & 2 & 0 \end{bmatrix}$.

For the 4×4 Haar analysis matrix the trend and detail blocks are

$$\mathbf{U} = \frac{1}{2}\begin{bmatrix} 1 & 1 & 0 & 0 \\ 0 & 0 & 1 & 1 \end{bmatrix} \quad \text{and} \quad \mathbf{V} = \frac{1}{2}\begin{bmatrix} 1 & -1 & 0 & 0 \\ 0 & 0 & 1 & -1 \end{bmatrix} .$$

We calculate that $\mathbf{Y_{ss}} = \begin{bmatrix} 0 & 1 \\ 1 & 1 \end{bmatrix}$, $\mathbf{Y_{sd}} = \begin{bmatrix} 0 & 0 \\ -1 & 0 \end{bmatrix}$, $\mathbf{Y_{ds}} = \begin{bmatrix} 0 & -1 \\ 0 & 0 \end{bmatrix}$, $\mathbf{Y_{dd}} = \begin{bmatrix} 0 & 0 \\ 0 & -1 \end{bmatrix}$.
The 4×4 Haar synthesis matrix has trend and detail blocks

$$\widetilde{\mathbf{U}} = \begin{bmatrix} 1 & 0 \\ 1 & 0 \\ 0 & 1 \\ 0 & 1 \end{bmatrix} \quad \text{and} \quad \widetilde{\mathbf{V}} = \begin{bmatrix} 1 & 0 \\ -1 & 0 \\ 0 & 1 \\ 0 & -1 \end{bmatrix} .$$

The four matrices that give the one-scale multiresolution representation of $\mathbf{X}$ are

$$\widetilde{\mathbf{U}}\begin{bmatrix} 0 & 1 \\ 1 & 1 \end{bmatrix}\widetilde{\mathbf{U}}^{\mathrm{T}}, \quad \widetilde{\mathbf{U}}\begin{bmatrix} 0 & 0 \\ -1 & 0 \end{bmatrix}\widetilde{\mathbf{V}}^{\mathrm{T}}, \quad \widetilde{\mathbf{V}}\begin{bmatrix} 0 & -1 \\ 0 & 0 \end{bmatrix}\widetilde{\mathbf{U}}^{\mathrm{T}}, \quad \widetilde{\mathbf{V}}\begin{bmatrix} 0 & 0 \\ 0 & -1 \end{bmatrix}\widetilde{\mathbf{V}}^{\mathrm{T}}.$$

Calculating the matrix products, we obtain

$$\mathbf{X_{ss}} = \begin{bmatrix} 0 & 0 & 1 & 1 \\ 0 & 0 & 1 & 1 \\ 1 & 1 & 1 & 1 \\ 1 & 1 & 1 & 1 \end{bmatrix}, \qquad \mathbf{X_{ds}} = \begin{bmatrix} 0 & 0 & -1 & -1 \\ 0 & 0 & 1 & 1 \\ 0 & 0 & 0 & 0 \\ 0 & 0 & 0 & 0 \end{bmatrix},$$

$$\mathbf{X_{sd}} = \begin{bmatrix} 0 & 0 & 0 & 0 \\ 0 & 0 & 0 & 0 \\ -1 & 1 & 0 & 0 \\ -1 & 1 & 0 & 0 \end{bmatrix}, \qquad \mathbf{X_{dd}} = \begin{bmatrix} 0 & 0 & 0 & 0 \\ 0 & 0 & 0 & 0 \\ 0 & 0 & -1 & 1 \\ 0 & 0 & 1 & -1 \end{bmatrix}.$$

Clearly these four matrices add up to $\mathbf{X}$. Comparing them with the original image, we see that $\mathbf{X_{ss}}$ gives the overall pattern (darker in the upper-left quadrant, lighter in the other three quadrants), $\mathbf{X_{sd}}$ shows the vertical line at the lower left, $\mathbf{X_{ds}}$ shows the horizontal line at the top right, and $\mathbf{X_{dd}}$ shows the diagonal line.   ∎

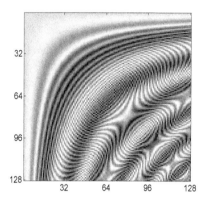

Fig. 3.20   Synthetic round wavy image

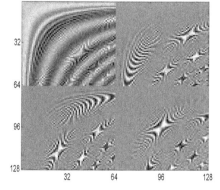

Fig. 3.21   One-scale CDF$(2,2)$ transform

Example 3.19. Figure 3.20 shows an interesting 128×128 synthetic image from the Uvi_Wave toolbox, and Figure 3.21 shows its one-scale CDF$(2,2)$ transform. The upper left 64×64 $\mathbf{Y_{ss}}$ block in the wavelet transform gives a lower-resolution blurred version of the original image. The other three blocks encode the different details (horizontal, vertical, diagonal) of the image.   ∎

3.6.3   *Multiscale 2D wavelet transform*

The multiscale 2D wavelet transform is obtained by the same pyramid algorithm used for one-dimensional signals: the three submatrices containing horizontal, vertical, or diagonal detail information are saved, and only the pure trend submatrix is subjected to another wavelet transform.[12]

---

[12]For a two-scale 2D wavelet transform, we do *not* use the two-scale 1D wavelet matrices $\mathbf{W_a}^{(2)}$ from Section 3.3. If we multiply the image matrix $\mathbf{X}$ on the left by $\mathbf{W_a}^{(2)}$ and on the right by

Let the image matrix $\mathbf{X}$ be of size $N \times N$, where now N is a multiple of 4, and let

$$\mathbf{Y}^{(1)} = \begin{bmatrix} \mathbf{Y}_{ss}^{(1)} & \mathbf{Y}_{sd}^{(1)} \\ \mathbf{Y}_{ds}^{(1)} & \mathbf{Y}_{dd}^{(1)} \end{bmatrix}$$

be the one-scale transform (3.55) of $\mathbf{X}$. Use the same notation $\mathbf{W_a}$ for the $N/2 \times N/2$ one-scale wavelet analysis matrix and write

$$\mathbf{W_a}\mathbf{Y}_{ss}^{(1)}(\mathbf{W_a})^{\mathrm{T}} = \begin{bmatrix} \mathbf{Y}_{ss}^{(2)} & \mathbf{Y}_{sd}^{(2)} \\ \mathbf{Y}_{ds}^{(2)} & \mathbf{Y}_{dd}^{(2)} \end{bmatrix},$$

where each block matrix is of size $N/4 \times N/4$. The two-scale transform $\mathbf{Y}^{(2)}$ of $\mathbf{X}$ is obtained by replacing the block $\mathbf{Y}_{ss}^{(1)}$ in $\mathbf{Y}^{(1)}$ by this block matrix form of $\mathbf{W_a}\mathbf{Y}_{ss}^{(1)}(\mathbf{W_a})^{\mathrm{T}}$:

$$\mathbf{Y}^{(2)} = \begin{bmatrix} \begin{bmatrix} \mathbf{Y}_{ss}^{(2)} & \mathbf{Y}_{sd}^{(2)} \\ \mathbf{Y}_{ds}^{(2)} & \mathbf{Y}_{dd}^{(2)} \end{bmatrix} & \mathbf{Y}_{sd}^{(1)} \\ \mathbf{Y}_{ds}^{(1)} & \mathbf{Y}_{dd}^{(1)} \end{bmatrix}.$$

The inverse transformation begins with

$$\mathbf{Y}_{ss}^{(1)} = \mathbf{W_s} \begin{bmatrix} \mathbf{Y}_{ss}^{(2)} & \mathbf{Y}_{sd}^{(2)} \\ \mathbf{Y}_{ds}^{(2)} & \mathbf{Y}_{dd}^{(2)} \end{bmatrix} \mathbf{W_s^{\mathrm{T}}},$$

where $\mathbf{W_s} = (\mathbf{W_a})^{-1}$ denotes the $N/2 \times N/2$ one-scale synthesis matrix. Then $\mathbf{X}$ is reconstructed from the one-scale transform $\mathbf{Y}^{(1)}$ using the $N \times N$ synthesis matrix as before.

Example 3.20. Figure 3.22 shows the two-scale transform of the round-wavy synthetic image from Figure 3.20. The 32×32 horizontal detail block (in rows 33–64 and columns 1–32), vertical detail block (in rows 1–32 and columns 33–64), and diagonal detail block (in rows 33–64 and columns 33–64) are the same as the corresponding 64×64 one-scale blocks.

This process can be continued to a three-scale multiresolution transform by applying a 2D wavelet transform to the 32×32 upper left-hand corner $\mathbf{Y}_{ss}^{(2)}$. ■

Remark 3.10. As we saw in Example 3.12, the arithmetic processing with the CDF$(2,2)$ transform only uses integer multiplication and division by powers of 2 (binary bit shifts). Since image matrices have integer entries, no round-off error

---

$(\mathbf{W_a}^{(2)})^{\mathrm{T}}$, then three of the four blocks in the one-scale transform must be recomputed (only the block $\mathbf{Y}_{dd}^{(1)}$ is left unchanged). This is too computationally expensive.

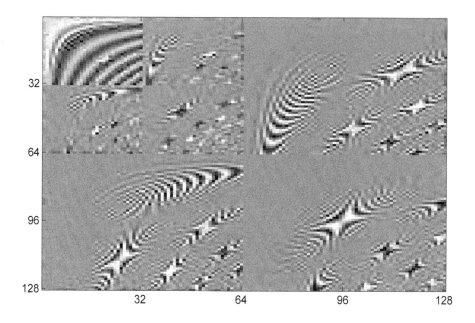

Fig. 3.22   Two-scale CDF$(2, 2)$ transform of synthetic round wavy image

occurs in the matrix multiplications and the calculations are fast. This is one of the reasons that this wavelet transform (also called the LeGall 5/3 transform) is used in the new JPEG 2000 image algorithm for lossless image encoding—see [Van Fleet (2008), §12.2]. Wavelet transforms have other advantages over the older JPEG discrete cosine transform (DCT) algorithm (based on the discrete Fourier transform). In reconstructing the image from its multiresolution representation the trend projection can be displayed first, and then the detail projections added. For example, a four-scale multiresolution wavelet transform on a 256×256 image uses the coefficients in the 16×16 matrix $\mathbf{Y}_{\mathbf{ss}}^{(4)}$ to obtain the trend projection, which can be displayed rapidly and shows basic image features. This is much more visually satisfying than the older JPEG algorithm, in which the image is divided into a grid of 8×8 blocks on which the DCT is used. The subdivision into small blocks is artificial and unrelated to the overall image, but these blocks are visible when the image is reconstructed.

3.6.4   *Image compression using wavelet transforms*

Many entries in the image matrix for Figure 3.19 don't differ very much from adjacent entries—see the data in (3.54), for example. By contrast, when there is a significant numerical change from one entry to a nearby entry, some significant detail of the image is located at this point. Wavelet methods allow us to process the numbers in an image matrix to take advantage of this phenomenon. Multiscale wavelet transforms separate a signal into a basic trend and several levels of details, so they are an effective mathematical tool for 2D *image compression* by the same method as for signals in Example 3.3: we replace some percentage of the smallest coefficients in the analysis transform of the image matrix by zero and then reconstruct an approximate image using the synthesis transform.

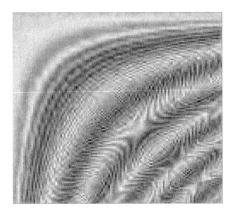 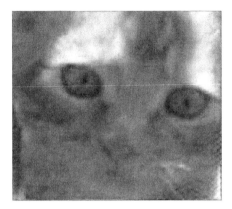

Fig. 3.23　Two-scale wavelet 10:1 compression of digital images

Figure 3.23 shows the compression of the synthetic round-wavy image (Fig. 3.20) and the kitten image (Fig. 3.19) using only the largest 10% of the two-scale $CDF(2, 2)$ wavelet analysis transform coefficients. This gives a 10:1 compression of each image, in the sense that from $(1/10) \times 128^2$ numbers taken from the analysis transform of the image we can reconstruct an approximate image matrix (which has 128^2 entries) using the synthesis transform (this algorithm doesn't depend on the particular image, of course). So if we are transmitting the compressed image over the internet, for example, we only need to send $(1/10) \times 128^2$ numbers, together with their locations in the 128×128 two-scale analysis matrix and the type of wavelet transform used.

The images in Figure 3.23 seem to be quite close to the original images. To obtain a quantitative measurement, let $\mathbf{X}$ be the image matrix and $\mathbf{X}_c$ the matrix for the compressed image. We define the *mean square error*

$$\text{MSE} = \|\mathbf{X} - \mathbf{X}_c\|^2 / N^2$$

for an $N \times N$ image matrix. For comparing two eight-bit images the largest possible MSE is 255^2, since each of the N^2 entries in $\mathbf{X} - \mathbf{X}_c$ is at most ± 255. In Figure 3.23 we have $N = 128$; for the round-wavy image we calculate that MSE $= 0.008$, while for the kitten image MSE $= 0.695$. The MSE for the compressed kitten image is about 85 times larger than the MSE for the compressed synthetic image, even though the small errors in the compressed synthetic image are more obvious to the eye than in the kitten image.[13] By comparison, if all entries in $\mathbf{X} - \mathbf{X}_c$ were ± 1,

---

[13]This is an advantage of using synthetic images for calibration purposes.

then MSE $= 1/128 = 0.008$.

In terms of visual perception a better measure of error is the *peak signal to noise ratio* PSNR, which is defined for eight-bit images as

$$\text{PSNR} = 10 \, \log_{10}(255^2/\text{MSE}).$$

This is on a logarithmic scale in units of *decibels*. In general PSNR ≥ 0 since MSE $\leq 255^2$. Small values of MSE give large values of PSNR. As a rule of thumb, if PSNR > 40 then the two images being compared are considered indistinguishable. For the round-wavy image we calculate that PSNR $= 69.0$, while for the kitten image PSNR $= 49.7$.

3.7   Computer Explorations

3.7.1   *Haar transform*

You will use MATLAB to explore the results in Section 3.3.

(a) Haar multiresolution basis

This basis for $\mathbb{R}^8$ is built from the *scaling vector* $\mathbf{h}_0$ (all components 1), and *wavelet vectors* at three levels. Write the following function m-file to generate the wavelet vectors at various levels.

```
% haar wavelet of length 2^n and level k
function h = hwavelet(n,k)
% create vector of length 2^n
h = zeros(2^n, 1);
h(1:2^(n-k),1) = 1;
h(2^(n-k)+1:2^(n-k+1),1) = -1;
```

Save this file as `hwavelet.m`. Then use it to create the Haar multiresolution basis for $\mathbb{R}^8$: First generate the *scaling vector* $\mathbf{h}_0$ and the basic wavelet vectors $\mathbf{h}_1$, $\mathbf{h}_2$, and $\mathbf{h}_4$ at levels 1, 2, and 3:

```
h0 = ones(2^3, 1)
h1 = hwavelet(3,1)
h2 = hwavelet(3,2)
h4 = hwavelet(3,3)
```

The other wavelet vectors at level 2 and 3 are obtained by shifting the vectors h_2 and h_4. You can carry out this operation easily in MATLAB. Use the `shift.m` function from Section 2.7.3 to generate the 8×8 shift matrix S and then apply this matrix to h_2 and h_4.

```
S = shift(8)
h3 = S^4 * h2
h5 = S^2 * h4
```

```
h6 = S^4*h4
h7 = S^6*h4
```

(b) Haar transform matrices

Make the Haar basis vectors for $\mathbb{R}^8$ that you generated in *(a)* into the columns of the $2^3 \times 2^3$ three-scale *Haar synthesis matrix* $\mathbf{W}_s^{(3)}$:

```
Ws = [h0  h1  h2  h3  h4  h5  h6  h7]
```

Calculate Ws'*Ws. What does this tell you about the orthogonality properties and lengths of the Haar basis vectors?

The Matlab command `diag(v)` creates a matrix with the entries of the column vector v on the diagonal and all other entries zero. Use this command to create a diagonal matrix D such that `Wa = D^(-1)*Ws'` is the *Haar analysis matrix* $W_a^{(3)}$ in equation (3.8). (HINT: Remember that *left* multiplication by a diagonal matrix multiplies each *row* by the corresponding diagonal entry.) Use MATLAB to verify that Wa is the inverse matrix to Ws.

(c) Data compression

Generate a random digital signal u of length 8 and plot it as a solid red line (consider u as a sample of an analog signal and use linear interpolation between the data points in the graph):

```
u = round(50*rand(8,1))
plot(u, 'r-'), axis([1 8 0 60]), hold on
```

Calculate the *Haar transform* v = Wa*u. Then create a *compressed* vector vc5 by replacing any entries in v whose *absolute value* is ≤ 5 by 0. For example, if v(2) = -3.5, then vc5(2) = 0. This is the *data compression step*. This is a *nonlinear* transformation, just like quantization. The compressed vector can be stored (or transmitted) using the *locations* of the nonzero entries and the values in the nonzero entries (for vectors with millions of entries that are 16 or 32 binary digits this is a significant saving). Calculate the *compression ratio* (the number of nonzero entries in the vector v divided by the number of nonzero entries in the compressed vector vc5).

Calculate the inverse transform uc5 of the compressed vector and plot the graph of uc5 on the same axes as the original signal u (using a dashed green line):

```
uc5 = Ws*vc5
plot(uc5, 'g--')
```

You should get a graph similar to Figure 3.5 (but your random signal is not the same as the signal in Figure 3.5, so your graph will differ in details). Notice that the graph of uc5 is fairly close to the graph of your uncompressed signal u. Calculate

```
100*norm(u - uc5)^2/norm(u)^2
```

to measure the relative percentage difference in energy between the compressed and uncompressed signals.

Finally, perform a more drastic compression on the signal **u**. Create another compressed Haar transform vector `vc10` by replacing all the entries in the Haar transform **v** whose absolute value is ≤ 10 by 0. (*Caution:* You must modify the *transform vector* **v**, not the original signal **u**.) Calculate the new compression ratio for `vc10`.

Calculate the inverse transform `uc10` of `vc10` and plot the graph of `uc10` on the same axes as the original signal **u** (using a dotted blue line):

```
uc10 = Ws*vc10
plot(uc10, 'b:')
```

The new graph will not be close to the original graph as the previous compression (unless your random transform vector **v** had no entries with absolute value ≤ 5). However, this graph follows the general trends of the original graph (this is the advantage of the wavelet method of compression). Calculate `100*norm(u - uc10)^2/norm(u)^2` to measure the relative percentage difference in energy between the compressed and uncompressed signals.

Insert arrows and labels on all the graphs in your MATLAB figure, indicating the original signal, the compression thresholds (5 and 10) of the two compressed signals, and the relative differences in energies of the two compressed signals. Insert the title Haar Wavelet Compression of Length 8 Signal.

3.7.2   *CDF*(2, 2) *wavelet transform*

You will use MATLAB to explore the results in Sections 3.4.1 and 3.5.

(a) CDF(2, 2) *analysis matrix*

Implement the CDF(2, 2) one-scale wavelet transform in matrix form for periodic signals **s** of even length $N \geq 4$, where **s** is a column vector, as follows. The $N \times N$ CDF(2, 2) analysis matrix is obtained by the lifting method as the product

$$\mathbf{T}_a = D\,U\,P\,\boxed{\text{split}}\,. \tag{3.58}$$

Here $\boxed{\text{split}}$ is the downsampling matrix that separates the signal **s** into subsignals s_{even} and s_{odd} of length $N/2$. After downsampling, the first lifting operation is a prediction step that is given by the matrix P in equation (3.12). Next, there is an update step that is given by the matrix U in equation (3.16). Finally, there is a normalization step that is performed by the diagonal matrix D in equation (3.17).

Write the following function m-file to generate the matrix $\mathbf{T}_a$. This m-file uses the MATLAB function m-file `shift.m` from Section 2.7.3 and the function m-file `downsamp.m` from Section 2.7.4.

```
% Generates N x N analysis matrix for
% one-scale CDF(2, 2) transform
% N >= 4 must be even
function Ta = cdfamat(N)
% generate special matrices
I = eye(N/2); S = shift(N/2); Z = zeros(N/2);
split2 = downsamp(N/2);
% prediction step
P = [I Z; -(1/2)*(I+S^(-1)) I];
% update step
U = [I (1/4)*(I + S); Z I];
% normalization step
D = [sqrt(2)*I Z; Z I/sqrt(2)];
% product of all steps
Ta = D*U*P*split2;
```

Save this file as `cdfamat.m`. Test it by generating the 8×8 CDF$(2,2)$ one-scale analysis matrix $\mathbf{T}_a$, a column vector $\mathbf{x}$ whose entries increase linearly, and the transform $\mathbf{y} = \mathbf{T}_a \mathbf{x}$:

```
Ta = cdfamat(8), x = [1:2:15]', y = Ta*x
```

You should get the same matrix as in Example 3.12. The first four entries in $\mathbf{y}$ are the *trend* $\mathbf{s}$ and the last four entries are the *detail* $\mathbf{d}$. Use the explicit form of the matrix $\mathbf{T}_a$ that you just generated to answer the following questions.

(i) How many entries of $\mathbf{x}$ are used to calculate a single entry in $\mathbf{s}$? (Don't count entries in $\mathbf{x}$ that are multiplied by a zero from a row of $\mathbf{T}_a$.)

(ii) How many entries of $\mathbf{x}$ are used to calculate a single entry in $\mathbf{d}$? (Don't count entries in $\mathbf{x}$ that are multiplied by a zero from a row of $\mathbf{T}_a$.)

(b) CDF(2, 2) synthesis matrix

The synthesis matrix $\mathbf{T}_s$ is the inverse of the analysis matrix. From (3.58)

$$\mathbf{T}_s = \boxed{\text{merge}}\ P^{-1} U^{-1} D^{-1}.$$

The matrix $\boxed{\text{merge}}$ is the inverse (= transpose) of the orthogonal permutation matrix $\boxed{\text{split}}$. Each of the other matrix inverses are also easy to calculate (this is one of the advantages of the lifting method). Use matrix multiplication in block form (hand calculation without MATLAB) to verify that

$$P^{-1} = \begin{bmatrix} I & 0 \\ \frac{1}{2}(I + S^{-1}) & I \end{bmatrix}, \quad U^{-1} = \begin{bmatrix} I & -\frac{1}{4}(I + S) \\ 0 & I \end{bmatrix}, \quad D^{-1} = \begin{bmatrix} \frac{1}{\sqrt{2}}I & 0 \\ 0 & \sqrt{2}I \end{bmatrix}.$$

Write the following MATLAB function m-file to generate the CDF$(2, 2)$ synthesis matrix $\mathbf{T}_s$. This m-file uses the function m-file `shift.m` from Section 2.7.3 and the function m-file `downsamp.m` from Section 2.7.4 and the formulas for P^{-1}, U^{-1}, and D^{-1} that you just verified.

```
% Generate N x N synthesis matrix for
% one-scale CDF(2, 2) transform
% N >= 4 must be even
function Ts = cdfsmat(N)
% generate special matrices
I = eye(N/2); S = shift(N/2); Z = zeros(N/2);
merge2 = downsamp(N/2)';
% prediction step
P = [I Z; (1/2)*(I+S^(-1)) I];
% update step
U = [I -(1/4)*(I + S); Z I];
% normalization step
D = [I/sqrt(2) Z; Z sqrt(2)*I];
% product of all steps
Ts = merge2*P*U*D;
```

Save this file as cdfsmat.m. Test it by generating the 8×8 CDF$(2,2)$ one-scale synthesis matrix and checking that it is the inverse of the analysis matrix:

```
Ts = cdfsmat(8), norm(Ts*Ta - eye(8))
```

You should get the same matrix as in Example 3.12.

(c) CDF$(2,2)$ basis vectors

In part *(b)* you generated the one-scale CDF$(2,2)$ synthesis matrix $\mathbf{T}_s$ of size 8×8. The columns of $\mathbf{T}_s$ give the CDF$(2,2)$ basis for $\mathbb{R}^8$ (see Section 3.5). Column 1 is the *scaling vector*. Observe that columns 2–4 of $\mathbf{T}_s$ are obtained from column 1 by applying S^2, S^4, and S^6. The trend vector in a one-scale CDF$(2,2)$ wavelet synthesis is a linear combination of these columns. Plot the columns by MATLAB:

```
figure, plot(Ts(:,1), 'r-'), axis([0  9  -.5  1.5]), hold on
plot(Ts(:,2), 'g-')
plot(Ts(:,3), 'b-')
plot(Ts(:,4), 'c-')
```

(see the left half of Figure 3.9). Use arrows to identify the plots as column 1, column 2, and so on; note that 'r-' plots red lines, 'g-' plots green, 'b-' plots blue, and 'c-' plots cyan. Insert the title CDF$(2,2)$ Trend Basis Vectors.

Column 5 of $\mathbf{T}_s$ is the *wavelet vector*, and columns 6–8 of $\mathbf{T}_s$ are obtained from column 5 by multiplying with S^2, S^4, and S^6, where S is the shift operator. The detail vector in a one-scale CDF$(2,2)$ wavelet synthesis is a linear combination of these columns. Plot the columns by MATLAB:

```
figure, plot(Ts(:,5), 'r-'), axis([0  9  -.5  1.5]), hold on
plot(Ts(:,6), 'g-')
plot(Ts(:,7), 'b-')
plot(Ts(:,8), 'c-')
```

(see the right half of Figure 3.9). Label the plots as column 5, column 6, and so on. Insert the title CDF(2, 2) Detail Basis Vectors in your MATLAB figure.

3.7.3   Daub4 wavelet transform

You will use MATLAB to explore the results in Sections 3.4.2 and 3.5.

(a) Daub4 analysis matrix

Implement the Daub4 one-scale wavelet transform in matrix form for periodic signals **s** of even length $N \geq 4$, viewing **s** as a column vector. The $N \times N$ analysis matrix is obtained by the lifting method as the product

$$\mathbf{T}_{\mathrm{a}} = D\, U_2\, P\, U_1\, \boxed{\text{split}} \qquad\qquad (3.59)$$

(see the flow chart in Figure 3.7). Here $\boxed{\text{split}}$ is the downsampling matrix that separates the signal **s** into subsignals $\mathbf{s}_{\mathrm{even}}$ and $\mathbf{s}_{\mathrm{odd}}$ of length $N/2$. After downsampling, the first lifting operation is an update step given by the matrix U_1 in equation (3.21). This is followed by a prediction step given by the matrix P in equation (3.24). Next, there is a second update step that is given by the matrix U_2 in equation (3.26). Finally, there is a normalization step that is performed by the diagonal matrix D in equation (3.27).

Write the following function m-file to generate the matrix $\mathbf{T}_{\mathrm{a}}$ This m-file uses the MATLAB function m-file `shift.m` from Section 2.7.3 and the function m-file `downsamp.m` from Section 2.7.4.

```
% Generates N x N analysis matrix for
% one-scale Daub4 transform
% N >= 4 must be even
function Ta = daub4mat(N)
% generate special matrices
I = eye(N/2); S = shift(N/2); Z = zeros(N/2);
split2 = downsamp(N/2);
% first update step
U1 = [I sqrt(3)*I; Z I];
% prediction step
P = [I Z; -(1/4)*(sqrt(3)*I+(sqrt(3)-2)*S) I];
% second update step
U2 = [I -S^(-1); Z I];
% normalization step
D = [(sqrt(3)-1)*I/sqrt(2) Z; Z (sqrt(3)+1)*I/sqrt(2)];
% product of all steps
Ta = D*U2*P*U1*split2;
```

Save this function m-file as `daub4mat.m`. Test it by generating the 8×8 Daub4 one-scale analysis matrix

```
Ta = daub4mat(8)
```

The matrix $\mathbf{T}_a$ is *orthogonal*; this is a special property of the Daub4 transform that is not true for the CDF$(2, 2)$ transform. Check this property by setting `Ts = Ta'` and calculating the distance `norm(Ts*Ta - eye(8))` between $\mathbf{T}_s\mathbf{T}_a$ and the identity matrix.

Suppose $\mathbf{x} \in \mathbb{R}^8$ and $\mathbf{y} = \mathbf{T}_a\mathbf{x}$. The first four entries in $\mathbf{y}$ are the *trend* vector $\mathbf{s}$, and the last four entries are the *detail* vector $\mathbf{d}$. Use the explicit form of the matrix $\mathbf{T}_a$ that you just generated to answer the following questions.

(i) How many entries of $\mathbf{x}$ are used to calculate a single entry in $\mathbf{s}$? (Don't count entries in $\mathbf{x}$ that are multiplied by a zero from a row of $\mathbf{T}_a$.)

(ii) How many entries of $\mathbf{x}$ are used to calculate a single entry in $\mathbf{d}$? (Don't count entries in $\mathbf{x}$ that are multiplied by a zero from a row of $\mathbf{T}_a$.)

(b) Two-scale Daub4 wavelet transform

Let $\mathbf{s}_3$ be a signal of length 2^3. To perform a two-scale Daub4 wavelet transform of $\mathbf{s}_3$, use the pyramid algorithm. First transform $\mathbf{s}_3$ into a *trend* $\mathbf{s}_2$ and a *detail* $\mathbf{d}_2$, each of length 2^2, using the 8×8 Daub4 analysis matrix. Then transform $\mathbf{s}_2$ into a trend $\mathbf{s}_1$ and a detail $\mathbf{d}_1$, each of length 2, using the 4×4 Daub4 analysis matrix. Stop at this point because the Daub4 transform needs input signals of length at least 4.

Generate the two-scale wavelet analysis matrix $\mathbf{W}_a^{(2)}$ for the Daub4 transform, which is denoted as `Wa` for MATLAB calculations, as follows (be sure to use the semicolons at the end of lines so that only the matrix `Wa` appears on screen).

```
T3 = daub4mat(8);
I = eye(4); Z = zeros(4,4); T2 = [daub4mat(4)  Z; Z  I];
Wa = T2*T3
```

Check that the matrix `Wa` is orthogonal by calculating `norm(Wa*Wa' - eye(8))`; the answer should be (essentially) zero.

Since the analysis matrix is orthogonal, you obtain the *two-scale* Daub4 *synthesis matrix* by

```
Ws = Wa'
```

(c) Two-scale Daub4 basis vectors

Let `Ws` be the 8×8 two-scale Daub4 synthesis matrix from part *(b)*. The columns of `Ws` give the *two-scale Daub4 basis* for $\mathbb{R}^8$ (see Section 3.5). Since `Ws` is an orthogonal matrix, this is an orthonormal basis. Column 1 is the *scaling vector*. Observe that column 2 of `Ws` is obtained from column 1 by multiplying by S^4, where S is the 8×8 shift matrix. The trend vector in a two-scale Daub4 wavelet synthesis is a linear combination of these columns. Plot the columns by MATLAB:

```
figure, plot(Ws(:,1), 'r-'), axis([0  9  -1  1]), hold on
plot(Ws(:,2), 'g-')
```

Insert arrows into the figure to label the plots column 1 and column 2. Notice the shift by 4 between the two plots. Insert the title Fig. 1: Daub4 Two-Scale Trend Basis Vectors.

Column 3 of Ws is the *coarse wavelet vector*. Observe that column 4 of Ws is obtained from column 3 by multiplying by S^4. The coarse detail vector in a two-scale Daub4 wavelet synthesis is a linear combination of these columns. Plot the columns by MATLAB:

```
figure, plot(Ws(:,3), 'r-'), axis([0 9 -1 1]), hold on
plot(Ws(:,4), 'g-')
```

Label the plots column 3 and column 4. Notice the shift by 4 between the two plots. Insert the title Daub4 Two-Scale Coarse Detail Basis Vectors in the MATLAB figure.

Columns 5 of Ws is the *fine wavelet vector*. Observe that columns 6–8 of Ws are obtained from column 5 by multiplying by S^2, S^4, and S^6. The fine detail vector in a two-scale Daub4 wavelet synthesis is a linear combination of these columns. Plot the columns by MATLAB:

```
figure, plot(Ws(:, 5), 'r-'), axis([0 9 -1 1]), hold on
plot(Ws(:, 6), 'g-')
plot(Ws(:, 7), 'b-')
plot(Ws(:, 8), 'c-')
```

Label the plots (column 5, column 6, and so on). Notice the shift by 2 between each plot. Insert the title Daub4 Two-Scale Fine Detail Basis Vectors in the MATLAB figure.

3.7.4 *Fast multiscale Haar transform*

The matrix formulation of the Haar transform and inverse Haar transform in Section 3.7.1 is helpful in understanding the theory of this transform and how it is similar to the discrete Fourier transform (it uses the Haar basis instead of the Fourier basis; both bases are orthogonal). For numerical calculation, however, the matrix formulation is impractical, since for a signal of length N direct matrix multiplication requires the order of N^2 arithmetic operations. Just as in the case of the Fourier transform, there is a fast implementation of the Haar transform through lifting. This algorithm only requires the order of N arithmetic operations.

(a) Implementing the one-scale Haar transform

Create the following function m-file

```
% one-scale discrete Haar wavelet transform
function T = dwthaar(Signal)
N = length(Signal); s = zeros(1, N/2); d = s;
```

```
for n=1:N/2
s(n) = 1/2*(Signal(2*n-1) + Signal(2*n));
d(n) = Signal(2*n-1) - s(n);
end
T = [s, d];
```

Save this file as dwthaar.m. Notice that now signals are *row vectors* rather than column vectors. Test the file on the signal

```
Signal = [56   40   8   24   48   48   40   16]
```

in Section 3.2, Table 3.1. Calculate

```
y = dwthaar(Signal)
```

The row vector y should be the same as the *second row* in Table 3.2.

(b) Inverting a one-scale Haar transform

Create the following function m-file:

```
% one-step discrete inverse Haar wavelet transform
function R = iwthaar(T)
N = length(T); R = zeros(1, N);
for n=1:N/2
R(2*n - 1) = T(n) + T(N/2 + n);
R(2*n) = T(n) - T(N/2 + n);
end
```

Save this file as iwthaar.m. Test it on the vector y by calculating

```
z = iwthaar(y)
```

You should obtain the row vector Signal.

(c) Complete Haar wavelet decomposition

To carry out a J-step Haar wavelet decomposition on a signal of length $N = 2^K$ (where $J \leq K$), you must apply the one-scale Haar transform J times to obtain a $J \times N$ matrix. The bottom row of the matrix will be the J-step Haar wavelet decomposition. Create the following function m-file to do this:

```
% Discrete Haar wavelet decomposition to level J
function T = haarwave(Signal, J)
N = size(Signal,2); T = zeros(J, N); L = N;
T(1,:)   = dwthaar(Signal);
for j=2:J
L = L/2;
T(j, 1:L) = dwthaar(T(j-1, 1:L));
T(j, L+1:N) = T(j-1, L+1:N);
end
```

Save this file as `haarwave.m`. Then test it by calculating

```
Table = haarwave(Signal, 3)
```

You should get a 3×8 matrix T whose bottom row is given in Table 3.5.

Remark 3.11. Notice that the first row of T is the one-scale Haar wavelet transform of the input signal y. The `for` loop implements a one-scale transform to obtain row j of T from row j-1. It applies the one-scale Haar wavelet transform to the initial segment of the row vector that is one-half the length of the segment used for the previous row.

(d) Inverting the level J Haar wavelet decomposition

Create the following function m-file:

```
% Inverse of level J Discrete Haar wavelet decomposition
function T = ihaarwave(y, J)
N = size(y,2); T = zeros(J, N); L = N/2^(J-1);
T(J,1:L) = iwthaar(y(1:L));
T(J, L+1:N) = y(L+1:N);
for k = J-1:-1:1
L = 2*L;
T(k, 1:L) = iwthaar( T(k+1, 1:L) );
T(k, L+1:N) = T(k+1, L+1:N);
end
```

Save this m-file as `ihaarwave.m`. Test it by entering

```
y = [35  -3  16   10   8  -8   0   12]
T = ihaarwave(y,3)
```

The first row of the 3×8 matrix T should be the original signal in Table 3.1.

Remark 3.12. Note that entries `1:L` of row J of T are the one-scale inverse Haar transform of y, where `L = N/2^(J-1)`. The remaining entries of row J are the same as the corresponding entries of y. The `for` loop repeats this inversion algorithm at each level from row J-1 down to row 1 of T, doubling the length L of the inverse transform vector at each step.

3.7.5   *Fast multiscale Daub4 transform*

The matrix formulation of the Daub4 transform and inverse transform in Section 3.7.3 is helpful in understanding the theory of this transform and how it is similar to the discrete Fourier transform and discrete Haar transform (it uses the Daub4 basis instead of the Fourier basis or the Haar basis; all these bases are orthogonal). For numerical calculation, however, the matrix formulation is impractical, just as for the discrete Fourier and Haar transforms. There is a fast implementation of the Daub4 transform through lifting. This algorithm only requires the order of N arithmetic operations.

(a) Implementing the one-scale Daub4 analysis transform

Create the following function m-file

```
% one-scale Daubechies 4 wavelet transform
% input signal S is a row vector of even length N >=4
function T = daub4wt(S)
N = length(S);
% allocate space in memory
s1 = zeros(1, N/2); d1 = zeros(1, N/2);
% first update for trend
s1 = S(1:2:N-1) + sqrt(3)*S(2:2:N);
% shift first trend with wrap-around s1shift = [s1(N/2)
s1(1:N/2-1)];
% first prediction for detail
d1 = S(2:2:N) - sqrt(3)/4*s1 - (sqrt(3)-2)/4*s1shift
% second update for trend; shift first detail
s2 = s1 - [d1(2:N/2)  d1(1)];
% normalization of trend
s = (sqrt(3)-1)/sqrt(2)*s2;
% normalization of detail  d = (sqrt(3)+1)/sqrt(2)*d1;
T = [s  d];
```

Save this file as daub4wt.m. Notice that signals are now *row vectors* rather than column vectors. To test this file, let S = [1 0 0 0], calculate T = daub4wt(S) and compare T with the first column of daub4mat(4).

(b) Implementing the one-scale Daub4 synthesis transform

Create the following function m-file that reverses the steps of the function in *(a)*.

```
% one-scale inverse Daubechies 4 wavelet transform
% input is a row vector T = [s  d] of even length N >=4
function S = daub4iwt(T)
N = length(T);
% Separate input into trend and detail
s = T(1:N/2); d = T(N/2+1:N);
% allocate space in memory
S = zeros(1, N);
% inverse normalization of trend and detail
s2 = (sqrt(3)+1)/sqrt(2)*s; d1 = (sqrt(3)-1)/sqrt(2)*d;
% inverse of second update for trend; shift first detail left
s1 = s2 + [d1(2:N/2)  d1(1)];
% inverse of first prediction for detail; shift first trend right
S(2:2:N) = d1 + sqrt(3)/4*s1 + (sqrt(3)-2)/4*[s1(N/2)  s1(1:N/2-1)];
% inverse of first update for trend
S(1:2:N-1) = s1 - sqrt(3)*S(2:2:N);
```

Save this file as daub4iwt.m. To test this file, let T = [1 0 0 0], calculate S = daub4iwt(T) and compare S with the first row of daub4mat(4).

(c) Multiscale Daub4 scaling and wavelet vectors

Write the following function m-file:

```
% N-2 scale Daub4 scaling function of length 2^N
function S = daub4scale(N)
T = zeros(1, 2^N); T(1) = 1;
for k = 1:(N-2)
T = daub4iwt(T);
end
S = T;
```

Save this file as daub4scale.m. Now write the following function m-file:

```
% N-3 scale Daub4 wavelet function of length 2^N
function S = daub4wave(N)
T = zeros(1, 2^N); T(5) = 1;
for k = 1:(N-3)
T = daub4iwt(T);
end
S = T;
```

Save this file as daub4wave.m. Generate and plot a ten-scale Daub4 scaling vector and a nine-scale wavelet vector (both of length $2^{12} = 4096$):

```
S = daub4scale(12); W = daub4wave(12);
figure, plot(S, 'r-'), hold on, plot(W, 'g-')
```

Label the plots as scaling function and wavelet. Insert the title Daub4 Nine-Scale Wavelet and Ten-Scale Scaling Function in the MATLAB figure.

Remark 3.13. The jagged graphs in these plots lead to the continuous but non-differentiable Daub4 *scaling function* and *wavelet function* as the number of scales goes to infinity (compare with Figure 3.13 which shows the four-scale scaling and wavelet vectors of length 32). These functions are very different from the sinusoidal functions of continuous Fourier analysis. Section 5.4 examines scaling and wavelet functions for analog signals in more detail.

3.7.6 *Signal processing with the multiscale Haar transform*

You will carry out some signal processing as in Section 3.3.2.

(a) Analyzing a synthetic signal

Consider the analog signal $s(t) = \sin(4\pi t)$. Sample this signal at $2^9 = 512$ equidistant points in $0 \le t \le 1$ to obtain a discrete signal called s_9:

```
s9 = sin([1:512]*4*pi/512);
figure, plot(s9), axis([0  550  -1.2  1.2])
```

Insert the title Sine Wave Signal, 512 samples in the MATLAB figure.

Now calculate a three-scale Haar wavelet transform of s_9 and plot the transform hws9 as a bar graph:

```
T = haarwave(s9, 3);
hws9 = T(3,:);
figure, bar(hws9, 0.1)
```

Insert the title Three-stage Haar transform of Sine Wave Signal, 512 samples in the MATLAB figure.

Remark 3.14. If you had created the 512×512 Haar analysis matrix $\mathbf{W}_a^{(9)}$ following the method of Section 3.7.1, then you could have obtained the column vector (hws9)' as the product of the 512×512 matrix $\mathbf{W}_a^{(9)}$ and the vector s_9. This is very inefficient, however, compared to the calculation using the fast Haar transform, which uses the 3×512 matrix haarwave(s9, 3).

To interpret the graph of the Haar transform of the signal, consider the row vector hws9 as a function of the row index j (for $1 \leq j \leq 512$). It is made up of a four parts:

Level-3 trend: $\mathbf{s}_6$ for $1 \leq j \leq 64$ (length 2^6)
Level-3 detail: $\mathbf{d}_6$ for $65 \leq j \leq 128$ (length 2^6)
Level-2 detail: $\mathbf{d}_7$ for $129 \leq j \leq 256$ (length 2^7)
Level-1 detail: $\mathbf{d}_8$ for $257 \leq j \leq 512$ (length 2^8)

Create row vectors in MATLAB for each of these parts of the transform and plot the trend $\mathbf{s}_6$ as a bar graph:

```
s6 = hws9(1:64); d6 = hws9(65:128);
d7 = hws9(129:256); d8 = hws9(257:512);
figure, bar(s6, 0.1)
```

Notice that the level-3 trend $\mathbf{s}_6$ is a coarse version of the graph of the original signal (with a sampling rate of $64 = 512/2^3$). Insert the title Level-3 Trend of Signal in the MATLAB figure.

(b) Approximating a signal

Create an *approximate version* fhws9 of the Haar transform using only the level-3 trend $\mathbf{s}_6$ and level-3 detail $\mathbf{d}_6$ of the signal (zero out the level-1 and level-2 details). Then take the inverse Haar transform fs9 of fhws9 and plot it:

```
fhws9 = hws9; fhws9(129:512) = 0;
fs9 = ihaarwave(fhws9, 3);
```

```
figure, plot(fs9(1,:)), axis([0  550  -1.2  1.2])
```

The graph should be a rough version of the original signal. The first row `fs9(1,:)` of the 3×512 matrix `fs9` is the projection of `s9` onto the subspace spanned by the first 128 Haar wavelet basis vectors (out of the total of 512). Although these are only one-fourth of the total basis vectors, they capture most of the information in a smooth signal such as a low-frequency sine wave. Calculate the relative approximation error

```
error = norm(s9 - fs9(1,:))/norm(s9)
```

Insert the title Approximation of Signal Using Level-3 Trend and Detail in the MATLAB figure.

(c) Compressing a noisy signal

Now add some *pops* to the signal s_9 that indicate some event such as a loud noise whose timing you want to determine. Do this by first generating a random two-component vector `pop` whose entries are integers between 100 and 400:

```
pop = round(100 + 300*rand(2,1))
```

Create a signal with pops by adding 1 to the two components of s_9 whose indices are given by the numbers in `pop`:

```
ps9 = s9;
ps9(pop(1)) = s9(pop(1)) + 1;
ps9(pop(2)) = s9(pop(2)) + 1;
```

Finally, add some normally-distributed *random static noise* and plot the noisy version of the original signal:

```
nps9 = ps9 + 0.1*randn(1, 512);
figure, plot(nps9), axis([0  550  -1.5  1.5])
```

The location of the two pops should be clear in the plot. Insert the title Signal with Pops and Static in the MATLAB figure.

Use the Haar transform to compress the signal while still retaining the information about time location of the pops. This is something that is not possible with the discrete Fourier transform (recall the train whistle in Example 2.10).

Calculate a three-scale Haar wavelet transform of the noisy signal with pops nps_9:

```
T = haarwave(nps9, 3);
hwnps9 = T(3,:);
```

As before, create row vectors in MATLAB for each of the four parts of the transform:

```
s6 = hwnps9(1:64); d6 = hwnps9(65:128);
d7 = hwnps9(129:256); d8 = hwnps9(257:512);
```

Plot bar graphs of the trend vector s_6 and the detail vectors d_6, d_7, and d_8 in four windows of the same figure:

```
figure, subplot(2,2,1), bar(s6, 0.1)
subplot(2,2,2), bar(d6, 0.1)
subplot(2,2,3), bar(d7, 0.1)
subplot(2,2,4), bar(d8, 0.1)
```

Notice that the pops are mostly hidden in the level-3 trend s_6, and may or may not show in the level-3 detail d_6 because of the random noise. However, they are the very clear in the details d_7 and d_8. Put title Trend and Details of Signal with Noise and Pops at the top of the MATLAB figure, and then put titles s6, d6, d7, d8 to the right of each of the plot windows (s6 is the upper left plot, d6 is the upper right, d7 is the lower left, and d8 is the lower right).

In the Haar transform most of the coefficients in the detail vectors d_7 and d_8 are less than 0.1 in absolute value, and are largely due to the random noise. The only significant coefficients come from the two pops. Compress and remove noise from the signal by setting to zero all entries in d_7 and d_8 that are less than 0.1 in absolute value (do not change the trend vector s_6 or the detail vector d_6). You did this by hand in Section 3.7.1 *(c)*. Now you must modify a vector with $512 - 128 = 384$ entries. To do this easily, create the following function m-file:

```
% threshold truncation function
% Zeros out all components of row vector x smaller than
% the level parameter
function y = threshold(x, level)
y = x.*(abs(x) >= level);
```

Save this file as `threshold.m`. Notice the period before * in the formula for y; this makes the product an entry-by-entry multiplication. The factor in parentheses after .* is a *logic vector*: its entries are 1 when the inequality is true for the corresponding entry of the vector `abs(x)` of absolute values of x, and 0 otherwise. Test this m-file by typing

```
Signal
threshold(Signal, 10)
```

(where `Signal` is the row vector from Section 3.7.4 (a)). The answer should be a vector in which each component of `Signal` whose absolute value is less than 10 has been replaced by 0.

Execute the following code. It creates a row vector of length 384 by concatenating the detail vectors d_7 and d_8, removes the small components using the `threshold` function, and then concatenates this compressed row vector with $\mathbf{s}_6$ and $\mathbf{d}_6$ to obtain the compressed Haar wavelet transform of $\mathbf{s}_9$:

```
detail = [d7  d8];
cdetail = threshold(detail, 0.1);
chws9 = [s6  d6  cdetail];
```

To calculate the compression ratio in this case, count the number of nonzero coefficients in the vector `cdetail` by

```
count = ones(1, 384).*(abs(cdetail)>0);
ratio = 512/(128 + sum(count))
```

Finally, take the three-scale inverse transform of the compressed Haar wavelet transform and plot the compressed version $\mathbf{cs}_9$ of the noisy signal $\mathbf{s}_9$:

```
T = ihaarwave(chws9, 3);
cs9 = T(1,:);
figure, plot(cs9), axis([0  550  -1.5  1.5])
```

Note that the graph of $\mathbf{cs}_9$ shows the main features of the signal with the two pops, namely the slow oscillation of the main signal and the location of the pops. Some of the static noise has been removed (other wavelet transforms are more efficient at removing such noise). Insert the title Compressed Signal with Pops and Less Noise in the MATLAB figure. Insert labels on the graph indicating the compression ratio and the locations of the pops.

3.8 Exercises

(1) Let $\mathbf{x} = \begin{bmatrix} 2\ 2\ 4\ 6\ 8\ 8\ 12\ 10 \end{bmatrix}^{\mathrm{T}}$. In this exercise you will carry out a three-scale Haar multiresolution analysis of this signal of length 8 using the Haar analysis matrices $\mathbf{T}_{\mathrm{a}}^{(k)}$ from equation (3.3).

(a) Obtain the vector $\mathbf{y} = \begin{bmatrix} \mathbf{s}_0\ \mathbf{d}_0\ \mathbf{d}_1\ \mathbf{d}_2 \end{bmatrix}^{\mathrm{T}}$ for $\mathbf{x}$ as follows: Calculate the level-two trend $\mathbf{s}_2$ and detail $\mathbf{d}_2$ of $\mathbf{x}$ by applying $\mathbf{T}_{\mathrm{a}}^{(3)}$ to $\mathbf{x}$. Then calculate the level-one trend $\mathbf{s}_1$ and detail $\mathbf{d}_1$ by applying $\mathbf{T}_{\mathrm{a}}^{(2)}$ to $\mathbf{s}_2$. Finally, calculate the level-zero trend $\mathbf{s}_0$ and detail $\mathbf{d}_0$ by applying $\mathbf{T}_{\mathrm{a}}^{(1)}$ to $\mathbf{s}_1$ (see Diagram 3.4).

(b) Check your answer in (a) by applying the 8×8 three-scale Haar analysis matrix $\mathbf{W}_{\mathrm{a}}^{(3)}$ to $\mathbf{x}$ (see equation (3.8)).

(c) Use $\mathbf{y}$ to write $\mathbf{x}$ as a linear combination of the columns of $\mathbf{W}_{\mathrm{s}}^{(3)}$, as in equation (3.9).

(d) Obtain a 2:1 compression $\tilde{\mathbf{x}} = \mathbf{W}_{\mathrm{s}}^{(3)} \begin{bmatrix} \mathbf{s}_0 \\ \mathbf{d}_0 \\ \mathbf{d}_1 \\ 0 \end{bmatrix}$ of $\mathbf{x}$ by setting the detail vector $\mathbf{d}_2$ to zero and calculating the inverse transform.

(e) Calculate the relative compression error $\|\mathbf{x} - \tilde{\mathbf{x}}\|^2/\|\mathbf{x}\|^2$ and draw the graphs of $\mathbf{x}$ and of $\tilde{\mathbf{x}}$ as piecewise linear functions on the interval $[0,7]$, following the style of Figure 3.5.

(2) The 6×6 one-scale Haar analysis matrix $\mathbf{T_a}$ and synthesis matrix $\mathbf{T_s}$ are given in 2×2 block form by equations (3.1) and (3.2). For $\mathbf{x} \in \mathbb{R}^6$ write $\mathbf{T_a x} = \begin{bmatrix} \mathbf{s} \\ \mathbf{d} \end{bmatrix}$, where the trend $\mathbf{s}$ and detail $\mathbf{d}$ are vectors in $\mathbb{R}^3$.

(a) Find the 6×6 permutation matrix P that carries out the $\boxed{\text{split}}$ transformation. Then use P in equation (3.1) to obtain the entries in $\mathbf{T_a}$. How are rows 2 and 3 of $\mathbf{T_a}$ obtained from row 1? How are rows 5 and 6 of $\mathbf{T_a}$ obtained from row 4?

(b) The inverse matrix P^T carries out the $\boxed{\text{merge}}$ transformation. Use P^T in (3.2) to obtain the entries in $\mathbf{T_s}$. How are columns 2 and 3 of $\mathbf{T_s}$ obtained from column 1? How are columns 5 and 6 of $\mathbf{T_s}$ obtained from column 4?

(c) Calculate the vectors $\mathbf{s}$ and $\mathbf{d}$ when $\mathbf{x} = \begin{bmatrix} 1\,2\,2\,3\,3\,3 \end{bmatrix}^\mathrm{T}$. Then use the inversion formula $\mathbf{x} = \mathbf{T_s} \begin{bmatrix} \mathbf{s} \\ \mathbf{d} \end{bmatrix}$ to write $\mathbf{x}$ as a linear combination of the columns of $\mathbf{T_s}$.

(d) Obtain a 2:1 compression $\tilde{\mathbf{x}} = \mathbf{T_s} \begin{bmatrix} \mathbf{s} \\ 0 \end{bmatrix}$ of $\mathbf{x}$ by setting the detail vector to zero and calculating the inverse transform.

(e) Calculate the relative compression error $\|\mathbf{x} - \tilde{\mathbf{x}}\|^2/\|\mathbf{x}\|^2$ and draw the graphs of $\mathbf{x}$ and of $\tilde{\mathbf{x}}$ as piecewise linear functions on the interval $[0,5]$, following the style of Figure 3.5.

(3) The two-scale Haar multiresolution analysis of a vector $\mathbf{x} \in \mathbb{R}^4$ is the vector $\begin{bmatrix} s_0 \\ d_0 \\ d_1 \end{bmatrix} = \mathbf{W_a^{(2)}} \mathbf{x}$, where $\mathbf{W_a^{(2)}} = \begin{bmatrix} \mathbf{T_a^{(1)}} & 0 \\ 0 & I \end{bmatrix} \mathbf{T_a^{(2)}} \mathbf{x}$. Here d_1 is in $\mathbb{R}^2$ while s_0 and d_0 are in $\mathbb{R}$. The Haar analysis matrices $\mathbf{T_a^{(k)}}$ are from equation (3.3) and I is the 2×2 identity matrix.

(a) Calculate the entries in the 4×4 matrix $\mathbf{W_a^{(2)}}$ and compare it to the three-scale analysis matrix $\mathbf{W_a^{(3)}}$ in equation (3.8).

(b) Give formulas for s_0, d_0 and d_1 when $\mathbf{x} = \begin{bmatrix} a\,b\,c\,d \end{bmatrix}^\mathrm{T}$ is an arbitrary vector in $\mathbb{R}^4$.

(4) This exercise leads to a proof of property (3.14) of the CDF$(2,2)$ wavelet transform. For a signal $\mathbf{x}$ of even length N let the detail $\mathbf{d}$ be defined by (3.10) and the trend $\mathbf{s}$ be defined by (3.13).

(a) Show that
$$\mathbf{d}[n]+\mathbf{d}[n-1] = \mathbf{x}_\mathrm{odd}[n]+\mathbf{x}_\mathrm{odd}[n-1]-\tfrac{1}{2}\mathbf{x}_\mathrm{even}[n-1]-\mathbf{x}_\mathrm{even}[n]-\tfrac{1}{2}\mathbf{x}_\mathrm{even}[n+1].$$

(b) Use (a) to show that
$$4\mathbf{s}[n] = \mathbf{x}_{\text{odd}}[n] + \mathbf{x}_{\text{odd}}[n-1] - \tfrac{1}{2}\mathbf{x}_{\text{even}}[n-1] + 3\mathbf{x}_{\text{even}}[n] - \tfrac{1}{2}\mathbf{x}_{\text{even}}[n+1].$$

(c) Suppose f is a function on the integers that is periodic of period $N/2$. Show that $\sum_{n=0}^{(N/2)-1} f[n+k] = \sum_{n=0}^{(N/2)-1} f[n]$ for all integers k.

(d) Use the results in (b) and (c) to prove (3.14) by showing
$$4\sum_{n=0}^{(N/2)-1} \mathbf{s}[n] = 2\sum_{n=0}^{(N/2)-1} \mathbf{x}_{\text{even}}[n] + 2\sum_{n=0}^{(N/2)-1} \mathbf{x}_{\text{odd}}[n] = 2\sum_{n=0}^{N-1} \mathbf{x}[n].$$

(5) Let $\mathbf{x}$ be a real-valued function on $\{0,1,2,3\}$. Extend $\mathbf{x}$ to be a periodic function on the integers of period 4. Define a *trend* function $\mathbf{s}$ and a *detail* function $\mathbf{d}$ by the following *lifting step* formulas for $n = 0,1$:

Prediction P: $\mathbf{d}[n] = \mathbf{x}[2n+1] - \mathbf{x}[2n] - 2\mathbf{x}[2n+2]$

Update U: $\mathbf{s}[n] = \mathbf{x}[2n] + \mathbf{d}[n] + 3\mathbf{d}[n-1]$

(a) Suppose $\mathbf{x}[0] = 4$, $\mathbf{x}[1] = 7$, $\mathbf{x}[2] = 0$, and $\mathbf{x}[3] = 3$. Calculate $\mathbf{d}[0]$, $\mathbf{d}[1]$, $\mathbf{s}[0]$, and $\mathbf{s}[1]$.

(b) Let $\mathbf{x}_{\text{even}} = \begin{bmatrix} \mathbf{x}[0] \\ \mathbf{x}[2] \end{bmatrix}$ and $\mathbf{x}_{\text{odd}} = \begin{bmatrix} \mathbf{x}[1] \\ \mathbf{x}[3] \end{bmatrix}$. Identify $\mathbf{s}$ and $\mathbf{d}$ with column vectors in $\mathbb{R}^2$ as usual. Let P be the *prediction* linear transformation: $P\begin{bmatrix} \mathbf{x}_{\text{even}} \\ \mathbf{x}_{\text{odd}} \end{bmatrix} = \begin{bmatrix} \mathbf{x}_{\text{even}} \\ \mathbf{d} \end{bmatrix}$. Write down the matrix for P. First give the matrix in 2×2 block form using the shift matrix S, and then give the 4×4 numerical matrix.

(c) Let U be the *update* linear transformation: $U\begin{bmatrix} \mathbf{x}_{\text{even}} \\ \mathbf{d} \end{bmatrix} = \begin{bmatrix} \mathbf{s} \\ \mathbf{d} \end{bmatrix}$. Write down the matrix for U. First give the matrix in 2×2 block form using the shift matrix S, and then give the 4×4 numerical matrix.

(6) Consider the Daub4 transform $\mathbf{T_a}$ from Example 3.7.

(a) Show that the coefficients in the transform satisfy $a + b + c + d = 8$ and $a + c = b + d$.

(b) Use the result in (a) to calculate $\mathbf{T_a x}$ when $\mathbf{x} = \begin{bmatrix} 1 & 1 & 1 & 1 \end{bmatrix}^{\mathrm{T}}$.

(c) Let $\mathbf{s}$ be the trend vector and $\mathbf{d}$ the detail vector for the signal $\mathbf{x}$ in (b). Check that $\|\mathbf{x}\|^2 = \|\mathbf{s}\|^2 + \|\mathbf{d}\|^2$.

(7) The CDF$(2,4)$ wavelet transform of a vector $\mathbf{x}$ of length N (even) consists of the following lifting steps in the order given:

Prediction P: $\mathbf{d}^{(1)}[n] = \mathbf{x}_{\text{odd}}[n] - (1/2)\big(\mathbf{x}_{\text{even}}[n] + \mathbf{x}_{\text{even}}[n+1]\big).$

Update U: $\mathbf{s}^{(1)}[n] = \mathbf{x}_{\text{even}}[n]$
$- (1/64)\big(3\,\mathbf{d}^{(1)}[n-2] - 19\,\mathbf{d}^{(1)}[n-1] - 19\,\mathbf{d}^{(1)}[n] + 3\,\mathbf{d}^{(1)}[n+1]\big).$

Normalization D: $\mathbf{s}[n] = \sqrt{2}\,\mathbf{s}^{(1)}[n]$, $\mathbf{d}[n] = (1/\sqrt{2})\,\mathbf{d}^{(1)}[n]$.

Note that P and D are the same as for the CDF$(2,2)$ transform.

(a) Give the equation for $\mathbf{s}^{(1)}$ in terms of $\mathbf{x}_{\text{even}}$, $\mathbf{d}^{(1)}$, and the $N/2 \times N/2$ shift matrix S, following the pattern for the CDF$(2,2)$ transformation in (3.15).

(b) Let U be the update linear transformation $U \begin{bmatrix} \mathbf{s}^{(1)} \\ \mathbf{d}^{(1)} \end{bmatrix} = \begin{bmatrix} \mathbf{x}_{\text{even}} \\ \mathbf{d}^{(1)} \end{bmatrix}$. Use (a)
to obtain the matrix for U in 2×2 block form.

(c) Use formula (3.12) for P and your formula from (b) to find polynomials
$p_{ij}(S)$ such that $DUP = \begin{bmatrix} p_{00}(S) & p_{01}(S) \\ p_{10}(S) & p_{11}(S) \end{bmatrix}$, where $D = \begin{bmatrix} \sqrt{2} & 0 \\ 0 & 1/\sqrt{2} \end{bmatrix}$. This
is the polyphase analysis matrix for the CDF$(2,4)$ transform. (For the
calculations use the symbolic toolbox of MATLAB and `format rational`,
with S and $\sqrt{2}$ symbolic variables and $I = 1$.)

(8) The CDF$(3,3)$ wavelet transform of a vector $\mathbf{x}$ of length N (even) consists of
the following lifting steps in the order given (omitting the final normalization
step):
First update U_1: $\mathbf{s}^{(1)}[n] = \mathbf{x}_{\text{even}}[n] - (1/3)\mathbf{x}_{\text{odd}}[n-1]$.
Prediction P: $\mathbf{d}^{(1)}[n] = \mathbf{x}_{\text{odd}}[n] - (1/8)\left(9\mathbf{s}^{(1)}[n] + 3\mathbf{s}^{(1)}[n+1]\right)$.
Second update U_2: $\mathbf{s}^{(2)}[n] = \mathbf{s}^{(1)}[n]$
$$+ (1/36)\left(3\mathbf{d}^{(1)}[n-1] + 16\mathbf{d}^{(1)}[n] - 3\mathbf{d}^{(1)}[n+1]\right).$$
Normalization D: $\mathbf{s}[n] = (3/\sqrt{2})\,\mathbf{s}^{(2)}[n]$, $\mathbf{d}[n] = (\sqrt{2}/3)\,\mathbf{d}^{(1)}[n]$.
Note that U_1, P, and D are the same as for the CDF$(3,1)$ transform.

(a) Give the equation for $\mathbf{s}^{(2)}$ in terms of $\mathbf{s}^{(1)}$, $\mathbf{d}^{(1)}$, and the $N/2 \times N/2$ shift
matrix S.

(b) Let U_2 be the second update linear transformation: $U_2 \begin{bmatrix} \mathbf{s}^{(1)} \\ \mathbf{d}^{(1)} \end{bmatrix} = \begin{bmatrix} \mathbf{s}^{(2)} \\ \mathbf{d}^{(1)} \end{bmatrix}$.
Write down the matrix for U_2 in 2×2 block form.

(c) Use the matrices U_1 and P from Example 3.11 and your formula
for U_2 from (b) to find polynomials $p_{ij}(S)$ such that $DU_2PU_1 = \begin{bmatrix} p_{00}(S) & p_{01}(S) \\ p_{10}(S) & p_{11}(S) \end{bmatrix}$, where $D = \begin{bmatrix} (3/\sqrt{2})I & 0 \\ 0 & (\sqrt{2}/3)I \end{bmatrix}$. This is the polyphase
analysis matrix for the CDF$(3,3)$ transform. (For the calculations use
the symbolic toolbox of Matlab and `format rational`, with S and $\sqrt{2}$
symbolic variables and $I = 1$.)

(9) Let $N = 2m$ be even. Suppose $\mathbf{u}_0$, $\mathbf{v}_0$ are $1 \times N$ real row vectors and $\tilde{\mathbf{u}}_0$, $\tilde{\mathbf{v}}_0$
are $N \times 1$ real column vectors that satisfy equations (3.42). Define $\mathbf{u}_k$, $\mathbf{v}_k$, $\tilde{\mathbf{u}}_k$,
$\tilde{\mathbf{v}}_k$ for $k = 1, \ldots, m-1$ as in Theorem 3.2. Prove that these vectors satisfy
the biorthogonality equations (3.36).

(10) This exercise examines the trend subspace for a one-scale wavelet transform
on $\mathbb{R}^N$ for any even N.

(a) Consider the one-scale CDF$(2,2)$ wavelet transform when $N = 2m \geq 6$.
Let $\mathbf{x} = \begin{bmatrix} \mathbf{x}[0] & \mathbf{x}[1] & \cdots & \mathbf{x}[N-1] \end{bmatrix}^\mathsf{T}$. Give m homogeneous linear equations
in the variables $\mathbf{x}[0], \ldots, \mathbf{x}[N-1]$ whose solutions give the trend subspace
for this transform.

(b) Let $\mathbf{T_a} = \begin{bmatrix} \mathbf{U} \\ \mathbf{V} \end{bmatrix}$ be any one-scale $N \times N$ wavelet analysis matrix as in Section 3.5. Prove that the trend subspace for $\mathbf{T_a}$ is the null space of $\mathbf{V}$.

(11) Consider the CDF$(3, 1)$ wavelet transform on $\mathbb{R}^6$ in Example 3.11.

(a) Fill in the missing entries in the one-scale analysis matrix (the normaliza-tion step has been put into the synthesis matrix, as in Example 3.12):

$$\mathbf{T_a} = \begin{bmatrix} 6 & 6 & -2 & 0 & 0 & -2 \\ \rule{1.5em}{0.4pt} & \rule{1.5em}{0.4pt} & \rule{1.5em}{0.4pt} & \rule{1.5em}{0.4pt} & \rule{1.5em}{0.4pt} & \rule{1.5em}{0.4pt} \\ -3 & 3 & -1 & 0 & 0 & 1 \\ \rule{1.5em}{0.4pt} & \rule{1.5em}{0.4pt} & \rule{1.5em}{0.4pt} & \rule{1.5em}{0.4pt} & \rule{1.5em}{0.4pt} & \rule{1.5em}{0.4pt} \\ \rule{1.5em}{0.4pt} & \rule{1.5em}{0.4pt} & \rule{1.5em}{0.4pt} & \rule{1.5em}{0.4pt} & \rule{1.5em}{0.4pt} & \rule{1.5em}{0.4pt} \end{bmatrix} .$$

(b) Fill in the missing entries in the one-scale synthesis matrix:

$$\mathbf{T_s} = \frac{1}{32} \begin{bmatrix} 3 & \rule{1.5em}{0.4pt} & \rule{1.5em}{0.4pt} & -6 & \rule{1.5em}{0.4pt} & \rule{1.5em}{0.4pt} \\ 3 & \rule{1.5em}{0.4pt} & \rule{1.5em}{0.4pt} & 6 & \rule{1.5em}{0.4pt} & \rule{1.5em}{0.4pt} \\ 1 & \rule{1.5em}{0.4pt} & \rule{1.5em}{0.4pt} & 2 & \rule{1.5em}{0.4pt} & \rule{1.5em}{0.4pt} \\ 0 & \rule{1.5em}{0.4pt} & \rule{1.5em}{0.4pt} & 0 & \rule{1.5em}{0.4pt} & \rule{1.5em}{0.4pt} \\ 0 & \rule{1.5em}{0.4pt} & \rule{1.5em}{0.4pt} & 0 & \rule{1.5em}{0.4pt} & \rule{1.5em}{0.4pt} \\ 1 & \rule{1.5em}{0.4pt} & \rule{1.5em}{0.4pt} & -2 & \rule{1.5em}{0.4pt} & \rule{1.5em}{0.4pt} \end{bmatrix} .$$

(c) Let $\tilde{\mathbf{u}}_0 = (1/32) \begin{bmatrix} 3 & 3 & 1 & 0 & 0 & 1 \end{bmatrix}^{\mathrm{T}}$ and $\tilde{\mathbf{v}}_0 = (1/32) \begin{bmatrix} -6 & 6 & 2 & 0 & 0 & -2 \end{bmatrix}^{\mathrm{T}}$ (the trend and detail vectors for this transform). Let $\mathbf{x} = \begin{bmatrix} 0 & 4 & 6 & 6 & 4 & 0 \end{bmatrix}^{\mathrm{T}}$. Use the matrices from (a) and (b) and `format rational` in MATLAB to calculate the rational numbers a_0, a_1, a_2 and b_0, b_1, b_2 such that $\mathbf{x}_s = a_0 \tilde{\mathbf{u}}_0 + a_1 S^2 \tilde{\mathbf{u}}_0 + a_2 S^4 \tilde{\mathbf{u}}_0$ is the projection of $\mathbf{x}$ onto the trend sub-space and $\mathbf{x}_d = b_0 \tilde{\mathbf{v}}_0 + b_1 S^2 \tilde{\mathbf{v}}_0 + b_2 S^4 \tilde{\mathbf{v}}_0$ is the projection of $\mathbf{x}$ onto the detail subspace for this transform (here S is the 6×6 shift matrix). Check that $\mathbf{x} = \mathbf{x}_s + \mathbf{x}_d$. Calculate the relative compression error $\|\mathbf{x}_d\|^2 / \|\mathbf{x}\|^2$ when $\mathbf{x}$ is approximated by its trend projection.

(12) Suppose A, B, C, D are $m \times m$ circulant matrices that satisfy $AD - BC = zI$, where $z \neq 0$ is a complex number.

(a) By carrying out block multiplication, show that the $2m \times 2m$ matrix $\begin{bmatrix} A & B \\ C & D \end{bmatrix}$ is invertible with inverse $z^{-1} \begin{bmatrix} D & -B \\ -C & A \end{bmatrix}$. (HINT: Remember that all cir-culant matrices of the same size mutually commute, so you can do block matrix multiplication as if A, B, C, D were scalars.)

(b) Use the result from (a) to obtain a formula for the CDF$(3, 1)$ polyphase synthesis matrix from the polyphase analysis matrix (see Example 3.11 and formula (3.33)).

(13) Suppose $\mathbf{u}_0 = \begin{bmatrix} a & b & c & d & 0 & 0 \end{bmatrix}$ and $\mathbf{v}_0 = \begin{bmatrix} d & -c & b & -a & 0 & 0 \end{bmatrix}$, where a, b, c, d are real numbers.

(a) Find the equations that a, b, c, d must satisfy so that $\mathbf{u}_0$ is the scaling vector and $\mathbf{v}_0$ is the wavelet vector for a 6×6 one-scale orthogonal wavelet analysis matrix $\mathbf{T_a}$.

(b) Suppose the real numbers a, b, c, d satisfy the equations you found in part (a). Concatenate an even number $2m$ of zeros at the ends of $\mathbf{u}_0$ and $\mathbf{v}_0$ to obtain row vectors with $6 + 2m$ entries. Will these new row vectors be the scaling vector and wavelet vector for a $(6 + 2m) \times (6 + 2m)$ one-scale orthogonal wavelet analysis matrix?

(14) The unnormalized 2×2 Haar analysis matrix is $\mathbf{W_a} = \dfrac{1}{2} \begin{bmatrix} 1 & 1 \\ 1 & -1 \end{bmatrix}$.

(a) Calculate the one-scale Haar wavelet transform $\mathbf{Y}$ of the matrix $\mathbf{X} = \begin{bmatrix} 2 & 4 \\ 0 & 8 \end{bmatrix}$.

(b) Calculate the *multiresolution representation* $\mathbf{X} = \mathbf{X_{ss}} + \mathbf{X_{sd}} + \mathbf{X_{ds}} + \mathbf{X_{dd}}$.

(15) Let $\mathbf{X}$ be the 4×4 matrix and $\mathbf{Y}^{(1)}$ its one-scale Haar transform from Example 3.18.

(a) Calculate the Haar analysis transform of the one-scale trend submatrix $\mathbf{Y}_{ss}^{(1)}$.

(b) Find the two-scale 2D Haar transform $\mathbf{Y}^{(2)}$ of $\mathbf{X}$ using the result from (a) and the calculations in Example 3.18.

(c) Use the 4 entries of largest magnitude in $\mathbf{Y}^{(2)}$ and the two-scale inverse Haar transform to reconstruct a 4:1 compression $\mathbf{X_c}$ of $\mathbf{X}$.

(d) Calculate the mean-square compression error $\text{MSE} = \|\mathbf{X} - \mathbf{X_c}\|^2 / 16$ and the peak signal-to-noise ratio $\text{PSNR} = 10 * \log_{10}(255^2 / \text{MSE})$.

Chapter 4

Wavelet Transforms from Filter Banks

4.1 Overview

In Chapter 3 we introduced several discrete wavelet transforms by the lifting process and saw how to use them to process signals and images. Now we investigate the mathematics behind discrete wavelet transforms and discover how they are designed to be effective for signal processing.

We will work in the vector space of digital signals of arbitrary finite length. This space is infinite dimensional, but it has a standard basis consisting of *unit impulses*. The signal processing tools we need are the shift operator, convolution, and the z-transform.

It is straightforward to carry over the one-scale wavelet transform constructions on periodic signals obtained by the lifting method in Chapter 3. To obtain the trend/detail wavelet decomposition of a non-periodic signal we pass it through the *lazy filter bank* to split the signal into two subsignals (the *polyphase decomposition*). This pair of subsignals is then multiplied by the 2×2 time-domain polyphase analysis matrix from Section 3.5, whose entries are Laurent polynomials in the shift operator. We can continue this process to obtain multiscale wavelet transforms of non-periodic signals from one-scale transforms by the pyramid algorithm, as in Chapter 3.

We have introduced wavelet transforms (based on slow/fast time scales for trend and detail) as an alternative to the Fourier transform (based on low/high frequency separation). But frequency analysis plays a crucial role in the construction of wavelet transforms by means of *two-channel filter banks*. However, for non-periodic digital signals (with time measured in unit steps), the frequency variable ω is *continuous* and *periodic* (of period 2π). To obtain a wavelet analysis transform, we convolve the signal with a pair of *finite impulse response* (FIR) filters to separate it into low-frequency (ω near 0) and high-frequency (ω near π) parts, with some overlap in the mid-range frequencies (ω near $\pm\pi/2$). Then we *downsample* each part so that the transformed signal is approximately the same length as the original signal (filtering spreads out the signal).

To understand the effect of these operations that we have performed in the time domain, we move to the frequency domain by the z-transform (where $z = \mathrm{e}^{\mathrm{i}\omega}$).

The z-transform of a FIR filter is a Laurent polynomial, and convolution with the filter becomes pointwise multiplication by this polynomial. Building a good wavelet transform thus turns into an algebraic problem of finding a pair of Laurent polynomials that interact suitably on the mid-range frequencies while also separating the low and high frequency parts of the signal. We construct a family of such polynomials and use them to obtain the filters for the CDF and Daubechies family of wavelet transforms.

In the filter-bank picture, the wavelet transform determined by a pair of FIR filters is expressed in terms of the z-transform by multiplication by a 2×2 *modulation matrix*, whose entries are Laurent polynomials in z determined by the filters. To translate into the lifting-step picture, we replace the modulation matrix by the corresponding polyphase matrix. We then show how to factor the polyphase matrix into a product of elementary matrices (these give the lifting steps in the time domain). This explains how the formulas for wavelet transforms such as the CDF$(2,2)$ and Daub4 in Chapter 3 are obtained and why these transforms work well for signal processing.

The frequency domain analysis (using modulation matrices) is the most effective way to design good *perfect reconstruction* (PR) wavelet transforms using FIR filters, but the signal processing computations are usually done in the time domain. Returning to the basic wavelet ideas of Chapter 3, we study the *trend-detail decomposition* in the time domain for a PR filter bank. The PR property corresponds to having a *biorthogonal wavelet basis* for the trend-detail decomposition.

The special case of an *orthogonal wavelet basis*, which was first constructed in [Daubechies (1988)], is more difficult. We analyze this case using *spectral factorization*, and obtain the Daub4 filters from Chapter 3 explicitly. We describe the computational steps needed to construct the Daub$2K$ filters for any positive integer K.

4.2   Filtering, Downsampling, and Upsampling

Let $\mathbf{x}$ be a real-valued function on $\mathbb{Z}$. We use the notation $\mathbf{x}[n]$ for the value of $\mathbf{x}$ at n and call $\mathbf{x}$ a *digital signal*. We say $\mathbf{x}$ is a *finite signal* if it has *finite support*; this means that there are integers $p \le q$ so that $\mathbf{x}[n] = 0$ when $n < p$ or $n > q$. Of course, p and q depend on $\mathbf{x}$. A real linear combination of finite signals is again a finite signal.

Definition 4.1. The real vector space consisting of all digital signals is denoted by $\ell(\mathbb{Z})$. The subspace of $\ell(\mathbb{Z})$ consisting of all finite signals is denoted[1] by $\ell_0(\mathbb{Z})$. For each positive integer N the subspace of N-periodic signals is denoted[2] by $\ell(\mathbb{Z}/N\mathbb{Z})$.

---

[1] Here the subscript 0 is used to indicate that for each $\mathbf{x} \in \ell_0(\mathbb{Z})$ the values $\mathbf{x}[n]$ are zero when n is outside some finite set.

[2] An N-periodic function on $\mathbb{Z}$ corresponds to a function on the additive group $\mathbb{Z}/N\mathbb{Z}$ of integers modulo N.

Define the *length* of a finite signal as follows.

Definition 4.2. The zero signal has length 0. If $\mathbf{x}$ is a nonzero finite signal, let $p \leq q$ be the integers such that $\mathbf{x}[p] \neq 0$, $\mathbf{x}[q] \neq 0$, and all the nonzero values of $\mathbf{x}[k]$ occur in the range $p \leq k \leq q$. Then $\mathsf{length}(\mathbf{x}) = q - p + 1$.

Let $k \in \mathbb{Z}$. Define the signal δ_k by

$$\delta_k[n] = \begin{cases} 1 & \text{if } n = k, \\ 0 & \text{if } n \neq k. \end{cases}$$

We call this signal the *unit impulse* at k. Every finite signal $\mathbf{x}$ can be written uniquely as a finite linear combination of unit impulses:

$$\mathbf{x} = \sum_{p \leq k \leq q} \mathbf{x}[k]\delta_k,$$

where p, q are from Definition 4.2. Thus the set $\{\delta_k : k \in \mathbb{Z}\}$ is a basis for $\ell_0(\mathbb{Z})$. In particular, $\ell_0(\mathbb{Z})$ is an *infinite-dimensional* vector space.

Given finite signals $\mathbf{x}$ and $\mathbf{y}$, we define their *inner product* to be

$$\langle \mathbf{x}, \mathbf{y} \rangle = \sum_{n \in \mathbb{Z}} \mathbf{x}[n]\,\mathbf{y}[n]$$

(by the finite support condition, there are only finitely many nonzero terms in the sum). Define the *total energy* of a finite signal to be the square of the norm:

$$\|\mathbf{x}\|^2 = \langle \mathbf{x}, \mathbf{x} \rangle = \sum_{n \in \mathbb{Z}} \mathbf{x}[n]^2.$$

Remark 4.1. The inner product space $\ell_0(\mathbb{Z})$ is a subspace of the finite-energy space $\ell^2(\mathbb{Z})$ from Example 1.24. This larger space is a natural framework for signal processing, as we will see in Chapter 5.

4.2.1 *Signals and z-transforms*

Let $\mathbf{x}$ be a finite signal. We define the *z-transform* of $\mathbf{x}$ to be the Laurent polynomial

$$X(z) = \sum_{n \in \mathbb{Z}} \mathbf{x}[n]\, z^{-n}.$$

Since $\mathbf{x}$ has finite support, there are only a finite number of nonzero terms in the sum: if $\mathbf{x}[n] = 0$ for $n < p$ and for $n > q$, then

$$X(z) = \mathbf{x}[p]\, z^{-p} + \mathbf{x}[p+1]\, z^{-p-1} + \cdots + \mathbf{x}[q]\, z^{-q}.$$

The transformation from a finite signal $\mathbf{x}$ to the Laurent polynomial $X(z)$ is linear: adding signals $\mathbf{x}$ and $\mathbf{y}$ or multiplying $\mathbf{x}$ by a real number corresponds to performing the same operations on the z-transforms $X(z)$ and $Y(z)$. Furthermore, any Laurent polynomial with real coefficients is the z-transform of a unique finite signal.

When ω is real, then $z = \mathrm{e}^{\mathrm{i}\omega}$ has absolute value 1 and $z^{-n} = \mathrm{e}^{-\mathrm{i}n\omega}$. Thus

$$X(\mathrm{e}^{\mathrm{i}\omega}) = \sum_{n \in \mathbb{Z}} \mathbf{x}[n]\,\mathrm{e}^{-\mathrm{i}n\omega}$$

is a trigonometric polynomial. In signal-processing language ω is the *continuous frequency* variable, whereas n is the *discrete time* variable. We can go from the z-transform back to the signal by integration:

$$\mathbf{x}[n] = \frac{1}{2\pi} \int_{-\pi}^{\pi} X(\mathrm{e}^{\mathrm{i}\omega})\,\mathrm{e}^{\mathrm{i}n\omega} d\omega\,. \tag{4.1}$$

Here we use formula (1.31) for Fourier coefficients with $k = -n$ and $x = \omega$.

The nonnegative function $P(\omega) = |X(\mathrm{e}^{\mathrm{i}\omega})|^2$ is called the *power spectral density* of the signal $\mathbf{x}$. Notice that $X(\mathrm{e}^{-\mathrm{i}\omega}) = \overline{X(\mathrm{e}^{\mathrm{i}\omega})}$ (complex conjugate) since the coefficients $\mathbf{x}[n]$ are real. Hence the functions $|X(\mathrm{e}^{-\mathrm{i}\omega})|$ and $P(\omega)$ are even functions of ω. By Parseval's formula (1.39) the *total energy* (norm) $\|\mathbf{x}\|^2$ of $\mathbf{x}$ is the average of the power spectral density over the frequency range $-\pi \le \omega \le \pi$:

$$\sum_{n \in \mathbb{Z}} \mathbf{x}[n]^2 = \frac{1}{2\pi} \int_{-\pi}^{\pi} P(\omega)\, d\omega\,. \tag{4.2}$$

Example 4.1. Suppose the nonzero values of $\mathbf{x}$ are $\mathbf{x}[-1] = 2$, $\mathbf{x}[0] = 3$, and $\mathbf{x}[2] = 4$. Then

$$\mathbf{x} = 2\delta_{-1} + 3\delta_0 + 4\delta_2\,,$$

and $\mathbf{x}$ is a finite signal of length 4, since the first nonzero value is at $n = -1$ and the last nonzero value is at $n = 2$. The z-transform of $\mathbf{x}$ is the Laurent polynomial $X(z) = 2z + 3 + 4z^{-2}$. When $z = \mathrm{e}^{\mathrm{i}\omega}$ then this z-transform is the trigonometric polynomial

$$X(\mathrm{e}^{\mathrm{i}\omega}) = 2\mathrm{e}^{\mathrm{i}\omega} + 3 + 4\mathrm{e}^{-2\mathrm{i}\omega}\,.$$

The graph of the power spectral density function for this signal is shown in Figure 4.1 on the range $-\pi \le \omega \le \pi$. As expected, the graph is symmetric around the point $\omega = 0$. By equation (4.2) the average value of $P(\omega)$ is the total energy $\|\mathbf{x}\|^2 = 2^2 + 3^2 + 4^2 = 29$ of $\mathbf{x}$.   ∎

4.2.2 *Convolution*

If $\mathbf{x}$ and $\mathbf{y}$ are signals and at least one of them is finite, we define their *convolution* $\mathbf{x} \star \mathbf{y}$ as the signal with values

$$(\mathbf{x} \star \mathbf{y})[n] = \sum_{j+k=n} \mathbf{x}[j]\,\mathbf{y}[k] \qquad \text{for } n \in \mathbb{Z}\,. \tag{4.3}$$

This formula for $\mathbf{x} \star \mathbf{y}$ can be described graphically as follows: Take the function $\mathbf{f}[j, k] = \mathbf{x}[j]\mathbf{y}[k]$ of the two (discrete) variables (j, k). If $\mathbf{x}$ has finite support, then $\mathbf{f}$ is supported on some vertical strip $V = \{(j, k) : p \le j \le q\}$ in $\mathbb{Z}^2$. To obtain

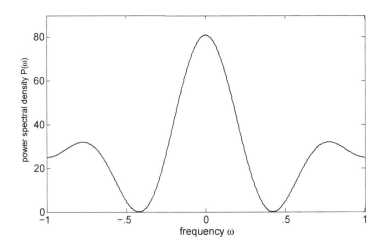

Fig. 4.1   Power spectral density function (ω in multiples of π)

the value $(\mathbf{x} \star \mathbf{y})[n]$, add the values of $\mathbf{f}$ at all integer points along the diagonal line $j + k = n$. For each n this line intersects L in a finite number of points, so the sum in equation (4.3) only has a finite number of nonzero terms. Likewise, if $\mathbf{y}$ has finite support, then $\mathbf{f}$ is supported on a horizontal strip H and the same argument applies. If both $\mathbf{x}$ and $\mathbf{y}$ have finite support, then $\mathbf{f}$ is supported on a rectangle $R = H \cap V$. When $|n|$ is sufficiently large the diagonal line $j + k = n$ does not intersect R. Hence $\mathbf{x} \star \mathbf{y}$ is a finite signal in this case.

The operation of convolution is *linear*: if $\mathbf{x}$ is a finite signal, $\mathbf{v}$ and $\mathbf{w}$ are signals, and $c \in \mathbb{R}$, then $\mathbf{x} \star (\mathbf{v} + c\mathbf{w}) = \mathbf{x} \star \mathbf{v} + c\mathbf{x} \star \mathbf{w}$. The formula (4.3) for convolution can be written in two other ways:

$$(\mathbf{x} \star \mathbf{y})[n] = \sum_{k \in \mathbb{Z}} \mathbf{x}[n - k]\, \mathbf{y}[k] = \sum_{j \in \mathbb{Z}} \mathbf{x}[j]\, \mathbf{y}[n - j]\,.$$

These alternate expressions for convolution show that $\mathbf{x} \star \mathbf{y} = \mathbf{y} \star \mathbf{x}$. In particular, if $\mathbf{x}[j] = 0$ for $j < p$ or $j > q$ (with $p \le q$), then

$$(\mathbf{x} \star \mathbf{y})[n] = \sum_{j=p}^{j=q} \mathbf{x}[j]\, \mathbf{y}[n - j]\,. \tag{4.4}$$

Thus for a given finite signal $\mathbf{x}$, the value of $\mathbf{x} \star \mathbf{y}$ at any point n depends only on the values of $\mathbf{y}$ between $n - p$ and $n - q$. Note that there is no wraparound in this formula, however, and it is not the same as the circular convolution of periodic extensions of $\mathbf{x}$ and $\mathbf{y}$ in Definition 2.4.

If $\mathbf{x} = \delta_p$ and $\mathbf{y} = \delta_q$ are unit impulses, then the function $\mathbf{x}[j]\mathbf{y}[k]$ is zero except when $j = p$ and $k = q$. Hence we get

$$\delta_p \star \delta_q = \delta_{p+q} \tag{4.5}$$

in this case. In general, let $\mathbf{x}$ and $\mathbf{y}$ be finite signals. When we write them as linear combinations of unit impulses and use linearity of convolution, we conclude from (4.5) that

$$\mathbf{x} \star \mathbf{y} = \sum_{p \in \mathbb{Z}} \sum_{q \in \mathbb{Z}} \mathbf{x}[p] \, \mathbf{y}[q] \, \delta_{p+q} \,. \tag{4.6}$$

Here is one of the most important properties of convolution.

Theorem 4.1. *Let $\mathbf{x}$ and $\mathbf{y}$ be finite signals. Then the z-transform of the finite signal $\mathbf{x} \star \mathbf{y}$ is the pointwise product $X(z)Y(z)$ of the z-transforms.*

Proof. Since the z-transform of the unit impulse δ_{p+q} is $z^{-p-q} = z^{-p}z^{-q}$, we see from (4.6) that the z-transform of $\mathbf{x} \star \mathbf{y}$ is

$$\sum_{p \in \mathbb{Z}} \sum_{q \in \mathbb{Z}} \mathbf{x}[p] \, \mathbf{y}[q] \, z^{-p-q} = X(z)Y(z) \,,$$

since taking the z-transform is a linear operation.   □

The shift operator S acts on a signal $\mathbf{x}$ by

$$(S\mathbf{x})[n] = \mathbf{x}[n-1] \quad \text{for } n \in \mathbb{Z}.$$

Note that for any signal $\mathbf{x}$,

$$(\delta_1 \star \mathbf{x})[n] = \sum_{j+k=n} \delta_1[j] \, \mathbf{x}[k] = \mathbf{x}[n-1] = (S\mathbf{x})[n] \,.$$

Combining this formula for $S\mathbf{x}$ with the unit impulse convolution formula (4.6), we can express any power of the shift operator in terms of convolution:

$$S^k \mathbf{x} = \delta_k \star \mathbf{x} \qquad \text{for all } k \in \mathbb{Z} \,. \tag{4.7}$$

If $\mathbf{x}$ is a finite signal, then $S\mathbf{x}$ is also a finite signal.

Theorem 4.2. *Let $\mathbf{x}$ be a finite signal. Then the z-transform of $S\mathbf{x}$ is $z^{-1}X(z)$.*

Proof. Since $S\mathbf{x} = \delta_1 \star \mathbf{x}$ by (4.7) and δ_1 has z-transform z^{-1}, the formula for the z-transform of $S\mathbf{x}$ follows from Theorem 4.1.   □

4.2.3   *Linear shift-invariant filters*

The most basic operation in signal processing is *linear filtering*. Given a finite signal $\mathbf{h} = \sum_k \mathbf{h}[k]\delta_k$, we use it to define a linear transformation $\mathbf{H}$ on the vector space of signals by

$$\mathbf{H}\mathbf{x} = \mathbf{h} \star \mathbf{x} \,.$$

If $\mathbf{h}$ is supported on the set $\{k : p \le k \le q\}$, then

$$\mathbf{H}\mathbf{x}[n] = \sum_{p \le k \le q} \mathbf{h}[k] \, \mathbf{x}[n-k] = \sum_{p \le k \le q} \mathbf{h}[k] \, S^k \mathbf{x}[n] \,.$$

This shows that $\mathbf{Hx}$ is a finite linear combination of shifts of $\mathbf{x}$, and we can write

$$\mathbf{H} = \sum_{p \leq k \leq q} \mathbf{h}[k]\, S^k \tag{4.8}$$

as a finite linear combination of powers of the shift operator. In particular, if $\mathbf{x}$ is a finite signal, then $T\mathbf{x}$ is also a finite signal.

If we take the input signal $\mathbf{x} = \delta_0$, then

$$\mathbf{H}\delta_0 = \sum_{p \leq k \leq q} \mathbf{h}[k]\, S^k \delta_0 = \sum_{p \leq k \leq q} \mathbf{h}[k]\, \delta_k = \mathbf{h}\,.$$

Thus $\mathbf{h}$ is uniquely determined by $\mathbf{H}$. We call the transformation $\mathbf{H}$ a *finite impulse response filter* (FIR filter) and $\mathbf{h}$ the *impulse response function* of the filter. We often will refer to $\mathbf{h}$ itself as the filter when considering the transformation $\mathbf{H}$. In particular, the *length* of the filter is defined as $\mathsf{length}(\mathbf{h})$.

If we apply the shift operator to $\mathbf{x}$ before applying $\mathbf{H}$, then we get

$$\mathbf{H}S\mathbf{x} = \sum_k \mathbf{h}[k]\, S^k S\mathbf{x} = \sum_k \mathbf{h}[k]\, SS^k \mathbf{x} = S\mathbf{H}\mathbf{x}\,.$$

In general, a linear transformation T on the vector space of signals is called *shift-invariant* if $ST\mathbf{x} = TS\mathbf{x}$ for all signals $\mathbf{x}$.

Theorem 4.3. *Suppose T is a shift-invariant linear transformation on the vector space of finite signals (so $T\mathbf{x}$ is a finite signal for all finite signals $\mathbf{x}$). Set $\mathbf{h} = T\delta_0$. Then $\mathbf{h}$ has finite support and $T\mathbf{x} = \mathbf{h} \star \mathbf{x}$ for all finite signals $\mathbf{x}$.*

Proof. The argument is essentially the same as in Theorem 2.2. Since T is linear and the unit impulses δ_k give a basis for the vector space of finite signals, it suffices to check that $T\delta_k = \mathbf{h} \star \delta_k$ for all integers k. To do this, note that $\delta_k = S^k \delta_0$, so that

$$T\delta_k = TS^k \delta_0 = S^k T\delta_0 = S^k \mathbf{h} = \delta_k \star \mathbf{h} = \mathbf{h} \star \delta_k\,.$$

Here we have used (4.7) and the commutative property of convolution. $\qquad\square$

Thus FIR filters are the linear transformations of finite signals that are analogous to $N \times N$ circulant matrices acting on N-periodic signals.

Example 4.2. Take $\mathbf{h}[n] = 1/3$ for $n = -1, 0, 1$ and zero otherwise. Then $\mathbf{h}$ is a filter of length 3, and

$$(\mathbf{h} \star \mathbf{x})[n] = (1/3)(\mathbf{x}[n-1] + \mathbf{x}[n] + \mathbf{x}[n+1])$$

can be described as a *moving average*. If $\mathbf{x}[n] = 0$ for $n < p$ and $n > q$, for example, then $(\mathbf{h} \star \mathbf{x})[n] = 0$ for $n < p-1$ and $n > q+1$. However $(\mathbf{h} \star \mathbf{x})[p-1] = (1/3)(\mathbf{x}[p] + \mathbf{x}[p+1])$ and $(\mathbf{h} \star \mathbf{x})[q+1] = (1/3)(\mathbf{x}[q-1] + \mathbf{x}[q])$, which are nonzero in general, even though $\mathbf{x}[p-1] = 0$ and $\mathbf{x}[q+1] = 0$. Thus filtering spreads out the support of $\mathbf{x}$. ∎

Define the *time-reversal* of a FIR filter $\mathbf{h}$ by $\mathbf{h}^{\mathrm{T}}[k] = \mathbf{h}[-k]$. Then $\mathbf{h}^{\mathrm{T}}$ is also a FIR filter. If $\mathbf{x}$ and $\mathbf{y}$ are finite signals, then

$$\langle \mathbf{h} \star \mathbf{x}, \mathbf{y} \rangle = \langle \mathbf{x}, \mathbf{h}^{\mathrm{T}} \star \mathbf{y} \rangle. \tag{4.9}$$

This is easy to verify this using (4.8) and the unitarity property $\langle S\mathbf{x}, \mathbf{y} \rangle = \langle \mathbf{x}, S^{-1}\mathbf{y} \rangle$ of the shift operator:

$$\langle \mathbf{h} \star \mathbf{x}, \mathbf{y} \rangle = \sum_k \mathbf{h}[k] \langle S^k \mathbf{x}, \mathbf{y} \rangle = \sum_k \mathbf{h}[k] \langle \mathbf{x}, S^{-k}\mathbf{y} \rangle$$

$$= \sum_k \mathbf{h}[-k] \langle \mathbf{x}, S^k \mathbf{y} \rangle = \langle \mathbf{x}, \mathbf{h}^{\mathrm{T}} \star \mathbf{y} \rangle.$$

We can write (4.9) in terms of the transformation $\mathbf{H}$ as

$$\langle \mathbf{H}\mathbf{x}, \mathbf{y} \rangle = \langle \mathbf{x}, \mathbf{H}^{\mathrm{T}} \star \mathbf{y} \rangle, \tag{4.10}$$

where $\mathbf{H}^{\mathrm{T}}$ is the filter with impulse response function $\mathbf{h}^{\mathrm{T}}$.

Definition 4.3. A FIR filter $\mathbf{h}$ is *causal* if $\mathbf{h}[k] = 0$ for all $k < 0$.

When $\mathbf{h}$ is causal and $\mathbf{x}$ is a signal, then the calculation of the filtered signal $(\mathbf{h} \star \mathbf{x})[k]$ at time k only uses the current and past values of $\mathbf{x}$:

$$(\mathbf{h} \star \mathbf{x})[k] = \sum_{n=0}^{L-1} \mathbf{h}[n]\mathbf{x}[k-n],$$

where $L = \mathsf{length}(\mathbf{h})$.

The filter $\mathbf{h} = (1/3)(\delta_{-1} + \delta_0 + \delta_1)$ in Example 4.2 is not causal, since it uses the value of $\mathbf{x}$ at $n+1$ to calculate the filtered value at n. However, if we insert a time shift, then we obtain a causal filter:

$$(\mathbf{h} \star \delta_1) \star \mathbf{x} = (1/3)(\mathbf{x}[n-2] + \mathbf{x}[n-1] + \mathbf{x}[n]).$$

The same shift method applies to any FIR filter.

4.2.4 *Downsampling and upsampling*

Let $\mathbf{x}$ be a finite signal. The *downsampling* by 2 of $\mathbf{x}$ is the signal

$$\mathbf{x}_{2\downarrow}[n] = \mathbf{x}[2n] \quad \text{for } n \in \mathbb{Z}.$$

This operation is the non-periodic version of $\mathbf{x} \to \mathbf{x}_{\text{even}}$ on periodic signals. It defines a linear transformation $\boxed{2\downarrow}\,\mathbf{x} = \mathbf{x}_{2\downarrow}$ on the vector space of finite signals. When $\mathbf{x} = \delta_n$ is a unit impulse, then

$$\boxed{2\downarrow}\,\delta_n = \begin{cases} \delta_{n/2} & \text{if } n \text{ is even}, \\ 0 & \text{if } n \text{ is odd}. \end{cases}$$

Example 4.3. Suppose the nonzero values of $\mathbf{x}$ are $\mathbf{x}[-2] = -1$, $\mathbf{x}[-1] = 2$, $\mathbf{x}[0] = 3$, $\mathbf{x}[1] = 4$, and $\mathbf{x}[2] = -2$. Then the nonzero values of $\mathbf{x}_{2\downarrow}$ are $\mathbf{x}_{2\downarrow}[-1] = -1$, $\mathbf{x}_{2\downarrow}[0] = 3$, and $\mathbf{x}_{2\downarrow}[1] = -2$. In terms of unit impulses, $\mathbf{x} = -\delta_{-2} + 2\delta_{-1} + 3\delta_0 + 4\delta_1 - 2\delta_2$ and $\boxed{2\downarrow}\,\mathbf{x} = -\delta_{-1} + 3\delta_0 - 2\delta_1$ (see Figure 4.2). ∎

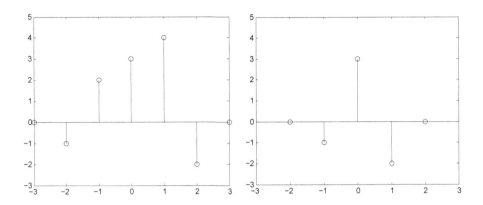

Fig. 4.2   Signal (left) and downsampled signal (right)

We denote the z-transform of $\mathbf{x}_{2\downarrow}$ by $X_{2\downarrow}(z)$. It is given by the formula

$$X_{2\downarrow}(z) = \sum_n \mathbf{x}[2n]\, z^{-n}\,. \tag{4.11}$$

The z-transforms in Example 4.3 are

$$X(z) = -z^2 + 2z + 3 + 4z^{-1} - 2z^{-2} \quad \text{and} \quad X_{2\downarrow}(z) = -z + 3 - 2z^{-1}\,.$$

The *upsampling* by 2 of $\mathbf{x}$ is the signal

$$\mathbf{x}_{2\uparrow}[n] = \begin{cases} \mathbf{x}[n/2] & \text{if } n \text{ is even}\,, \\ 0 & \text{if } n \text{ is odd}\,. \end{cases}$$

This operation gives a linear transformation $\boxed{2\uparrow}\,\mathbf{x} = \mathbf{x}_{2\uparrow}$ on the vector space of finite signals. When $\mathbf{x}$ is a unit impulse, then from the definition just given we see that

$$\boxed{2\uparrow}\,\delta_n = \delta_{2n}\,.$$

We denote the z-transform of $\mathbf{x}_{2\uparrow}$ by $X_{2\uparrow}(z)$. It is given by the formula

$$X_{2\uparrow}(z) = \sum_n \mathbf{x}[n]\, z^{-2n} = X(z^2)\,. \tag{4.12}$$

Example 4.4. Let $\mathbf{x}$ be the signal in Example 4.3. If we set $\mathbf{y} = \mathbf{x}_{2\downarrow}$, then the nonzero values of $\mathbf{y}_{2\uparrow}$ are $\mathbf{y}_{2\uparrow}[-2] = -1$, $\mathbf{y}_{2\uparrow}[0] = 3$, and $\mathbf{y}_{2\uparrow}[2] = -2$. In terms of unit impulses, $\boxed{2\uparrow}\,\mathbf{y} = -\delta_{-2} + 3\delta_0 - 2\delta_2$. Thus the z-transform is

$$Y_{2\uparrow}(z) = -z^2 + 3 - 2z^{-2} = Y(z^2)$$

(the even terms in $X(z)$). ∎

For any finite signal $\mathbf{x}$ the upsampled signal $\mathbf{x}_{2\uparrow}$ is a stretched version of $\mathbf{x}$ (with zeros interlaced); we recover $\mathbf{x}$ from $\mathbf{x}_{2\uparrow}$ by downsampling:

$$\boxed{2\downarrow}\ \boxed{2\uparrow}\,\mathbf{x} = \mathbf{x}\,.$$

By linearity, it suffices to prove this property when $\mathbf{x} = \delta_n$ is a unit impulse; in this case

$$\boxed{2\downarrow}\ \boxed{2\uparrow}\,\delta_n = \boxed{2\downarrow}\,\delta_{2n} = \delta_n\,.$$

Thus the downsampling transformation $\boxed{2\downarrow}$ is a *left inverse* to the upsampling transformation $\boxed{2\uparrow}$. If we apply these transformations in the opposite order, we obtain

$$\left(\,\boxed{2\uparrow}\ \boxed{2\downarrow}\,\mathbf{x}\right)[n] = \begin{cases} \mathbf{x}[n] & \text{if } n \text{ is even}\,, \\ 0 & \text{if } n \text{ is odd}\,. \end{cases} \tag{4.13}$$

Hence $\boxed{2\uparrow}\ \boxed{2\downarrow}\,\mathbf{x}$ is the projection of $\mathbf{x}$ onto the subspace of finite signals that are zero at all odd integers.

4.2.5 *Periodic signals*

Fix a positive integer N. Given a finite signal $\mathbf{x}$, define a function $\mathbf{x}_{\mathrm{per},N}$ on $\mathbb{Z}$ by

$$\mathbf{x}_{\mathrm{per},N}[k] = \sum_{j\in\mathbb{Z}} \mathbf{x}[k + jN]\,. \tag{4.14}$$

There are only a finite number of nonzero terms in the sum (4.14), since the length of $\mathbf{x}$ is finite. Note that $\mathbf{x}_{\mathrm{per},N}$ is not a finite signal unless $\mathbf{x} = \mathbf{0}$, since the nonzero values of $\mathbf{x}_{\mathrm{per},N}$ repeat indefinitely.

The function $\mathbf{x}_{\mathrm{per},N}$ is periodic of period N:

$$\mathbf{x}_{\mathrm{per},N}[k + N] = \sum_{j\in\mathbb{Z}} \mathbf{x}[k + (j+1)N] = \sum_{\ell\in\mathbb{Z}} \mathbf{x}[k + \ell N] = \mathbf{x}_{\mathrm{per},N}[k]\,.$$

In particular, if $\mathsf{length}(\mathbf{x}) \leq N$, then for each k there is at most one nonzero term in the sum (4.14). In this case we call $\mathbf{x}_{\mathrm{per},N}$ the N-*periodic extension* of $\mathbf{x}$.

The transformation $\mathbf{x} \to \mathbf{x}_{\mathrm{per},N}$ from the vector space $\ell_0(\mathbb{Z})$ of finite signals to the vector space $\ell(\mathbb{Z}/N\mathbb{Z})$ of N-periodic signals is linear and compatible with the shift operator S on finite signals:

$$(S\mathbf{x})_{\mathrm{per},N} = S(\mathbf{x}_{\mathrm{per},N})\,. \tag{4.15}$$

To verify (4.15), note from formula (4.14) that

$$(S\mathbf{x})_{\mathrm{per},N}[k] = \sum_{j\in\mathbb{Z}} (S\mathbf{x})[k + jN] = \sum_{j\in\mathbb{Z}} \mathbf{x}[k - 1 + jN] = \mathbf{x}_{\mathrm{per},N}[k - 1]\,.$$

If $\mathbf{y}$ is any signal, define

$$P_N\mathbf{y} = \begin{bmatrix} \mathbf{y}[0] \\ \mathbf{y}[1] \\ \vdots \\ \mathbf{y}[N-1] \end{bmatrix} \in \mathbb{R}^N\,. \tag{4.16}$$

As in Section 2.4.1, we will use the linear transformation P_N to identify the real vector space $\ell(\mathbb{Z}/N\mathbb{Z})$ with column vectors in $\mathbb{R}^N$. If $\mathbf{y}$ is N-periodic and S_N is the $N \times N$ shift matrix, then

$$S_N P_N \mathbf{y} = \begin{bmatrix} \mathbf{y}[N-1] \\ \mathbf{y}[0] \\ \vdots \\ \mathbf{y}[N-2] \end{bmatrix} = \begin{bmatrix} S\mathbf{y}[0] \\ S\mathbf{y}[1] \\ \vdots \\ S\mathbf{y}[N-1] \end{bmatrix} = P_N S\mathbf{y}. \tag{4.17}$$

Using equations (4.15) and (4.17), with $\mathbf{y} = \mathbf{x}_{\mathrm{per},N}$, we see that

$$P_N\big((S\mathbf{x})_{\mathrm{per},N}\big) = S_N\big(P_N(\mathbf{x}_{\mathrm{per},N})\big) \quad \text{for all finite signals } \mathbf{x}. \tag{4.18}$$

Example 4.5. The signal $\mathbf{x} = 3\delta_3 + 4\delta_4 + 5\delta_5 + 6\delta_6$ has length 4. If we take $N = 4$, then $P_4(\mathbf{x}_{\mathrm{per},4}) = \begin{bmatrix} 4\ 5\ 6\ 3 \end{bmatrix}^{\mathrm{T}}$ (the positions of the components of $P_4(\mathbf{x}_{\mathrm{per},4})$ are determined by reading the subscripts on the unit impulse functions modulo 4). The shifted (non-periodic) signal $S\mathbf{x} = 3\delta_4 + 4\delta_5 + 5\delta_6 + 6\delta_7$ also has length 4 and $P_4(S\mathbf{x})_{\mathrm{per},4} = \begin{bmatrix} 3\ 4\ 5\ 6 \end{bmatrix}^{\mathrm{T}} = S_4 P_4(\mathbf{x}_{\mathrm{per},4})$ (observe the wraparound). But note that $P_4(\mathbf{x}_{\mathrm{per},4}) \neq P_4(\mathbf{x}) = \begin{bmatrix} 0\ 0\ 0\ 3 \end{bmatrix}^{\mathrm{T}}$. ∎

4.2.6 *Filtering and downsampling of periodic signals*

The filter bank approach to discrete wavelet transforms, which we will introduce in Section 4.4, combines filtering and downsampling. In the present section we show that these operations, when applied to periodic signals, give the trend and detail blocks of one-scale wavelet matrices, as in Theorem 3.2.

Let $\mathbf{h} = \mathbf{h}[0]\delta_0 + \cdots + \mathbf{h}[L-1]\delta_{L-1}$ be a causal[3] FIR filter of length L. Fix an even integer $M \geq L$. If $\mathbf{x}$ is a periodic signal of period M, define

$$T\mathbf{x} = \boxed{2\downarrow}(\mathbf{h} \star \mathbf{x}).$$

Because $L \leq M$ the values of the signal $T\mathbf{x}$ are

$$T\mathbf{x}[k] = \sum_{j=0}^{M-1} \mathbf{h}[j]\,\mathbf{x}[2k-j] \quad \text{for all integers } k. \tag{4.19}$$

From this formula it is clear that $T\mathbf{x}$ is a periodic signal of period $M/2$. Hence T is a linear transformation from the vector space of M-periodic signals to the vector space of $M/2$-periodic signals. Thus there is a unique $(M/2) \times M$ matrix $\mathbf{U}$ such that

$$P_{M/2}T\mathbf{x} = \mathbf{U}P_M\mathbf{x} \quad \text{for all } M\text{-periodic signals } \mathbf{x}. \tag{4.20}$$

This matrix has the following form.

---

[3] Any FIR filter can be shifted to become causal; this is convenient for Proposition 4.1.

Proposition 4.1. *Let* $\mathbf{u} = \begin{bmatrix} \mathbf{h}[0] \ \mathbf{h}[1] \ \cdots \ \mathbf{h}[L-1] \ 0 \ \cdots \ 0 \end{bmatrix}$ *be the* $1 \times M$ *row vector consisting of the filter coefficients padded by zeros. Then the matrix* $\mathbf{U}$ *in (4.20) is*

$$
\mathbf{U} = \begin{bmatrix} \mathbf{u} \\ \mathbf{u}(S_M)^{-2} \\ \vdots \\ \mathbf{u}(S_M)^{-M+2} \end{bmatrix},
$$

where S_M *is the* $M \times M$ *shift matrix.*

Proof. This follows immediately from (4.19), as in the proof of Theorem 3.2. □

Example 4.6. Take a causal filter $\mathbf{h}$ of length $L = 4$ acting on periodic signals of period $M = 8$ by convolution followed by downsampling. Then this linear transformation from $\mathbb{R}^8$ to $\mathbb{R}^4$ has matrix

$$
\mathbf{U} = \begin{bmatrix} \mathbf{h}[0] & \mathbf{h}[1] & \mathbf{h}[2] & \mathbf{h}[3] & 0 & 0 & 0 & 0 \\ 0 & 0 & \mathbf{h}[0] & \mathbf{h}[1] & \mathbf{h}[2] & \mathbf{h}[3] & 0 & 0 \\ 0 & 0 & 0 & 0 & \mathbf{h}[0] & \mathbf{h}[1] & \mathbf{h}[2] & \mathbf{h}[3] \\ \mathbf{h}[2] & \mathbf{h}[3] & 0 & 0 & 0 & 0 & \mathbf{h}[0] & \mathbf{h}[1] \end{bmatrix}.
$$

Notice the wrap-around that occurs on the last row.   ■

4.2.7   *Discrete Fourier transform and z-transform*

Theorem 4.4. *Let* $\mathbf{x}$ *be a finite signal with* z*-transform* $X(z)$*. The discrete Fourier transform of* $\mathbf{x}_{\mathrm{per},N}$ *is obtained by sampling* $X(z)$ *at the* N*th roots of unity (going counterclockwise around the unit circle):*

$$
\widehat{\mathbf{x}}_{\mathrm{per},N}[k] = X(\omega_N^k) \quad \text{for } k = 0, 1, \ldots, N-1, \text{ where } \omega_N = e^{2\pi i k/N}.
$$

Proof. Every integer n can be written uniquely as $n = j + mN$, where m and j are integers and $0 \le j < N$. Hence from (4.14) we can write

$$
X(\omega_N^k) = \sum_{n \in \mathbb{Z}} \mathbf{x}[n]\, \omega_N^{-kn} = \sum_{j=0}^{N-1} \left\{ \sum_{m \in \mathbb{Z}} \mathbf{x}[j + mN] \right\} \omega_N^{-(j+mN)k}
$$

$$
= \sum_{j=0}^{N-1} \mathbf{x}_{\mathrm{per},N}[j]\, \omega_N^{-jk} = \widehat{\mathbf{x}}_{\mathrm{per},N}[k].
$$

Here we have used the properties

$$
\omega_N^{-(j+mN)k} = (\omega_N)^{-jk}(\omega_N)^{-Nmk} \quad \text{and} \quad (\omega_N)^N = 1.
$$

Note that the sums over $\mathbb{Z}$ in these formulas only have a finite number of nonzero terms, since $\mathbf{x}$ is a finite signal.   □

Example 4.7. Suppose $\mathbf{x} = 2\delta_{-1} + 3\delta_0 + 4\delta_2$ as in Example 4.1. Since $\mathbf{x}$ has length 4, we can make an N-periodic extension of $\mathbf{x}$ for any integer $N \geq 4$. In particular, the 4-periodic extension has values

$$\mathbf{x}_{\mathrm{per},4}[0] = 3, \quad \mathbf{x}_{\mathrm{per},4}[1] = 0, \quad \mathbf{x}_{\mathrm{per},4}[2] = 4, \quad \mathbf{x}_{\mathrm{per},4}[3] = 2$$

(the value at 3 is $\mathbf{x}[-1] = 2$ since $3 = -1 + 4$). Since $\omega_4 = e^{2\pi i/4} = i$, we have

$$\widehat{\mathbf{x}}_{\mathrm{per},4}[k] = X(i^k) \quad \text{for } k = 0, 1, 2, 3$$

in this case. For this example $X(z) = 2z + 3 + 4z^{-2}$; thus Theorem 4.4 shows that $\widehat{\mathbf{x}}_{\mathrm{per},4}$ corresponds to the column vector

$$\begin{bmatrix} X(1) \\ X(i) \\ X(-1) \\ X(-i) \end{bmatrix} = \begin{bmatrix} 9 \\ 2i + 3 - 4 \\ -2 + 3 + 4 \\ -2i + 3 - 4 \end{bmatrix} = \begin{bmatrix} 9 \\ -1 + 2i \\ 5 \\ -1 - 2i \end{bmatrix}.$$

We get the same column vector by matrix-vector multiplication using the 4×4 Fourier matrix:

$$F_4\,\mathbf{x}_{\mathrm{per},4} = \begin{bmatrix} 1 & 1 & 1 & 1 \\ 1 & -i & -1 & i \\ 1 & -1 & 1 & -1 \\ 1 & i & -1 & -i \end{bmatrix} \begin{bmatrix} 3 \\ 0 \\ 4 \\ 2 \end{bmatrix} = \begin{bmatrix} 9 \\ -1 + 2i \\ 5 \\ -1 - 2i \end{bmatrix}.$$

We can also define an 8-periodic extension of $\mathbf{x}$ by zero padding. In this case

$$\mathbf{x}_{\mathrm{per},8}[0] = 3, \quad \mathbf{x}_{\mathrm{per},8}[1] = 0, \quad \mathbf{x}_{\mathrm{per},8}[2] = 4, \quad \mathbf{x}_{\mathrm{per},8}[3] = 0,$$
$$\mathbf{x}_{\mathrm{per},8}[4] = \mathbf{x}_{\mathrm{per},8}[5] = \mathbf{x}_{\mathrm{per},8}[6] = 0, \quad \mathbf{x}_{\mathrm{per},8}[7] = 2,$$

since $7 \equiv -1 \pmod 8$. Note that $\mathbf{x}_{\mathrm{per},8}[3] \neq \mathbf{x}_{\mathrm{per},4}[3]$. The discrete Fourier transform of $\mathbf{x}_{\mathrm{per},8}$ has values $\widehat{\mathbf{x}}_{\mathrm{per},8}[k] = X(\omega_8^k)$ for $k = 0, 1, \ldots, 7$, where $\omega_8 = e^{\pi i/4} = (1 + i)/\sqrt{2}$. For $k = 0, 2, 4, 6$ these are the same as the values of $\widehat{\mathbf{x}}_{\mathrm{per},4}$ at $k/2$, since $(\omega_8)^2 = \omega_4$. ∎

Theorem 4.5. *If $\mathbf{x}$ is a finite signal then the discrete Fourier transform of $S\mathbf{x}_{\mathrm{per},N}$ is obtained by sampling $z^{-1}X(z)$ at the Nth roots of unity $1, \omega, \ldots, \omega^{N-1}$, where $\omega = \omega_N = e^{2\pi i/N}$.*

Proof. Set $\mathbf{y} = S\mathbf{x}$. Then $Y(z) = z^{-1}X(z)$ by Theorem 4.2, while from Theorem 4.4 we know that $\mathbf{x}_{\mathrm{per},N}$ and $\mathbf{y}_{\mathrm{per},N}$ have discrete Fourier transforms

$$\widehat{\mathbf{x}}_{\mathrm{per},N}[k] = X(\omega^k) \quad \text{and} \quad \widehat{\mathbf{y}}_{\mathrm{per},N}[k] = Y(\omega^k).$$

Hence $\widehat{\mathbf{y}}_{\mathrm{per},N}[k] = \omega^{-k} X(\omega^k) = \omega^{-k} \widehat{\mathbf{x}}_{\mathrm{per},N}[k]$. □

4.3   Filter Banks and Polyphase Matrices

Recall from Chapter 3 that for a one-scale wavelet transform of a periodic signal, the first analysis operation is to split the signal into even and odd subsignals. Then lifting operations (prediction, update, normalization) are applied to obtain the trend and detail signals. Signal processing (such as compression) is performed on these two signals. Then the synthesis transform applies lifting operations to the modified trend and detail, and finally the merge operation. When the trend and detail signals are not modified before synthesis, the final output is the same as the original signal. This property is called *perfect reconstruction*. We will now carry out similar constructions for finite non-periodic signals using a *filter bank*.

4.3.1   *Lazy filter bank*

Let $\mathbf{x}$ be a finite signal. We can split $\mathbf{x}$ into two subsignals by downsampling $\mathbf{x}$ and downsampling the left shift of $\mathbf{x}$. Define

$$\mathbf{x}_0 = \boxed{2\downarrow}\mathbf{x} \quad \text{and} \quad \mathbf{x}_1 = \boxed{2\downarrow}S^{-1}\mathbf{x}. \qquad (4.21)$$

This splitting is called the *polyphase decomposition* of the signal $\mathbf{x}$. It is the *analysis* part of the so-called *lazy filter bank* and can be described by the flow chart

$$\boxed{2\downarrow} \quad \longrightarrow \mathbf{x}_0$$
$$\nearrow$$
$$\mathbf{x}$$
$$\searrow$$
$$\boxed{S^{-1}} \rightarrow \boxed{2\downarrow} \rightarrow \mathbf{x}_1$$

In the opposite direction, given any two finite signals $\mathbf{x}_0$ and $\mathbf{x}_1$, we obtain another finite signal $\widetilde{\mathbf{x}}$ by

$$\widetilde{\mathbf{x}} = \boxed{2\uparrow}\mathbf{x}_0 + S\boxed{2\uparrow}\mathbf{x}_1. \qquad (4.22)$$

This way of combining two signals is the *synthesis* part of the lazy filter bank. It is described by the flow chart

$$\mathbf{x}_0 \longrightarrow \boxed{2\uparrow}$$
$$\searrow$$
$$\boxed{+} \longrightarrow \widetilde{\mathbf{x}}$$
$$\nearrow$$
$$\mathbf{x}_1 \rightarrow \boxed{2\uparrow} \rightarrow \boxed{S}$$

Proposition 4.2. *The lazy filter bank has the* perfect reconstruction *property: Let* $\mathbf{x}$ *be a finite signal, define* $\mathbf{x}_0$, $\mathbf{x}_1$ *by (4.21), and define* $\widetilde{\mathbf{x}}$ *by (4.22). Then* $\widetilde{\mathbf{x}} = \mathbf{x}$.

Proof. It is enough (by linearity) to check the case where $x = \delta_n$ is a unit impulse. If $n = 2m$ is even, then $\mathbf{x}_0 = \delta_m$ and $\mathbf{x}_1 = 0$. Hence

$$\boxed{2\uparrow}\mathbf{x}_0 + S\boxed{2\uparrow}\mathbf{x}_1 = \boxed{2\uparrow}\delta_m = \delta_{2m} = \mathbf{x}$$

in this case. If $n = 2m + 1$ is odd, then $\mathbf{x}_0 = 0$ and $\mathbf{x}_1 = \delta_m$. Hence

$$\boxed{2\uparrow}\,\mathbf{x}_0 + S\,\boxed{2\uparrow}\,\mathbf{x}_1 = S\,\boxed{2\uparrow}\,\delta_m = S\delta_{2m} = \delta_{2m+1} = \mathbf{x}$$

in this case also. $\qquad\square$

In graphical terms, we can visualize the stem graphs of $\boxed{2\uparrow}\,\mathbf{x}_0$ and $S\,\boxed{2\uparrow}\,\mathbf{x}_1$ fitting together like the teeth of a zipper (thanks to the shift S) to give the stem graph of $\mathbf{x}$. This is the *perfect reconstruction formula*

$$\mathbf{x} = \boxed{2\uparrow}\,\boxed{2\downarrow}\,\mathbf{x} + S\,\boxed{2\uparrow}\,\boxed{2\downarrow}\,S^{-1}\mathbf{x}. \tag{4.23}$$

Putting the analysis part together with the synthesis part, we obtain the flow chart of the lazy filter bank:

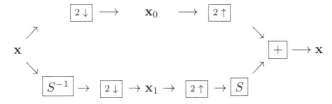

We obtain the $\boxed{\text{split}}$ and $\boxed{\text{merge}}$ operations on $\mathbb{R}^N$ when $N = 2m$ is an even integer by making the analysis and synthesis operations of the lazy filter bank periodic and then using the transformations P_N and P_m from (4.16).

Lemma 4.1. *Let $\mathbf{x}$ be a finite signal. Set $\mathbf{x}_0 = \boxed{2\downarrow}\,\mathbf{x}$ and $\mathbf{x}_1 = \boxed{2\downarrow}\,S^{-1}\mathbf{x}$ and let $N = 2m$ be even. Then*

$$\boxed{\text{split}}\,P_N\mathbf{x}_{\text{per},N} = \begin{bmatrix} P_m(\mathbf{x}_0)_{\text{per},m} \\ P_m(\mathbf{x}_1)_{\text{per},m} \end{bmatrix},$$

$$P_N\mathbf{x}_{\text{per},N} = \boxed{\text{merge}}\begin{bmatrix} P_m(\mathbf{x}_0)_{\text{per},m} \\ P_m(\mathbf{x}_1)_{\text{per},m} \end{bmatrix}. \tag{4.24}$$

Proof. Equations (4.24) are linear in $\mathbf{x}$, and the second equation follows from the first since $\boxed{\text{merge}}$ is the inverse of $\boxed{\text{split}}$ on $\mathbb{R}^N$. So it is enough to verify the first equation when $\mathbf{x} = \delta_k$ is a unit impulse. This follows directly from the formulas for $\mathbf{x}_0$ and $\mathbf{x}_1$ by taking $k = 2n$ or $k = 2n + 1$ with $0 \le n \le m - 1$. $\qquad\square$

We now describe the lazy filter bank in terms of z-transforms. The z-transform of $\mathbf{x}_0$ is

$$X_0(z) = \sum_{n\in\mathbb{Z}} \mathbf{x}[2n]z^{-n}.$$

Since $S^{-1}\mathbf{x} = \sum_{k\in\mathbb{Z}} \mathbf{x}[k]\delta_{k-1} = \sum_{k\in\mathbb{Z}} \mathbf{x}[k+1]\delta_k$, the z-transform of $\mathbf{x}_1$ is obtained by taking the terms in $S^{-1}\mathbf{x}$ with $k = 2n$ even:

$$X_1(z) = \sum_{n\in\mathbb{Z}} \mathbf{x}[2n+1]z^{-n}.$$

The z-transform version of the perfect reconstruction formula (4.23) is

$$X(z) = X_0(z)_{2\uparrow} + z^{-1}\left(X_1(z)_{2\uparrow}\right) = X_0(z^2) + z^{-1}X_1(z^2). \qquad (4.25)$$

Notice in (4.25) that we multiply by z^{-1} *after* upsampling.

Example 4.8. Consider the signal $\mathbf{x} = -\delta_{-2} + 2\delta_{-1} + 3\delta_0 + 4\delta_1 - 2\delta_2$ from Example 4.3 with z-transform $X(z) = -z^2 + 2z + 3 + 4z^{-1} - 2z^{-2}$. In this case we have

$$X_0(z) = X_{2\downarrow}(z) = -z + 3 - 2z^{-1},$$

$$X_1(z) = (zX(z))_{2\downarrow} = \left(-z^3 + 2z^2 + 3z + 4 - 2z^{-1}\right)_{2\downarrow} = 2z + 4$$

(notice the shift of the exponents). We recover $X(z)$ by (4.25):

$$X_0(z^2) + z^{-1}X_1(z^2) = (-z^2 + 3 - 2z^{-2}) + z^{-1}(2z^2 + 4) = X(z).$$

See how the even and odd powers of z zip together to reconstruct $X(z)$. ∎

4.3.2 Filter banks from lifting

Recall that the wavelet transforms of discrete periodic signals in Section 3.4 use the operators $\boxed{\text{split}}$ and $\boxed{\text{merge}}$, together with the 2×2 block-form polyphase analysis and synthesis matrices, whose entries are Laurent polynomials in the shift operator (see Definition 3.1).

To obtain wavelet transforms of finite nonperiodic signals we replace $\boxed{\text{split}}$ by the analysis part of the lazy filter bank, and we replace $\boxed{\text{merge}}$ by the synthesis part of the lazy filter bank, as suggested by Lemma 4.1. We then use the same 2×2 block form for the polyphase analysis and synthesis matrices, but now acting on the space $\ell_0(\mathbb{Z}) \oplus \ell_0(\mathbb{Z})$ (see Sections 1.5.2 and 1.5.3). The entry in each block is a Laurent polynomial in the shift operator acting on $\ell_0(\mathbb{Z})$. The polyphase synthesis matrix is the inverse of the polyphase analysis matrix; hence the resulting analysis and synthesis transforms on $\ell_0(\mathbb{Z})$ have the perfect reconstruction property by Proposition 4.2.

Example 4.9. The polyphase matrix for the CDF$(2,2)$ transform is the product of a prediction step, followed by an update step and a normalization. It acts in the time domain by

$$\begin{bmatrix} \mathbf{y}_0 \\ \mathbf{y}_1 \end{bmatrix} = \frac{1}{4\sqrt{2}} \begin{bmatrix} (-S^{-1} + 6I - S) & (2I + 2S) \\ -(2S^{-1} + 2I) & 4I \end{bmatrix} \begin{bmatrix} \mathbf{x}_0 \\ \mathbf{x}_1 \end{bmatrix}$$

(see Example 3.10). In terms of z-transforms, the analysis transform becomes matrix multiplication

$$\begin{bmatrix} Y_0(z) \\ Y_1(z) \end{bmatrix} = \mathbf{H}_p(z) \begin{bmatrix} X_0(z) \\ X_1(z) \end{bmatrix}$$

on the vector of z-transforms of the polyphase components $\mathbf{x}_0$ and $\mathbf{x}_1$ of $\mathbf{x}$. Here the polyphase analysis matrix is

$$\mathbf{H}_p(z) = \frac{1}{4\sqrt{2}} \begin{bmatrix} (-z + 6 - z^{-1}) & (2 + 2z^{-1}) \\ -(2 + 2z) & 4 \end{bmatrix},$$

since the shift operator S becomes the operator of multiplication by z^{-1} when we use z-transforms. The inverse transform is obtained by inverting the individual lifting steps (see Example 3.10):

$$\begin{bmatrix} \mathbf{x}_0 \\ \mathbf{x}_1 \end{bmatrix} = \frac{1}{4\sqrt{2}} \begin{bmatrix} 4I & -(2I+2S) \\ (2S^{-1}+2I) & (-S^{-1}+6I-S) \end{bmatrix} \begin{bmatrix} \mathbf{y}_0 \\ \mathbf{y}_1 \end{bmatrix}.$$

We can then reconstruct $\mathbf{x}$ using the synthesis part of the lazy filter bank:

$$\mathbf{x} = \boxed{2\uparrow}\mathbf{x}_0 + S\,\boxed{2\uparrow}\mathbf{x}_1\,.$$

In terms of z-transforms, the inverse transformation becomes

$$\begin{bmatrix} X_0(z) \\ X_1(z) \end{bmatrix} = \mathbf{G}_p(z) \begin{bmatrix} Y_0(z) \\ Y_1(z) \end{bmatrix},$$

where the *polyphase synthesis matrix* is

$$\mathbf{G}_p(z) = \frac{1}{4\sqrt{2}} \begin{bmatrix} 4 & -(2+2z^{-1}) \\ (2+2z) & (-z+6-z^{-1}) \end{bmatrix}.$$

Since $\mathbf{G}_p(z)$ is the inverse matrix to $\mathbf{H}_p(z)$ for every nonzero complex number z, the formula for $\mathbf{G}_p(z)$ can be obtained directly from the entries in $\mathbf{H}_p(z)$ using *Cramer's Rule*:

$$\begin{bmatrix} a & b \\ c & d \end{bmatrix}^{-1} = \frac{1}{\delta} \begin{bmatrix} d & -b \\ -c & a \end{bmatrix} \quad \text{when } \delta = ad - bc \neq 0. \tag{4.26}$$

The lifting-step factorization of $\mathbf{H}_p(z)$ shows it has determinant 1, so $\delta = 1$ in this application of Cramer's rule. ∎

The general one-scale wavelet analysis transform $\mathbf{T_a}$ is constructed by the lifting process in the same way as the CDF$(2,2)$ transform. The signal $\mathbf{x}$ is passed through the analysis part of the lazy filter bank to get $\mathbf{x}_0$ and $\mathbf{x}_1$. Then this pair of signals is transformed into the pair $\mathbf{y}_0$ and $\mathbf{y}_1$ by a succession of lifting steps (predictions, updates). Each of these lifting steps acts by a 2×2 block matrix with diagonal entries I and one nonzero off-diagonal term that is a Laurent polynomial in S. Then there is a final normalization by a diagonal matrix of determinant 1. The product of these matrices is the *polyphase analysis matrix* of the wavelet transform.

We now express this process in terms of the z-transforms of $\mathbf{x}_0$, $\mathbf{x}_1$, $\mathbf{y}_0$, and $\mathbf{y}_1$. The polyphase analysis matrix becomes a product

$$\mathbf{H}_p(z) = \begin{bmatrix} \kappa & 0 \\ 0 & 1/\kappa \end{bmatrix} \begin{bmatrix} 1 & F_1(z) \\ 0 & 1 \end{bmatrix} \cdots \begin{bmatrix} 1 & 0 \\ F_k(z) & 1 \end{bmatrix} = \begin{bmatrix} H_{00}(z) & H_{01}(z) \\ H_{10}(z) & H_{11}(z) \end{bmatrix}, \tag{4.27}$$

where $F_1(z),\ldots,F_k(z)$ and $H_{ij}(z)$ are Laurent polynomials and $\kappa \neq 0$ (there may be several prediction and update factors), and

$$\begin{bmatrix} Y_0(z) \\ Y_1(z) \end{bmatrix} = \mathbf{H}_p(z) \begin{bmatrix} X_0(z) \\ X_1(z) \end{bmatrix}.$$

This transformation is described by the flow chart

Here $\boxed{z}$ denotes the operation of multiplication by z that corresponds to the left shift S^{-1} in the time domain. Notice that the polyphase matrix acts on the signal *after* it has passed through the analysis part of the lazy filter bank.

The inverse transform is obtained using the *polyphase synthesis matrix*:

$$\mathbf{G}_p(z) = \mathbf{H}_p(z)^{-1} = \begin{bmatrix} 1 & 0 \\ -F_k(z) & 1 \end{bmatrix} \cdots \begin{bmatrix} 1 & -F_1(z) \\ 0 & 1 \end{bmatrix} \begin{bmatrix} 1/\kappa & 0 \\ 0 & \kappa \end{bmatrix}$$

$$= \begin{bmatrix} H_{11}(z) & -H_{01}(z) \\ -H_{10}(z) & H_{00}(z) \end{bmatrix}. \qquad (4.28)$$

We have obtained $\mathbf{G}_p(z)$ directly from the four Laurent polynomials in the matrix $\mathbf{H}_p(z)$ using Cramer's rule (note that the normalization matrix has been chosen to ensure $\det \mathbf{H}_p(z) = 1$, so no division is needed). The flow chart for the inverse transform is

Here $\boxed{z^{-1}}$ denotes the operation of multiplication by z that corresponds to the right shift S in the time domain. In algebraic terms,

$$\begin{bmatrix} X_0(z) \\ X_1(z) \end{bmatrix} = \mathbf{G}_p(z) \begin{bmatrix} Y_0(z) \\ Y_1(z) \end{bmatrix}.$$

Finally, we reconstruct $X(z)$ by applying the synthesis part of the lazy filter bank (equation (4.25)).

Remark 4.2. If $\mathbf{x}$ is a finite signal and $N = 2m$ is even, then $(\mathbf{T_a x})_{\text{per},N}$ is the periodic wavelet analysis transform calculated in Chapter 3 for the N-periodic signal $\mathbf{x}_{\text{per},N}$. To verify this, let $\mathbf{T}_{\mathbf{a},N}$ be the $N \times N$ wavelet analysis matrix obtained by setting $S = S_m$ in the time-domain polyphase analysis matrix. Let $\mathbf{u} = P_N \mathbf{x}_{\text{per},N} \in \mathbb{R}^N$. Then from Lemma 4.1 and the commutation relation (4.18) it follows that

$$\mathbf{T_a x} = \begin{bmatrix} \mathbf{y}_0 \\ \mathbf{y}_1 \end{bmatrix} \quad \text{and} \quad \mathbf{T}_{\mathbf{a},N} \mathbf{u} = \begin{bmatrix} P_m(\mathbf{y}_0)_{\text{per},m} \\ P_m(\mathbf{y}_1)_{\text{per},m} \end{bmatrix}. \qquad (4.29)$$

This makes the connection between wavelet transforms of periodic signals and wavelet transforms of finite, non-periodic signals. But we cannot recover $\mathbf{T_a x}$ from $\mathbf{T}_{\mathbf{a},N} \mathbf{x}_{\text{per},N}$ in general because of the aliasing that occurs in periodization, as the following example illustrates.

Example 4.10. For the CDF$(2, 2)$ transform in Example 4.9, take $\mathbf{x} = \delta_0$. Then $\mathbf{x}_0 = \delta_0$ and $\mathbf{x}_1 = \mathbf{0}$, so

$$\mathbf{T}_a \mathbf{x} = \frac{1}{4\sqrt{2}} \begin{bmatrix} (-S^{-1} + 6I - S) & (2I + 2S) \\ -(2I + 2S^{-1}) & 4I \end{bmatrix} \begin{bmatrix} \delta_0 \\ \mathbf{0} \end{bmatrix} = \frac{1}{4\sqrt{2}} \begin{bmatrix} -\delta_{-1} + 6\delta_0 - \delta_1 \\ -2\delta_{-1} - 2\delta_0 \end{bmatrix}.$$

Thus $(4\sqrt{2})\mathbf{y}_0 = -\delta_{-1} + 6\delta_0 - \delta_1$ and $(4\sqrt{2})\mathbf{y}_1 = -2\delta_{-1} - 2\delta_0$ for this signal. If we take $N = 4$, so that $m = 2$, then $P_4 \mathbf{x}_{per,4} = \begin{bmatrix} 1 & 0 & 0 & 0 \end{bmatrix}^T$, whereas

$$P_2(\mathbf{y}_0)_{per,2} = \frac{1}{2\sqrt{2}} \begin{bmatrix} 3 \\ -1 \end{bmatrix} \quad \text{and} \quad P_2(\mathbf{y}_1)_{per,2} = \frac{1}{2\sqrt{2}} \begin{bmatrix} -1 \\ -1 \end{bmatrix}.$$

From Example 3.5 we have

$$\mathbf{T}_{a,4} P_4 \mathbf{x}_{per,4} = \frac{1}{2\sqrt{2}} \begin{bmatrix} 3 & 1 & -1 & 1 \\ -1 & 1 & 3 & 1 \\ -1 & 2 & -1 & 0 \\ -1 & 0 & -1 & 2 \end{bmatrix} \begin{bmatrix} 1 \\ 0 \\ 0 \\ 0 \end{bmatrix} = \frac{1}{2\sqrt{2}} \begin{bmatrix} 3 \\ -1 \\ -1 \\ -1 \end{bmatrix} = \begin{bmatrix} P_2(\mathbf{y}_0)_{per,2} \\ P_2(\mathbf{y}_1)_{per,2} \end{bmatrix},$$

as predicted by (4.29). But the signals $\mathbf{y}_0$ and $\mathbf{y}_1$ have length 3, and we cannot recover them from knowing $P_2(\mathbf{y}_0)_{per,2}$ and $P_2(\mathbf{y}_1)_{per,2}$. ∎

4.4 Filter Banks and Modulation Matrices

The definition of wavelet transforms in Section 4.3 using the polyphase method assures that the perfect reconstruction property always holds, since each step of the lifting process (prediction, update, normalization) uses an invertible triangular matrix, and the lazy filter bank has perfect reconstruction. However, this approach doesn't explain how the lifting steps are chosen to obtain desirable properties in the wavelet transform, such as separation of trend and detail. To understand this aspect, we need an alternate description of wavelet transforms using ideas from signal processing (lowpass and highpass filters).

4.4.1 *Lowpass and highpass filters*

Let T be a FIR filter with impulse response function $\mathbf{h}$ and let $\mathbf{x}$ be a finite signal. Write $H(z)$ for the z-transform of $\mathbf{h}$. Then the z-transform of $T\mathbf{x}$ is $H(z)X(z)$ by Theorem 4.1. Thus the action of the filter on the z-transform of $\mathbf{x}$ emphasizes the range of frequencies ω in $\mathbf{x}$ where $|H(e^{\omega i})|$ is large, and suppresses the range of frequencies in $\mathbf{x}$ where $|H(e^{\omega i})|$ is small. We illustrate the effect this has in the time domain in the next example.

Example 4.11. Consider the filters T_0 and T_1 with impulse response functions

$$\mathbf{h}_0 = \frac{1}{4}(\delta_{-1} + 2\delta_0 + \delta_1) \quad \text{and} \quad \mathbf{h}_1 = \frac{1}{4}(-\delta_{-1} + 2\delta_0 - \delta_1), \tag{4.30}$$

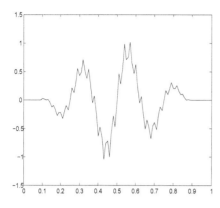
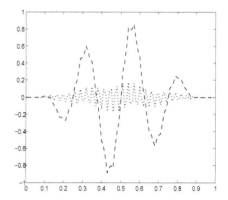

Fig. 4.3   Signal   Fig. 4.4   Lowpass/highpass filtered signals

respectively. The filtered signals $\mathbf{y}_0 = \mathbf{h}_0 \star \mathbf{x}$ and $\mathbf{y}_1 = \mathbf{h}_1 \star \mathbf{x}$ use three adjacent values of the input signal $\mathbf{x}$:

$$\mathbf{y}_0[n] = \frac{1}{4}\big(\mathbf{x}[n-1] + 2\mathbf{x}[n] + \mathbf{x}[n+1]\big),$$

$$\mathbf{y}_1[n] = \frac{1}{4}\big(-\mathbf{x}[n-1] + 2\mathbf{x}[n] - \mathbf{x}[n+1]\big).$$

Evidently $\mathbf{x} = \mathbf{y}_0 + \mathbf{y}_1$, so we can recover the original signal $\mathbf{x}$ as the sum of the two filtered signals $\mathbf{y}_0$ and $\mathbf{y}_1$.

To compare the effects of the filters T_0 and T_1, we create an analog signal that is the sum of two sine waves:

$$f(t) = \sin(8\pi t) + 0.25\sin(64\pi t)$$

(a fundamental oscillation at a frequency of 4 Hz and the eighth overtone at 32 Hz with smaller amplitude). After multiplying $f(t)$ by a *window function* that is zero outside the interval $0.1 \le t \le 0.9$ and making an analog to digital conversion by sampling, we obtain the signal $\mathbf{x}$ in Figure 4.3. In Figure 4.4 we see the signals $\mathbf{y}_0$ (dashed line) and $\mathbf{y}_1$ (dotted line). The filter T_0 has picked out the low-frequency part of $\mathbf{x}$; by contrast, the filter T_1 has picked out the high frequency part.   ∎

The term *lowpass* or *highpass* for a filter describes its frequency response. An *ideal* pair of filters $\mathbf{h}_0$ (lowpass) and $\mathbf{h}_1$ (highpass) for frequency band separation would have z-transforms

$$H_0(\mathrm{e}^{\mathrm{i}\omega}) = \begin{cases} 1 & \text{when } 0 \le \omega < L, \\ 0 & \text{when } L \le \omega \le \pi, \end{cases} \qquad H_1(\mathrm{e}^{\mathrm{i}\omega}) = \begin{cases} 0 & \text{when } 0 \le \omega < L, \\ 1 & \text{when } L \le \omega \le \pi, \end{cases} \qquad (4.31)$$

where L is some number between 0 and π that is called the *crossover frequency* between the two filters.[4] For such a pair of filters, the filtered signal $\mathbf{h}_0 \star \mathbf{x}$ only has

----

[4] A real-valued digital filter $\mathbf{h}$ has a z-transform H that satisfies $H(\mathrm{e}^{-\mathrm{i}\omega}) = \overline{H(\mathrm{e}^{\mathrm{i}\omega})}$. Since the function $\omega \longrightarrow H(\mathrm{e}^{\mathrm{i}\omega})$ is periodic of period 2π, we only need to specify it for $0 \le \omega \le \pi$.

frequencies in the low range $|\omega| < L$, since its Fourier transform $H_0(e^{i\omega})X(e^{i\omega})$ is zero when $L \le |\omega| < \pi$. Likewise, the filtered signal $\mathbf{h}_1 \star \mathbf{x}$ only has frequencies in the high range $L \le |\omega| \le \pi$.

However, we cannot find finite impulse response filters that satisfy (4.31). For example, if $L = \pi/2$, then for $n \ne 0$

$$\mathbf{h}_0[n] = \frac{1}{2\pi} \int_{-\pi}^{\pi} H_0(e^{i\omega})e^{in\omega}\, d\omega = \frac{1}{2\pi} \int_{-\pi/2}^{\pi/2} e^{in\omega}\, d\omega = \frac{\sin(n\pi/2)}{n\pi}.$$

But $\sin(n\pi/2) = (-1)^m \ne 0$ when $n = 2m+1$ is odd, so $\mathbf{h}_0$ is not a FIR filter. This means that the filtered output $\mathbf{h}_0 \star \mathbf{x}$ from a finite input $\mathbf{x}$ has an infinitely long slowly decaying tail.

To obtain FIR filters we will modify (4.31) to allow some overlap between the low and high frequency bands, while still using convolution with $\mathbf{h}_1$ to remove most of the energy in the low frequencies ($|\omega|$ near zero) and convolution with $\mathbf{h}_0$ to remove most of the energy in the high frequencies ($|\omega|$ near π), as in Example 4.12 and Figure 4.4.

Definition 4.4. The pair $\mathbf{h}_0$, $\mathbf{h}_1$ of FIR filters satisfies the *lowpass/highpass condition* if $\mathbf{h}_0 \ne \mathbf{0}$, $\mathbf{h}_1 \ne \mathbf{0}$, and the z-transforms satisfy $H_0(-1) = 0$, $H_1(1) = 0$.

In this definition the value $z = -1$ corresponds to the highest frequency $\omega = \pi$, and the value $z = 1$ corresponds to the lowest frequency $\omega = 0$. If $\mathbf{h}_0$, $\mathbf{h}_1$ satisfy the lowpass/highpass condition, then the nonzero Laurent polynomial $H_0(z)$ is divisible by $(1 + z)^q$ for some integer $q > 0$, and the nonzero Laurent polynomial $H_1(z)$ is divisible by $(1 - z)^p$ for some integer $p > 0$. Thus

$$H_0(z) = (1+z)^q \varphi(z) \quad \text{and} \quad H_1(z) = (1-z)^p \psi(z), \tag{4.32}$$

where $\varphi(z)$ and $\psi(z)$ are Laurent polynomials with $\varphi(-1) \ne 0$ and $\psi(1) \ne 0$. A large value of q means that $H_0(e^{i\omega})$ stays very close to zero when ω is near π, while a large value of p means that $H_1(e^{i\omega})$ stays very close to zero when ω is near 0. Thus $\mathbf{h}_0$ will remove most of the high frequency part of a signal, and $\mathbf{h}_1$ will remove most of the low frequency part of a signal.

Example 4.12. Consider the filters T_0 and T_1 in Example 4.11. From (4.30) the z-transforms of the impulse response functions are $H_0(z) = \frac{1}{4}(z + 2 + z^{-1})$ and $H_1(z) = \frac{1}{4}(-z + 2 - z^{-1})$. In terms of the frequency ω, we can write

$$H_0(e^{i\omega}) = \left(\frac{e^{i\omega/2} + e^{-i\omega/2}}{2}\right)^2 = \cos^2(\omega/2),$$

$$H_1(e^{i\omega}) = \left(\frac{e^{i\omega/2} - e^{-i\omega/2}}{2i}\right)^2 = \sin^2(\omega/2).$$

Notice that $H_0(e^{i\omega}) + H_1(e^{i\omega}) = 1$ since $\mathbf{h}_0 + \mathbf{h}_1 = \delta_0$. The graphs of $H_0(e^{i\omega})$ (solid line) and $H_0(e^{i\omega})$ (dotted line) for $-\pi \le \omega \le \pi$ are shown in Figure 4.5. Since $H_0(z)$ vanishes to order 2 at $z = -1$ ($\omega = \pi$), T_0 is a lowpass filter, whereas $H_1(z)$ vanishes to order 2 at $z = 1$ ($\omega = 0$) and T_1 is a highpass filter. This explains the frequency separation observed in Figure 4.4. ∎

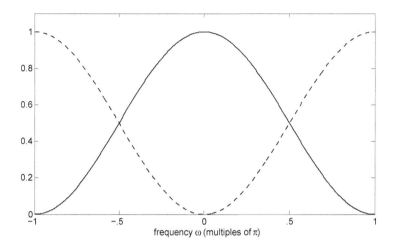

frequency ω (multiples of π)

Fig. 4.5   Graphs of $H_0(e^{i\omega})$ and $H_1(e^{i\omega})$ for Example 4.12

4.4.2   *Filter banks from filter pairs*

A *two-channel analysis filter bank* uses two FIR filters with impulse responses $\mathbf{h}_0$ (*lowpass*) and $\mathbf{h}_1$ (*highpass*) to transform the input signal $\mathbf{x}$ into

$$\mathbf{T_a x} = \begin{bmatrix} \mathbf{y}_0 \\ \mathbf{y}_1 \end{bmatrix}, \quad \text{where} \quad \mathbf{y}_0 = \boxed{2\downarrow}\,\mathbf{h}_0 \star \mathbf{x} \quad \text{and} \quad \mathbf{y}_1 = \boxed{2\downarrow}\,\mathbf{h}_1 \star \mathbf{x}. \quad (4.33)$$

Since convolution and downsampling are linear processes, this gives a linear transformation $\mathbf{T_a}$ whose output is the pair of signals $\mathbf{y}_0, \mathbf{y}_1$:

$$\boxed{\mathbf{h}_0\star} \longrightarrow \boxed{2\downarrow} \longrightarrow \mathbf{y}_0$$
$$\nearrow$$
$$\mathbf{T_a} : \mathbf{x}$$
$$\searrow$$
$$\boxed{\mathbf{h}_1\star} \longrightarrow \boxed{2\downarrow} \longrightarrow \mathbf{y}_1$$

Thus the entire signal first passes through each filter separately and then the two filtered signals are downsampled.

We can express the filter bank operations on finite signals in terms of z-transforms. Let $\mathbf{x}$ be any finite signal. Then

$$X_{2\downarrow}(z) = \frac{1}{2}\left\{X(z^{1/2}) + X(-z^{1/2})\right\}. \quad (4.34)$$

To verify the correctness of (4.34), observe that the right side of (4.34) is the sum over n of the terms $(1/2)\mathbf{x}[n]\big((z^{1/2})^{-n} + (-1)^n(z^{1/2})^{-n}\big)$. These terms are zero when n is odd and are $\mathbf{x}[2m]z^{-m}$ when $n = 2m$ is even.

Remark 4.3. The notation in (4.34) means that we substitute $z^{1/2}$ in z-transform of $\mathbf{x}$ and simplify by the usual algebraic rules for exponents:

$$(z^{1/2})^n = z^{n/2} \quad \text{and} \quad (-z^{1/2})^n = (-1)^n z^{n/2}\,.$$

Notice the placement of parentheses: $-z^{1/2} = (-1)z^{1/2}$. When we write $z = e^{i\omega}$ in terms of the frequency ω, then $(z^{1/2})^n = e^{in\omega/2}$, so the frequency is divided by two.

Since the z-transform of $\mathbf{h}_0 \star \mathbf{x}$ is $H_0(z)X(z)$ and the z-transform of $\mathbf{h}_1 \star \mathbf{x}$ is $H_1(z)X(z)$, we can use (4.34) to express the z-transforms of $\mathbf{y}_0$ and $\mathbf{y}_1$ as

$$Y_0(z) = \frac{1}{2}\left\{H_0(z^{1/2})X(z^{1/2}) + H_0(-z^{1/2})X(-z^{1/2})\right\} ,$$

$$Y_1(z) = \frac{1}{2}\left\{H_1(z^{1/2})X(z^{1/2}) + H_1(-z^{1/2})X(-z^{1/2})\right\} .$$

Define the 2×2 *analysis modulation matrix*

$$\mathbf{H}_m(z) = \begin{bmatrix} H_0(z) & H_0(-z) \\ H_1(z) & H_1(-z) \end{bmatrix} . \tag{4.35}$$

Then the formulas for $Y_0(z)$ and $Y_1(z)$ can be combined into a single vector-matrix equation

$$2\begin{bmatrix} Y_0(z) \\ Y_1(z) \end{bmatrix} = \mathbf{H}_m(z^{1/2})\begin{bmatrix} X(z^{1/2}) \\ X(-z^{1/2}) \end{bmatrix} . \tag{4.36}$$

The term *modulation* is used for $\mathbf{H}_m(z)$ because replacing z by $-z$ corresponds to a half-band *frequency shift*: when $z = e^{i\omega}$ then $-z = e^{i\pi}e^{i\omega} = e^{i(\omega+\pi)}$. Thus the entries $H_0(-z)$ and $H_1(-z)$ in the second column of $\mathbf{H}_m(z)$ are modulations by frequency shift $\omega \to \omega + \pi$ of the entries in the first column.

A *two-channel synthesis filter bank* uses two FIR filters with impulse responses $\mathbf{g}_0$ (*lowpass*) and $\mathbf{g}_1$ (*highpass*) to transform a pair of input signals $\mathbf{y}_0$, $\mathbf{y}_1$ into

$$\widetilde{\mathbf{x}} = \mathbf{T_s}\begin{bmatrix} \mathbf{y}_0 \\ \mathbf{y}_1 \end{bmatrix} = \mathbf{g}_0 \star (\boxed{2\uparrow}\mathbf{y}_0) + \mathbf{g}_1 \star (\boxed{2\uparrow}\mathbf{y}_1) . \tag{4.37}$$

Upsampling and convolution are linear processes, so this gives a linear transformation $\mathbf{T_s}$ whose output we have denoted as $\widetilde{\mathbf{x}}$:

$$\mathbf{T_s}:\quad \begin{array}{c} \mathbf{y}_0 \longrightarrow \boxed{2\uparrow} \longrightarrow \boxed{\mathbf{g}_0\star} \\ \\ \\ \mathbf{y}_1 \longrightarrow \boxed{2\uparrow} \longrightarrow \boxed{\mathbf{g}_1\star} \end{array} \begin{array}{c} \searrow \\ \boxed{+} \longrightarrow \widetilde{\mathbf{x}} \\ \nearrow \end{array}$$

Notice that each signal is first upsampled and then filtered. Since the z-transform of $\boxed{2\uparrow}\mathbf{y}_0$ is $Y_0(z^2)$ and the z-transform of $\boxed{2\uparrow}\mathbf{y}_1$ is $Y_1(z^2)$, it follows that the z-transform of $\widetilde{\mathbf{x}}$ is

$$\widetilde{X}(z) = G_0(z)Y_0(z^2) + G_1(z)Y_1(z^2) .$$

If we apply the analysis transform $\mathbf{T_a}$ to a signal $\mathbf{x}$ and then apply the synthesis transform $\mathbf{T_s}$ to $\mathbf{T_a}\mathbf{x}$, we obtain a signal $\widetilde{\mathbf{x}}$. We want to express the z-transform $\widetilde{X}(z)$ of the output in terms of the z-transform $X(z)$ of the input. Since $Y_0((-z)^2) =$

$Y_0(z^2)$ and $Y_1((-z)^2) = Y_1(z^2)$, the equation we just found for $\widetilde{X}(z)$ furnishes the pair of equations

$$\widetilde{X}(z) = G_0(z)Y_0(z^2) + G_1(z)Y_1(z^2),$$
$$\widetilde{X}(-z) = G_0(-z)Y_0(z^2) + G_1(-z)Y_1(z^2).$$

We can write these two equations in matrix form as

$$\begin{bmatrix} \widetilde{X}(z) \\ \widetilde{X}(-z) \end{bmatrix} = \mathbf{G}_m(z) \begin{bmatrix} Y_0(z^2) \\ Y_1(z^2) \end{bmatrix},$$

where

$$\mathbf{G}_m(z) = \begin{bmatrix} G_0(z) & G_1(z) \\ G_0(-z) & G_1(-z) \end{bmatrix} \tag{4.38}$$

is called the *synthesis modulation matrix*. Notice that the entries $G_0(-z)$ and $G_1(-z)$ in the second row of $\mathbf{G}_m(z)$ are the half-band frequency modulations of the entries in the first row.

From (4.36) we know that

$$2 \begin{bmatrix} Y_0(z^2) \\ Y_1(z^2) \end{bmatrix} = \mathbf{H}_m(z) \begin{bmatrix} X(z) \\ X(-z) \end{bmatrix}.$$

Hence the z-transforms of the input $\mathbf{x}$ and output $\widetilde{\mathbf{x}}$ of a two-channel analysis-synthesis filter bank are related by

$$2 \begin{bmatrix} \widetilde{X}(z) \\ \widetilde{X}(-z) \end{bmatrix} = \mathbf{G}_m(z)\mathbf{H}_m(z) \begin{bmatrix} X(z) \\ X(-z) \end{bmatrix}. \tag{4.39}$$

Definition 4.5. A filter bank has the *perfect reconstruction* (PR) property if the output signal $\widetilde{\mathbf{x}}$ is the same as the input signal $\mathbf{x}$ for all finite signals $\mathbf{x}$.

Theorem 4.6. *The perfect reconstruction property holds if and only if the modulation matrices satisfy* $\mathbf{G}_m(z)\mathbf{H}_m(z) = 2I$, *where I is the 2×2 identity matrix. This condition is the same as* $\mathbf{H}_m(z)\mathbf{G}_m(z) = 2I$.

Proof. Suppose $\mathbf{G}_m(z)\mathbf{H}_m(z) = 2I$. For $n \times n$ matrices A and B, if $AB = cI$ where c is a nonzero constant, then A is invertible and $B = cA^{-1}$ (see Example 1.13). Hence $\mathbf{G}_m(z) = 2\mathbf{H}_m(z)^{-1}$ and also $\mathbf{H}_m(z)\mathbf{G}_m(z) = 2I$. It follows from (4.39) that the PR property holds.

Conversely, if the PR property is satisfied, take $\mathbf{x} = \delta_0$. Then $X(z) = 1$, so (4.39) implies that

$$2 \begin{bmatrix} 1 \\ 1 \end{bmatrix} = \mathbf{G}_m(z)\mathbf{H}_m(z) \begin{bmatrix} 1 \\ 1 \end{bmatrix}. \tag{4.40}$$

Now take $\mathbf{x} = \delta_1$. Then $X(z) = z^{-1}$ and (4.39) implies that

$$2 \begin{bmatrix} z^{-1} \\ -z^{-1} \end{bmatrix} = \mathbf{G}_m(z)\mathbf{H}_m(z) \begin{bmatrix} z^{-1} \\ -z^{-1} \end{bmatrix}.$$

Multiply this equation by z to obtain a new equation

$$2 \begin{bmatrix} 1 \\ -1 \end{bmatrix} = \mathbf{G}_m(z)\mathbf{H}_m(z) \begin{bmatrix} 1 \\ -1 \end{bmatrix}. \tag{4.41}$$

Adding and subtracting equations (4.40) and (4.41), we find that

$$\mathbf{G}_m(z)\mathbf{H}_m(z) \begin{bmatrix} 1 \\ 0 \end{bmatrix} = 2 \begin{bmatrix} 1 \\ 0 \end{bmatrix} \quad \text{and} \quad \mathbf{G}_m(z)\mathbf{H}_m(z) \begin{bmatrix} 0 \\ 1 \end{bmatrix} = 2 \begin{bmatrix} 0 \\ 1 \end{bmatrix}.$$

Hence $\mathbf{G}_m(z)\mathbf{H}_m(z) = 2I$. □

To describe the possible FIR analysis filters in a PR filter bank, we need the following basic algebraic fact about Laurent polynomials.

Lemma 4.2. *Suppose $g(z)$ and $h(z)$ are Laurent polynomials such that the polynomial $g(z)h(z)$ is a nonzero constant. Then $g(z)$ and $h(z)$ are monomials.*

Proof. Since $g(z)$ is not the zero polynomial, we can write $g(z) = c_m z^m + \cdots + c_n z^n$ with $m \leq n$, $c_m \neq 0$, and $c_n \neq 0$. Likewise $h(z) = d_p z^p + \cdots + d_q z^q$ with $p \leq q$, $d_p \neq 0$, and $d_q \neq 0$. The product is

$$g(z)h(z) = c_m d_p z^{m+p} + \cdots + c_n d_q z^{n+q}.$$

By assumption the right side of this equation is a nonzero constant. Hence $m + p = n + q = 0$, $m = n$, and $p = q$. Thus $g(z) = c_n z^n$ and $h(z) = d_{-n} z^{-n}$. □

Theorem 4.7. *Suppose $\mathbf{h}_0$ and $\mathbf{h}_1$ are FIR filters. These filters are the analysis part of a two-channel FIR filter bank with perfect reconstruction if and only if the corresponding modulation matrix $\mathbf{H}_m(z)$ satisfies*

$$\det \mathbf{H}_m(z) = cz^{2K-1} \quad (c \neq 0 \text{ a constant and } K \text{ an integer}). \tag{4.42}$$

When (4.42) is satisfied then the synthesis filters $\mathbf{g}_0$ and $\mathbf{g}_1$ in the filter bank are uniquely determined by the analysis filters:

$$G_0(z) = \frac{2}{d(z)} H_1(-z) \quad \text{and} \quad G_1(z) = -\frac{2}{d(z)} H_0(-z), \tag{4.43}$$

where $d(z) = \det \mathbf{H}_m(z) = H_0(z)H_1(-z) - H_0(-z)H_1(z)$.

Proof. Suppose that there exist FIR filters $\mathbf{g}_0$ and $\mathbf{g}_1$ such that the filter bank with analysis/synthesis filters $\mathbf{h}_0$, $\mathbf{h}_1$ and $\mathbf{g}_0$, $\mathbf{g}_1$ has perfect reconstruction. Then the modulation matrices satisfy $\mathbf{G}_m(z)\mathbf{H}_m(z) = 2I$ by Theorem 4.6. Hence

$$\det \mathbf{G}_m(z) \det \mathbf{H}_m(z) = \det(2I) = 4.$$

Lemma 4.2 implies that $d(z) = \det \mathbf{H}_m(z)$ is a nonzero monomial. From its definition it is clear that $d(-z) = -d(z)$. Hence $d(z)$ must be a monomial of odd degree $2K - 1$. Furthermore, $\mathbf{G}_m(z) = 2\mathbf{H}_m(z)^{-1}$, so by Cramer's rule for the inverse of a 2×2 matrix

$$\begin{bmatrix} G_0(z) & G_1(z) \\ G_0(-z) & G_1(-z) \end{bmatrix} = \frac{2}{d(z)} \begin{bmatrix} H_1(-z) & -H_0(-z) \\ -H_1(z) & H_0(z) \end{bmatrix}. \tag{4.44}$$

We now obtain equations (4.43) by comparing entries on each side of (4.44).

Conversely, if $\det \mathbf{H}_m(z)$ is a monomial, then we can define Laurent polynomials $G_0(z)$ and $G_1(z)$ by (4.43). The synthesis modulation matrix is then given by (4.44) and the PR condition is satisfied.   □

Equations (4.43) show that the lowpass synthesis filter $\mathbf{g}_0$ is obtained from the highpass analysis filter $\mathbf{h}_1$ by the following operations:

- *half-band frequency shift*: when $z = e^{i\omega}$ then $-z = e^{i(\omega+\pi)}$,
- *time shift*: multiplication of z-transforms by z^{-2K+1} (time shift by $2K-1$ units),
- *rescaling*: multiplication by a nonzero constant.

The highpass synthesis filter $\mathbf{g}_1$ is obtained from the lowpass analysis filter $\mathbf{h}_0$ in the same way.

4.4.3 *Perfect reconstruction from analysis filters*

We have now reached the challenging *design problem* for two-channel PR filter banks with FIR filters; namely, to combine the PR property with the lowpass/highpass condition. The z-transforms of the filters $\mathbf{h}_0$ and $\mathbf{h}_1$ must factor as in equation (4.32) for some positive integers p and q. Furthermore, the Laurent polynomials $\varphi(z)$ and $\psi(z)$ in these factorizations must be chosen so that the PR condition (4.42) holds. To express the PR condition in terms of $\varphi(z)$ and $\psi(z)$, let

$$Q(z) = H_0(z)H_1(-z) = (1+z)^{p+q}\varphi(z)\psi(-z) .$$

Then $\det \mathbf{H}_m(z) = Q(z) - Q(-z)$ is twice the sum of the odd-degree terms in $Q(z)$. Thus the PR condition (4.42) is the same as

$$Q(z) \text{ contains exactly } one \text{ nonzero term of odd degree.} \qquad (4.45)$$

We use condition (4.45) to explore the PR design problem with some simple examples at this point. In Section 4.5 we will prove an algebraic result (Bezout's theorem) that will be the key to obtaining families of PR wavelet transforms with FIR filters.

Example 4.13. Take $\varphi(z) = \psi(z) = 1$ and $p = q = 1$ in (4.32). Then $H_0(z) = 1+z$ and $H_1(z) = 1 - z$; thus $Q(z) = (1+z)^2 = 1+2z+z^2$ in this case. Hence condition (4.45) is satisfied, and $\mathbf{h}_0$, $\mathbf{h}_1$ are the analysis filters for a PR filter bank. Referring to Example 4.19, we see that, except for a normalizing factor of $1/2$, these are the analysis filters of the Haar transform.   ■

Example 4.14. Take $\varphi(z) = \psi(z) = 1$ and $p + q > 2$ in (4.32). In this case

$$Q(z) = (1+z)^{p+q} = 1 + (p+q)z + \cdots + (p+q)z^{p+q-1} + z^{p+q} .$$

If $p + q$ is even, then $(p+q)z^{p+q-1}$ has odd degree, whereas if $p + q$ is odd, then z^{p+q} has odd degree. So when $p + q > 2$, the polynomial $Q(z)$ has two or more terms of odd degree. Hence condition (4.45) is *not* satisfied, and $\mathbf{h}_0$, $\mathbf{h}_1$ cannot be the analysis filters for a PR filter bank.   ■

Example 4.15. We modify Example 4.14 by taking $\varphi(z) = 1 + bz + cz^2$, $\psi(z) = 1$, and $p = 2$, $q = 2$, where b and c are real parameters to be determined. Then $H_0(z) = (1 + z)^2(1 + bz + cz^2)$ and $H_1(z) = (1 - z)^2$; thus

$$Q(z) = (1 + z)^4(1 + bz + cz^2) = (1 + 4z + 6z^2 + 4z^3 + z^4)(1 + bz + cz^2) \, .$$

The terms of odd degree in $Q(z)$ are $(4c + b)z^5 + (4 + 6b + 4c)z^3 + (4 + b)z$. So if we take $b = -4$ and $c = 1$, then condition (4.45) is satisfied, and $\mathbf{h}_0$, $\mathbf{h}_1$ are the analysis filters for a PR filter bank that satisfies the lowpass/highpass condition. We have

$$H_0(z) = (1 + z)^2(1 - 4z + z^2) = 1 - 2z - 6z^2 - 2z^3 + z^4 \, ,$$
$$H_1(z) = (1 - z)^2 = 1 - 2z + z^2 \, .$$

Now modify these filters by multiplying $H_0(z)$ by the normalizing factor and time shift $-z^{-2}\sqrt{2}/8$ and multiplying $H_1(z)$ by the normalizing factor $-\sqrt{2}/4$. For simplicity we continue to denote the new filters as $\mathbf{h}_0$ and $\mathbf{h}_1$. With these modifications (which don't change the PR property) we have found the CDF$(2, 2)$ filters with z-transforms (see Example 4.20):

$$H_0(z) = \frac{\sqrt{2}}{8}\left(-z^2 + 2z + 6 + 2z^{-1} - z^{-2}\right) \, ,$$
$$H_1(z) = \frac{\sqrt{2}}{4}\left(-z^2 + 2z - 1\right) \, .$$

Observe that with these modifications $H_0(z) = H_0(z^{-1})$ (so the filter $\mathbf{h}_0$ is symmetric around 0 in the time domain) and $|H_0(1)| = |H_1(-1)| = \sqrt{2}$.

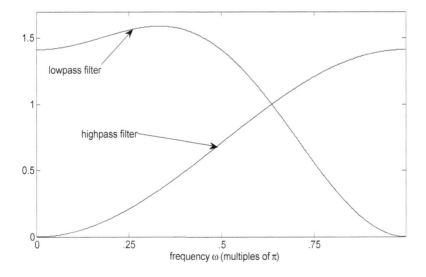

Fig. 4.6 Frequency response of CDF$(2, 2)$ analysis filters

It is instructive to determine the frequency response of these filters. Let $z = e^{\omega i}$ and for a real number a define $z^a = e^{a\omega i}$. Then $z^a + z^{-a} = 2\cos(a\omega)$. Hence

$$H_0(z) = \frac{\sqrt{2}}{4}\left(3 + 2\cos(\omega) - \cos(2\omega)\right),$$

$$H_1(z) = \frac{z\sqrt{2}}{4}(z - 2 + z^{-1}) = \frac{z\sqrt{2}}{2}\left(\cos(\omega) - 1\right).$$

Since $|z| = 1$, we obtain

$$|H_0(e^{\omega i})| = \frac{\sqrt{2}}{4}\left|3 + 2\cos(\omega) - \cos(2\omega)\right|,$$

$$|H_1(e^{\omega i})| = \frac{\sqrt{2}}{2}\left|\cos(\omega) - 1\right|.$$

The graphs of $|H_0(e^{\omega i})|$ and $|H_1(e^{\omega i})|$ are shown in Figure 4.6.   ■

4.5 Perfect Reconstruction Filter Pairs

To construct a two-channel FIR filter bank with the PR property, it suffices to specify two of the four filters. In Section 4.4 we used the lowpass and highpass analysis filters $\mathbf{h}_0$, $\mathbf{h}_1$, and expressed the PR condition as (4.45). Now we focus on the two lowpass filters.

4.5.1 *Perfect reconstruction from lowpass filters*

We can express the PR condition on a filter bank in terms of the lowpass analysis and synthesis filters as follows.

Theorem 4.8. *Suppose $\mathbf{h}_0$ and $\mathbf{g}_0$ are FIR filters. These filters are the low-pass part of a two-channel FIR filter bank with perfect reconstruction if and only if $H_0(-1) = 0$, $G_0(-1) = 0$, and*

$$H_0(z)G_0(z) + H_0(-z)G_0(-z) = 2 . \tag{4.46}$$

Conversely, if condition (4.46) is satisfied, define FIR filters $\mathbf{h}_1$ and $\mathbf{g}_1$ by

$$H_1(z) = zG_0(-z) \quad and \quad G_1(z) = z^{-1}H_0(-z) . \tag{4.47}$$

Then $\mathbf{h}_1$ and $\mathbf{g}_1$ are highpass filters. Furthermore, the set of filters $\mathbf{h}_0$, $\mathbf{h}_1$ (analysis) and $\mathbf{g}_0$, $\mathbf{g}_1$ (synthesis) give a PR filter bank.

Proof. The left side of (4.46) is the upper left entry in the matrix $\mathbf{H}_m(z)\mathbf{G}_m(z)$. If $\mathbf{h}_0$ and $\mathbf{g}_0$ are lowpass filters for a PR filter bank, then $\mathbf{H}_m(z)\mathbf{G}_m(z) = 2I$, and hence (4.46) holds. Conversely, assume condition (4.46) holds and define $H_1(z)$ and $G_1(z)$ by (4.47). Then $H_1(1) = G_0(-1) = 0$ and $G_1(1) = H_0(-1) = 0$, so $\mathbf{h}_1$ and $\mathbf{g}_1$ are highpass FIR filters. Write $F(z) = H_0(z)G_0(z)$. Then

$$\mathbf{H}_m(z)\mathbf{G}_m(z) = \begin{bmatrix} H_0(z) & H_0(-z) \\ zG_0(-z) & -zG_0(z) \end{bmatrix} \begin{bmatrix} G_0(z) & z^{-1}H_0(-z) \\ G_0(-z) & -z^{-1}H_0(z) \end{bmatrix}$$

$$= \begin{bmatrix} H_0(z)G_0(z) + H_0(-z)G_0(-z) & 0 \\ 0 & H_0(-z)G_0(-z) + H_0(z)G_0(z) \end{bmatrix} .$$

By Theorem 4.6 the condition for PR is $\mathbf{H}_m(z)\mathbf{G}_m(z) = 2I$. The calculation just made shows that the PR condition reduces to equation (4.46) in the present situation. □

4.5.2 *Lowpass filters and the Bezout polynomials*

Suppose that $\mathbf{h}_0$ and $\mathbf{g}_0$ are nonzero FIR lowpass filters. Since $H_0(-1) = 0$ and $G_0(-1) = 0$, we can write $H_0(z) = (1 + z)^q \varphi(z)$ and $G_0(z) = (1 + z)^p \psi(z)$, where p and q are positive integers, whereas $\varphi(z)$ and $\psi(z)$ are Laurent polynomials that do not vanish at $z = -1$. The PR condition (4.46) can be expressed as

$$(1 + z)^n f(z) + (1 - z)^n f(-z) = 2 , \tag{4.48}$$

where $n = p + q$ and $f(z) = \varphi(z)\psi(z)$ does not vanish at $z = -1$.

The algebraic calculations become much simpler if we make a quadratic change of variable

$$y = \frac{1}{4}\left(-z + 2 - z^{-1}\right) \tag{4.49}$$

(notice that y is unchanged when z is replaced by z^{-1}). To understand the choice of this transformation, we observe that when $z = e^{i\omega}$, then

$$y = \frac{1}{4}\left(-e^{i\omega} + 2 - e^{-i\omega}\right) = \left(\frac{e^{i\omega/2} - e^{-i\omega/2}}{2i}\right)^2 = \sin^2\frac{\omega}{2} . \tag{4.50}$$

Thus $1 - y = \cos^2(\omega/2)$. Note that the low and high frequency values $z = 1$ ($\omega = 0$) and $z = -1$ ($\omega = \pi$) correspond to $y = 0$ and $y = 1$, respectively. Furthermore,

$$1 - y = 1 + \frac{1}{4}\left(z - 2 + z^{-1}\right) = \frac{1}{4}\left(z + 2 + z^{-1}\right) . \tag{4.51}$$

Thus the replacement of z by $-z$ (frequency modulation by π) corresponds to replacing y by $1 - y$. We now prove the following key algebraic result:

Proposition 4.3 (Bezout's theorem). *For every integer $n \geq 1$ there is a unique Bezout polynomial $B_n(y)$ of degree $n - 1$ that satisfies*

$$y^n B_n(1 - y) + (1 - y)^n B_n(y) = 1 . \tag{4.52}$$

It is given by the formula

$$B_n(y) = \sum_{k=0}^{n-1} \binom{2n-1}{k} (1 - y)^{n-1-k} y^k , \tag{4.53}$$

where $\binom{m}{k} = \dfrac{m(m-1)\cdots(m-k+1)}{1 \cdot 2 \cdots k}$ is the binomial coefficient. It satisfies $B_n(y) > 1$ for all $y > 0$, and it can also be written as

$$B_n(y) = 1 + ny + \frac{n(n+1)}{1 \cdot 2} y^2 + \cdots + \binom{n+k-1}{k} y^k + \cdots + \binom{2n-2}{n-1} y^{n-1} . \tag{4.54}$$

Proof. For the moment let x and y be any variables. Consider the binomial expansions[5]

$$(x+y)^3 = x^3 + 3x^2 y + 3xy^2 + y^3$$
$$= x^2 (1+3y) + y^2 (1+3x),$$
$$(x+y)^5 = x^5 + 5x^4 y + 10x^3 y^2 + 10x^2 y^3 + 5xy^4 + y^5$$
$$= x^3 (1+5xy+10y^2) + y^3 (1+5xy+10x^2).$$

The pattern is the same for the binomial expansion of $(x+y)^{2n-1}$ with any odd exponent $2n-1$:

$$(x+y)^{2n-1} = \sum_{k=0}^{n-1} \binom{2n-1}{k} x^{2n-1-k} y^k + \sum_{k=0}^{n-1} \binom{2n-1}{2n-1-k} x^k y^{2n-1-k}.$$

There are $2n$ terms, but the second set of n binomial coefficients are the mirror symmetries of the first n coefficients:

$$\binom{2n-1}{k} = \frac{(2n-1)!}{k!(2n-1-k)!} = \binom{2n-1}{2n-1-k} \quad \text{for} \quad k=0,1,\ldots,n-1. \quad (4.55)$$

Using this symmetry, we can write the binomial expansion as

$$(x+y)^{2n-1} = \sum_{k=0}^{n-1} \binom{2n-1}{k} x^{2n-1-k} y^k + \sum_{k=0}^{n-1} \binom{2n-1}{k} y^{2n-1-k} x^k$$
$$= x^n P_n(x,y) + y^n P_n(y,x), \quad (4.56)$$

where we have factored x^n out of the first summation, factored y^n out of the second summation, and introduced the two-variable polynomial

$$P_n(x,y) = \sum_{k=0}^{n-1} \binom{2n-1}{k} x^{n-1-k} y^k.$$

This agrees with the calculation above in the cases $n=2$ and $n=3$.

To obtain the Bezout polynomial, set $x = 1-y$ and let $B_n(y) = P_n(1-y, y)$. Then $x+y = 1$ and the binomial expansion (4.56) becomes

$$1 = (1-y)^n P_n(1-y, y) + y^n P_n(y, 1-y) = (1-y)^n B_n(y) + y^n B_n(1-y),$$

which is the Bezout equation (4.52).

For the proof that $B_n(y)$ is the unique polynomial of degree $n-1$ that satisfies (4.52) and the verification of equation (4.54) see Exercises 4.12 #9. Since the coefficients in equation (4.54) are all positive, it follows that $B_n(y) > 1$ when $y > 0$. □

---

[5]This argument is from [Gundy (2007), §12], who points out that $(1-y)^n B_n(y)$ is the probability of one player winning in the famous *problem of the points* of Fermat from 1654.

4.5.3   CDF(p, q) filters

With the Bezout polynomials as our tool, it is easy to construct the filters for the CDF(p, q) family of wavelet transforms, where p and q are positive integers and $p + q = 2n$ is an even integer. Define y by the quadratic change of variable (4.49). We factor out the term z^{-1} in (4.49) and (4.51) to obtain the equivalent formulas

$$y = (-4z)^{-1} (1 - z)^2 \quad \text{and} \quad 1 - y = (4z)^{-1} (1 + z)^2 . \qquad (4.57)$$

Thus we can write the Bezout equation (4.52) in terms of z as

$$
(1 + z)^{2n} (4z)^{-n} B_n \left(\frac{-z + 2 - z^{-1}}{4} \right)
$$
$$
+ (1 - z)^{2n} (-4z)^{-n} B_n \left(\frac{z + 2 + z^{-1}}{4} \right) = 1. \qquad (4.58)
$$

Define[6]

$$G_0(z) = \frac{\sqrt{2}}{2^p} (1 + z)^p , \qquad (4.59)$$

$$H_0(z) = \frac{\sqrt{2}}{2^q} z^{-n} (1 + z)^q B_n \left(\frac{-z + 2 - z^{-1}}{4} \right) . \qquad (4.60)$$

Since $4^n = 2^{2n} = 2^p 2^q$, we see from (4.58) and the factors of $\sqrt{2}$ included in the definition that $H_0(z)G_0(z) + H_0(-z)G_0(-z) = 2$. Clearly $H_0(-1) = G_0(-1) = 0$ and $H_0(1) = G_0(1) = \sqrt{2}$. Thus the lowpass filters $\mathbf{h}_0$ and $\mathbf{g}_0$ satisfy the conditions of Theorem 4.8.

Following Theorem 4.8, we take the highpass filters to have z-transforms

$$H_1(z) = zG_0(-z) = \frac{\sqrt{2}}{2^p} z (1 - z)^p , \qquad (4.61)$$

$$G_1(z) = z^{-1}H_0(-z) = (-1)^n \frac{\sqrt{2}}{2^q} (1 - z)^q z^{-(n+1)} B_n \left(\frac{z + 2 + z^{-1}}{4} \right) . \qquad (4.62)$$

The parameters (p, q) give the orders of vanishing of $G_0(z)$ and $H_0(z)$ at $z = -1$, which is the same as the order of vanishing of $H_1(z)$ and $G_1(z)$ at $z = 1$.

Remark 4.4. In the formula for the CDF filters, both factors of $\sqrt{2}$ can be put on one of the lowpass filters; when this is done all the filter coefficients become rational numbers with denominators that are powers of 2 (bit shifts), since the polynomial $B_n(y)$ has integer coefficients.

In Chapter 3 we used the lifting process to construct several examples of the CDF family of wavelet transforms. The CDF$(1, 1)$ filters give the Haar transform, up to normalizing constants (Example 4.13); this is immediate from the definition, since $B_1(y) = 1$. From the equation $B_2(y) = 1 + 2y$ it is easy to check that the PR filters in Example 4.15 are the same as the filters just defined when $p = q = 2$, up

---

[6]The spitting of the term z^{-n} (which corresponds to a time shift) between the two filters can be done in other ways for alignment purposes; here we have simply put it into $H_0(z)$.

to powers of z (time shifts). We will return to this filter pair in Example 4.20. We now examine some other filter pairs in this family of wavelet transforms.

Example 4.16. For the CDF$(3,1)$ filters, we take $p = 3$ and $q = 1$ in equations (4.59) and (4.60). Then $n = (p+q)/2 = 2$ and $B_2(y) = 1 + 2y$, so we obtain

$$B_2\left(\frac{-z + 2 - z^{-1}}{4}\right) = 1 + \frac{1}{2}\left(-z + 2 - z^{-1}\right) = \frac{1}{2}\left(-z + 4 - z^{-1}\right).$$

Thus the lowpass filters have z-transforms

$$G_0(z) = \frac{\sqrt{2}}{8}(z+1)^3 = \frac{\sqrt{2}}{8}\left(z^3 + 3z^2 + 3z + 1\right),$$

$$H_0(z) = \frac{\sqrt{2}}{4}(z+1)z^{-2}\left(-z + 4 - z^{-1}\right) = \frac{\sqrt{2}}{4}\left(-1 + 3z^{-1} + 3z^{-2} - z^{-3}\right).$$

The highpass filters have z-transforms

$$G_1(z) = z^{-1}H_0(-z) = \frac{\sqrt{2}}{4}\left(-z^{-1} - 3z^{-2} + 3z^{-3} + z^{-4}\right),$$

$$H_1(z) = zG_0(-z) = \frac{\sqrt{2}}{8}\left(-z^4 + 3z^3 - 3z^2 + z\right).$$

The filters all have length four and are the following linear combinations of unit impulses:

$$\mathbf{g}_0 = \frac{\sqrt{2}}{8}\left(\delta_{-3} + 3\delta_{-2} + 3\delta_{-1} + \delta_0\right), \quad \mathbf{g}_1 = \frac{\sqrt{2}}{4}\left(-\delta_{-1} - 3\delta_{-2} + 3\delta_{-3} + \delta_{-4}\right),$$

$$\mathbf{h}_0 = \frac{\sqrt{2}}{4}\left(-\delta_0 + 3\delta_1 + 3\delta_2 - \delta_3\right), \quad \mathbf{h}_1 = \frac{\sqrt{2}}{8}\left(-\delta_{-4} + 3\delta_{-3} - 3\delta_{-2} + \delta_{-1}\right).$$

To examine the frequency responses of these filters, let $z = e^{\omega i}$. Then $2\cos(a\omega) = z^a + z^{-a}$ and $2i\sin(a\omega) = z^a - z^{-a}$, so

$$G_0(z) = (\sqrt{2}/8)z^{3/2}\left(z^{3/2} + 3z^{1/2} + 3z^{-1/2} + z^{-3/2}\right)$$

$$= (\sqrt{2}/4)z^{3/2}\left(\cos(3\omega/2) + 3\cos(\omega/2)\right).$$

Since $|z^{3/2}| = 1$, we see that

$$|G_0(e^{\omega i})| = (\sqrt{2}/4)\left|\cos(3\omega/2) + 3\cos(\omega/2)\right|.$$

In the same way we can write

$$G_1(z) = -(\sqrt{2}/4)z^{-5/2}\left(z^{3/2} + 3z^{1/2} - 3z^{-1/2} - z^{-3/2}\right)$$

$$= -(\sqrt{2}i/2)z^{-5/2}\left(\sin(3\omega/2) + 3\sin(\omega/2)\right).$$

Since $|z^{-5/2}| = 1$, we see that

$$|G_1(e^{\omega i})| = (\sqrt{2}/2)\left|\sin(3\omega/2) + 3\sin(\omega/2)\right|.$$

In the relations $H_0(z) = -zG_1(-z)$ and $H_1(z) = zG_0(-z)$, multiplication by z gives a time shift, whereas the interchange of z and $-z$ gives a frequency modulation

$\omega \longleftrightarrow \omega + \pi$ between low and high frequencies. Since $\cos(\omega + 3\pi/2) = \sin(\omega)$ and $\cos(\omega + \pi/2) = -\sin(\omega)$, the formulas we just obtained now imply that

$$|H_0(e^{\omega i})| = (\sqrt{2}/2)\big|\cos(3\omega/2) - 3\cos(\omega/2)\big|,$$
$$|H_1(e^{\omega i})| = (\sqrt{2}/4)\big|\sin(3\omega/2) - 3\sin(\omega/2)\big|.$$

Notice the *modulation* in the frequency responses of the pairs of filters: the lowpass analysis filter is shifted (modulated) by π in the frequency domain to become the highpass synthesis filter. The same shift changes the highpass analysis filter into the lowpass synthesis filter.

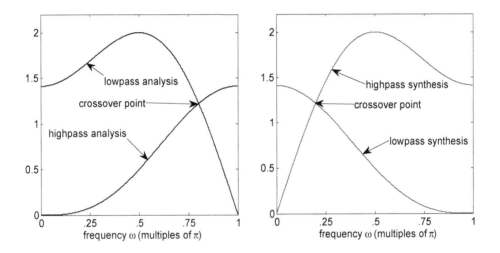

Fig. 4.7   Frequency response of CDF$(3, 1)$ filters

The construction of the CDF$(3, 1)$ filters makes the highpass analysis filter vanish to order three at $\omega = 0$ and the lowpass analysis filter vanish to order one at $\omega = \pi$. The *crossover point* of the pair of analysis filters, at which $|H_0(e^{\omega i})| = |H_1(e^{\omega i})|$, occurs at a relatively high frequency ω, as we see in Figure 4.7. Thus very little mid-range frequency energy (when $|\omega| \approx \pi/2$) in the input signal goes through the highpass analysis channel, whereas a large part of this mid-range frequency energy goes through the lowpass analysis channel of the filter bank. For the pair of synthesis filters the situation is reversed, both for the order of vanishing and the crossover point. The highpass synthesis filter then passes a large part of the mid-range energy coming from the lowpass channel to achieve perfect reconstruction of the input signal. ∎

Example 4.17 (CDF$(6, 2)$ filters). Consider the filter bank in the CDF family obtained by taking $p = 6$ and $q = 2$ in equations (4.59) and (4.60). The lowpass synthesis filter has z-transform given by the binomial expansion of $(1 + z)^6$ with a normalizing factor. We can write it in symmetric form with a power of z (time shift):

$$G_0(z) = cz^3 \left(z^{-3} + 6z^{-2} + 15z^{-1} + 20 + 15z + 6z^2 + z^3 \right),$$

where $c = \sqrt{2}/64$. In this case $n = (p+q)/2 = 4$ and $B_4(y) = 1 + 4y + 10y^2 + 20y^3$. With the aid of a computer algebra system, we calculate that the highpass synthesis filter has z-transform

$$G_1(z) = 16c(1-z)^2 z^{-5} B_4 \left(\frac{z + 2 + z^{-1}}{4} \right)$$
$$= cz^{-4} \left(5z^{-4} + 30z^{-3} + 56z^{-2} - 14z^{-1} - 154 - 14z + 56z^2 + 30z^3 + 5z^4 \right).$$

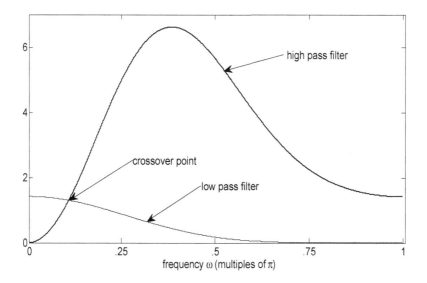

Fig. 4.8   Frequency response of CDF$(6, 2)$ synthesis filters

To determine the frequency response of these filters, we proceed as in Example 4.20. When $z = e^{\omega i}$, then from the symmetric form of $G_0(z)$ and $G_1(z)$ given above, together with the relation $z^a + z^{-a} = 2\cos(a\omega)$, we obtain

$$G_0(z) = 2cz^3 \left(\cos(3\omega) + 6\cos(2\omega) + 15\cos(\omega) + 10 \right),$$
$$G_1(z) = 2cz^{-4} \left(5\cos(4\omega) + 30\cos(3\omega) + 56\cos(2\omega) - 14\cos(\omega) - 77 \right).$$

The graphs of $|G_0(e^{\omega i})|$ and $|G_1(e^{\omega i})|$ are shown in Figure 4.8. Notice how far the crossover point has moved into the low-frequency range, and how much of the low to middle frequencies are emphasized by the high-frequency filter.   ∎

Example 4.18 (Daubechies 9/7 filters). The imbalance between the lowpass and the highpass CDF$(6, 2)$ filters that is evident in Figure 4.8 can be remedied by factoring the polynomial $B_4(y)$ that occurs in $H_0(z)$, shifting the linear factor with a real root from $H_0(z)$ to $G_0(z)$, and shifting a factor $(1+z)^2$ from $G_0(z)$ to $H_0(z)$. With these changes both lowpass filters vanish to the same order 4 at $z = -1$, and still have lengths 9 and 7. The resulting pair of filters, called the *Daubechies 9/7 filters*, are used in a five-level wavelet transform in the JPEG 2000 image algorithm for lossy image compression. They give excellent results, at the price of more computational cost and the elimination of rational coefficients. Since the roots of $B_4(y)$ are irrational numbers, high precision floating-point arithmetic operations would be needed for maximum accuracy; in practice, all arithmetic calculations are quantized to a fixed number of bits and then encoded, as mentioned in Section 1.2. See [Van Fleet (2008), Ch. 12] and [Walker (2008), §4.7] for details.

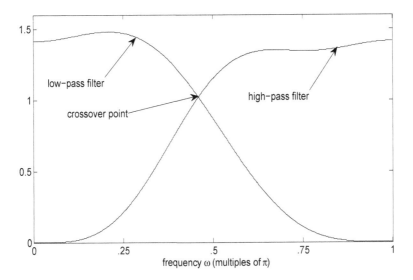

Fig. 4.9   Frequency response of Daub 9/7 synthesis filters

The graphs of $|G_0(e^{\omega i})|$ and $|G_1(e^{\omega i})|$ are shown in Figure 4.9. Notice how the crossover point is now almost at the mid-point, and how evenly the low and high frequency ranges are split between the two filters (see Exercise 4.12 #13 for more details).   ∎

4.6   Comparing Polyphase and Modulation Matrices

A one-scale discrete wavelet analysis transform can be implemented in two ways:

Lazy filter bank and polyphase matrix: A finite signal $\mathbf{x}$ is split by downsampling into $\mathbf{x}_0 = \boxed{2\downarrow}\,\mathbf{x}$ and $\mathbf{x}_1 = \boxed{2\downarrow}\,S^{-1}\mathbf{x}$ (this is the same as passing the signal through the analysis part of the lazy filter bank). Then lifting steps (predictions, updates, and a normalization) are applied to $\begin{bmatrix} \mathbf{x}_0 \\ \mathbf{x}_1 \end{bmatrix}$ to give an output $\begin{bmatrix} \mathbf{y}_0 \\ \mathbf{y}_1 \end{bmatrix}$ (see

Section 4.3.2). The z-transform of the output is

$$\begin{bmatrix} Y_0(z) \\ Y_1(z) \end{bmatrix} = \mathbf{H}_p(z) \begin{bmatrix} X_0(z) \\ X_1(z) \end{bmatrix} = \frac{1}{2}\mathbf{H}_p(z) \begin{bmatrix} X(z^{1/2}) + X(-z^{1/2}) \\ z^{1/2}X(z^{1/2}) - z^{1/2}X(-z^{1/2}) \end{bmatrix},$$

where $\mathbf{H}_p(z)$ is the polyphase analysis matrix from equation (4.27). Replacing z by z^2 in these equations, we get

$$\begin{bmatrix} Y_0(z^2) \\ Y_1(z^2) \end{bmatrix} = \frac{1}{2}\mathbf{H}_p(z^2) \begin{bmatrix} 1 & 1 \\ z & -z \end{bmatrix} \begin{bmatrix} X(z) \\ X(-z) \end{bmatrix}. \tag{4.63}$$

Two-channel filter bank: First the signal $\mathbf{x}$ is filtered by h_0 and h_1. Then the two filtered signals are downsampled to give the output

$$\begin{bmatrix} \mathbf{y}_0 \\ \mathbf{y}_1 \end{bmatrix} = \begin{bmatrix} \boxed{2\downarrow}(\mathbf{h}_0 * \mathbf{x}) \\ \boxed{2\downarrow}(\mathbf{h}_1 * \mathbf{x}) \end{bmatrix}$$

(see Section 4.4.2). The z-transform of the output is

$$\begin{bmatrix} Y_0(z) \\ Y_1(z) \end{bmatrix} = \frac{1}{2} \begin{bmatrix} H_0(z^{1/2})X(z^{1/2}) + H_0(-z^{1/2})X(-z^{1/2}) \\ H_1(z^{1/2})X(z^{1/2}) + H_1(-z^{1/2})X(-z^{1/2}) \end{bmatrix}.$$

Replacing z by z^2 in this equation, we get the relation

$$\begin{bmatrix} Y_0(z^2) \\ Y_1(z^2) \end{bmatrix} = \frac{1}{2}\mathbf{H}_m(z) \begin{bmatrix} X(z) \\ X(-z) \end{bmatrix}, \tag{4.64}$$

where $\mathbf{H}_m(z)$ is the modulation analysis matrix.

The four Laurent polynomials in the analysis modulation matrix are completely determined by the two polynomials in the first row, whereas the polyphase analysis matrix coming from the lifting steps has four different polynomial entries, subject only to the constraint that the determinant must be a nonzero constant. These two matrices are related as follows.

Theorem 4.9. *Let $\mathbf{H}_p(z)$ be the polyphase analysis matrix for a one-scale wavelet transform obtained by the lifting procedure. Define*

$$\mathbf{H}_m(z) = \mathbf{H}_p(z^2) \begin{bmatrix} 1 & 1 \\ z & -z \end{bmatrix}. \tag{4.65}$$

Then $\mathbf{H}_m(z)$ is the analysis modulation matrix for a two-channel filter bank with perfect reconstruction. The analysis filters for this filter bank have z-transforms

$$H_0(z) = H_{00}(z^2) + zH_{01}(z^2), \quad H_1(z) = H_{10}(z^2) + zH_{11}(z^2). \tag{4.66}$$

Proof. Carrying out the matrix multiplication in (4.65), we find that

$$\mathbf{H}_m(z) = \begin{bmatrix} H_{00}(z^2) + zH_{01}(z^2) & H_{00}(z^2) - zH_{01}(z^2) \\ H_{10}(z^2) + zH_{11}(z^2) & H_{10}(z^2) - zH_{11}(z^2) \end{bmatrix}.$$

This is the analysis modulation matrix for the filters defined by equations (4.66). Its determinant is $-2z \det \mathbf{H}_p(z)$. But a polyphase analysis matrix obtained by the lifting procedure is the product of upper and lower triangular matrices of determinant 1 and a diagonal normalization matrix whose determinant is a nonzero constant c. Now apply Theorem 4.7. $\qquad\square$

Example 4.19. The polyphase analysis matrix of the (unnormalized) Haar transform is $\mathbf{H}_p(z) = \frac{1}{2}\begin{bmatrix} 1 & 1 \\ 1 & -1 \end{bmatrix}$. Hence the analysis modulation matrix is

$$\mathbf{H}_m(z) = \frac{1}{2}\begin{bmatrix} 1 & 1 \\ 1 & -1 \end{bmatrix}\begin{bmatrix} 1 & 1 \\ z & -z \end{bmatrix} = \frac{1}{2}\begin{bmatrix} (1+z) & (1-z) \\ (1-z) & (1+z) \end{bmatrix}.$$

Thus $H_0(z) = \frac{1}{2}(1+z)$ and $H_1(z) = \frac{1}{2}(1-z)$. These are the z-transforms of the filters $\mathbf{h}_0 = \frac{1}{2}(\delta_0 + \delta_{-1})$ and $\mathbf{h}_1 = \frac{1}{2}(\delta_0 - \delta_{-1})$ that take averages and differences of adjacent signal values. Since $\det\mathbf{H}_p(z) = -1/2$, equations (4.43) give

$$G_0(z) = 2z^{-1}H_1(-z) = 1 + z^{-1} \quad \text{and} \quad G_1(z) = -2z^{-1}H_0(-z) = 1 - z^{-1}.$$

These are the z-transforms of the filters $\mathbf{g}_0 = \delta_0 + \delta_1$ and $\mathbf{g}_1 = \delta_0 - \delta_1$. If we compare these filters with those given by formulas (4.59) and (4.60) with $p = q = 1$, we find that they differ by factors of z and z^{-1} and $1/\sqrt{2}$. These factors correspond to a time shift in each channel and a normalization; they don't change the perfect reconstruction property. ∎

Example 4.20. The polyphase analysis matrix of the CDF$(2,2)$ transform is

$$\mathbf{H}_p(z) = \frac{\sqrt{2}}{8}\begin{bmatrix} (-z+6-z^{-1}) & (2+2z^{-1}) \\ -(2+2z) & 4 \end{bmatrix}$$

(see Example 4.9). Hence the analysis modulation matrix is

$$\mathbf{H}_m(z) = \frac{\sqrt{2}}{8}\begin{bmatrix} (-z^2+6-z^{-2}) & (2+2z^{-2}) \\ -(2+2z^2) & 4 \end{bmatrix}\begin{bmatrix} 1 & 1 \\ z & -z \end{bmatrix}$$

$$= \frac{\sqrt{2}}{8}\begin{bmatrix} (-z^2+2z+6+2z^{-1}-z^{-2}) & (-z^2-2z+6-2z^{-1}-z^{-2}) \\ (-2z^2+4z-2) & (-2z^2-4z-2) \end{bmatrix}.$$

Thus the analysis filters have z-transforms $H_0(z) = (\sqrt{2}/8)(-z^2+2z+6+2z^{-1}-z^{-2})$ and $H_1(z) = (\sqrt{2}/4)(-z^2+2z-1)$. These are the z-transforms of the filters

$$\mathbf{h}_0 = \frac{\sqrt{2}}{8}(-\delta_{-2} + 2\delta_{-1} + 6\delta_0 + 2\delta_1 - \delta_2), \quad \mathbf{h}_1 = \frac{\sqrt{2}}{4}(-\delta_{-2} + 2\delta_{-1} - \delta_0).$$

From the factored form of $\mathbf{H}_p(z)$ in Example 4.9 we see that $\det\mathbf{H}_p(z) = 1$. Hence equations (4.43) give

$$G_0(z) = -z^{-1}H_1(-z) = \frac{\sqrt{2}}{4}(z+2+z^{-1}),$$

$$G_1(z) = z^{-1}H_0(-z) = \frac{\sqrt{2}}{8}(-z-2+6z^{-1}-2z^{-2}-z^{-3}).$$

These are the z-transforms of the filters

$$\mathbf{g}_0 = \frac{\sqrt{2}}{4}(\delta_{-1} + 2\delta_0 + \delta_1), \quad \mathbf{g}_1 = \frac{\sqrt{2}}{8}(-\delta_{-1} - 2\delta_0 + 6\delta_1 - 2\delta_2 - \delta_3).$$

If we compare these filters with those given by formulas (4.59) and (4.60), we see that they differ by factors of z and z^{-1}. These factors correspond to time shifts in each channel, but don't affect perfect reconstruction. ∎

Equation (4.65) determines the z-transforms of the filters from the entries in the analysis modulation matrix. Relation (4.66) lets us go the other direction. Calculating the entries in the polyphase analysis matrix from the z-transforms of the filters amounts to writing a Laurent polynomial $P(z)$, such as $H_0(z)$ or $H_1(z)$, as

$$P(z) = Q(z^2) + zR(z^2).$$

Here $Q(z)$ is the unique Laurent polynomial such that $Q(z^2)$ is the sum of the even-degree terms in $P(z)$, and $R(z)$ is the unique Laurent polynomial such that $zR(z^2)$ is the sum of the odd-degree terms in $P(z)$. For example, if

$$P(z) = z^{-3} + 6z^{-2} + 15z^{-1} + 20 + 15z + 6z^2 + z^3$$

(as in Example 4.17), then

$$Q(z^2) = 6z^{-2} + 20 + 6z^2,$$
$$zR(z^2) = z^{-3} + 15z^{-1} + 15z + z^3 = z\left(z^{-4} + 15z^{-2} + 15 + z^2\right).$$

Hence $Q(z) = 6z^{-1} + 20 + 6z$ and $R(z) = z^{-2} + 15z^{-1} + 15 + z$ in this case.

Example 4.21. Consider the filter bank with filters

$$H_0(z) = \frac{1}{2}z^2 + z + \frac{1}{2} = \frac{1}{2}(z+1)^2 \quad \text{and} \quad H_1(z) = -\frac{3}{4} + \frac{1}{2}z^{-1} + \frac{1}{4}z^{-2}.$$

These filters satisfy the lowpass/highpass conditions $H_0(-1) = 0$ and $H_1(1) = 0$. Furthermore,

$$H_0(z)H_1(-z) = -\frac{3}{8}z^2 - z - \frac{3}{4} + \frac{1}{8}z^{-2}$$

has exactly one term of odd degree, so the PR condition (4.45) is satisfied. To obtain the entries in the polyphase analysis matrix, write

$$H_0(z) = \frac{1}{2}\left(z^2 + 1\right) + z = H_{00}(z^2) + zH_{01}(z^2),$$
$$H_1(z) = \frac{1}{4}\left(-3 + z^{-2}\right) + \frac{z}{2}\left(z^{-2}\right) = H_{10}(z^2) + zH_{11}(z^2).$$

By the even-odd splitting of a polynomial described above, we see that

$$\mathbf{H}_p(z) = \begin{bmatrix} \frac{1}{2}(z+1) & 1 \\ \frac{1}{4}\left(-3 + z^{-1}\right) & \frac{1}{2}z^{-1} \end{bmatrix}$$

is the polyphase analysis matrix for this filter bank. ∎

4.7   Lifting Step Factorization of Polyphase Matrices

We now describe the last step in the efficient implementation of two-channel filter banks in polyphase form. Given a two-channel filter bank with perfect reconstruction, we define the polyphase analysis matrix by equation (4.65). From the PR condition we know that $\det \mathbf{H}_m(z) = cz^{2k+1}$ for some constant $c \neq 0$ and some integer k (see Theorem 4.7). We change the filter $\mathbf{h}_1$ to $-(2/c)\delta_{2k} \star \mathbf{h}_1$ (a time shift and a rescaling). Then $H_1(z)$ is multiplied by $-(2/c)z^{-2k}$ and $\det \mathbf{H}_m(z)$ becomes $-2z$. Thus the modified filter bank still has the PR property and the highpass condition $H_1(1) = 0$ is still satisfied. From equation (4.65) we then have

$$\det \mathbf{H}_p(z^2) = -[1/(2z)] \det \mathbf{H}_m(z) = 1.$$

In the following we assume that the filters have been modified as just described to obtain $\det \mathbf{H}_p(z) = 1$ for all $z \neq 0$. To implement the lifting scheme we need to factor $\mathbf{H}_p(z)$ into a product of a diagonal matrix $\mathrm{diag}[c, \; c^{-1}]$ (for some constant $c \neq 0$) and upper-triangular or lower-triangular matrices with 1 in the diagonal positions and zero or a Laurent polynomial in the off-diagonal positions. Each of the matrix factors corresponds to a lifting step (prediction, update, or normalization). Here is the basic result about matrices with determinant 1 whose entries are Laurent polynomials.

Theorem 4.10.   Let $\mathbf{F}(z) = \begin{bmatrix} a(z) & b(z) \\ c(z) & d(z) \end{bmatrix}$ with $a(z)$, $b(z)$, $c(z)$, and $d(z)$ Laurent polynomials. Assume that $\det \mathbf{F}(z) = 1$ for all $z \neq 0$. Then $\mathbf{F}(z)$ can be factored into a product of matrices of the form $\mathbf{P}(z) = \begin{bmatrix} 1 & 0 \\ g(z) & 1 \end{bmatrix}$ and $\mathbf{U}(z) = \begin{bmatrix} 1 & h(z) \\ 0 & 1 \end{bmatrix}$.

Proof. Call a matrix of the form $\mathbf{P}(z)$ a *prediction matrix* and a matrix of the form $\mathbf{U}(z)$ an *update matrix*. When $c(z) = 0$ so that $\mathbf{F}(z)$ is an upper triangular matrix the proof is easy. The condition $\det \mathbf{F}(z) = 1$ means that $d(z) = 1/a(z)$ in this case. But then the Laurent polynomial $a(z)$ must be a monomial cz^k with some scalar $c \neq 0$ and integer exponent k in order for $1/a(z)$ also to be a Laurent polynomial (Lemma 4.2). Hence we can use an elementary row operation to change $\mathbf{F}(z)$ into a diagonal matrix:

$$\begin{bmatrix} 1 & -a(z)b(z) \\ 0 & 1 \end{bmatrix} \begin{bmatrix} a(z) & b(z) \\ 0 & 1/a(z) \end{bmatrix} = \begin{bmatrix} a(z) & 0 \\ 0 & 1/a(z) \end{bmatrix}.$$

Next we use the factorization

$$\begin{bmatrix} a(z) & 0 \\ 0 & 1/a(z) \end{bmatrix} = \begin{bmatrix} 1 & -a(z) \\ 0 & 1 \end{bmatrix} \begin{bmatrix} 1 & 0 \\ (1/a(z)) - 1 & 1 \end{bmatrix} \begin{bmatrix} 1 & 1 \\ 0 & 1 \end{bmatrix} \begin{bmatrix} 1 & 0 \\ (a(z)) - 1 & 1 \end{bmatrix} \tag{4.67}$$

(verification left as an exercise). The inverse of a unit upper triangular matrix is unit upper triangular:

$$\begin{bmatrix} 1 & -a(z)b(z) \\ 0 & 1 \end{bmatrix}^{-1} = \begin{bmatrix} 1 & a(z)b(z) \\ 0 & 1 \end{bmatrix}.$$

Also

$$\begin{bmatrix} 1 & a(z)b(z) \\ 0 & 1 \end{bmatrix} \begin{bmatrix} 1 & -a(z) \\ 0 & 1 \end{bmatrix} = \begin{bmatrix} 1 & a(z)(b(z)-1) \\ 0 & 1 \end{bmatrix}.$$

Thus we can write our original matrix $\mathbf{F}(z)$ as the product of four prediction and update matrices:

$$\begin{bmatrix} a(z) & b(z) \\ 0 & 1/a(z) \end{bmatrix} = \begin{bmatrix} 1 & a(z)b(z) \\ 0 & 1 \end{bmatrix} \begin{bmatrix} a(z) & 0 \\ 0 & 1/a(z) \end{bmatrix}$$

$$= \begin{bmatrix} 1 & a(z)(b(z)-1) \\ 0 & 1 \end{bmatrix} \begin{bmatrix} 1 & 0 \\ (1/a(z)-1) & 1 \end{bmatrix} \begin{bmatrix} 1 & 1 \\ 0 & 1 \end{bmatrix} \begin{bmatrix} 1 & 0 \\ (a(z)-1) & 1 \end{bmatrix}.$$

This proves the theorem for the special case of an upper-triangular matrix.

If $a(z) = 0$, then we can interchange the rows of $\mathbf{F}(z)$ as follows:

$$\begin{bmatrix} 0 & -1 \\ 1 & 0 \end{bmatrix} \begin{bmatrix} 0 & b(z) \\ c(z) & d(z) \end{bmatrix} = \begin{bmatrix} -c(z) & -d(z) \\ 0 & b(z) \end{bmatrix}.$$

We have already showed that the matrix on the right is a product of prediction and update matrices. Since

$$\begin{bmatrix} 0 & -1 \\ 1 & 0 \end{bmatrix} = \begin{bmatrix} 1 & -1 \\ 0 & 1 \end{bmatrix} \begin{bmatrix} 1 & 0 \\ 1 & 1 \end{bmatrix} \begin{bmatrix} 1 & -1 \\ 0 & 1 \end{bmatrix} \tag{4.68}$$

is also a product of prediction and update matrices, the same is true for the original matrix $\mathbf{F}(z)$.

In the general case when $a(z) \neq 0$ and $c(z) \neq 0$, we would like to multiply by prediction and update matrices to make $\mathbf{F}(z)$ an upper-triangular matrix. Just as in the case of row reduction of a matrix with scalar entries, we have

$$\begin{bmatrix} 1 & 0 \\ -c(z)/a(z) & 1 \end{bmatrix} \begin{bmatrix} a(z) & b(z) \\ c(z) & d(z) \end{bmatrix} = \begin{bmatrix} a(z) & b(z) \\ 0 & 1/a(z) \end{bmatrix} \tag{4.69}$$

(here we have used the condition $a(z)d(z) - b(z)c(z) = 1$ to obtain the entry $1/a(z)$ in the matrix on the right). But the rational functions $1/a(z)$ and $c(z)/a(z)$ are *not* Laurent polynomials, in general. So

$$\begin{bmatrix} 1 & 0 \\ -c(z)/a(z) & 1 \end{bmatrix}$$

is *not* a valid prediction matrix and the matrix on the right in (4.69) does not have Laurent polynomial entries.

The way around this obstacle is to use the *Euclidean division algorithm with remainder* for Laurent polynomials. We define the *degree* of a nonzero Laurent polynomial

$$f(z) = a_p z^p + \cdots + a_q z^q \quad \text{(where } p \leq q, \ a_p \neq 0, \text{ and } a_q \neq 0)$$

as $\deg f(z) = q - p$. Then $1/f(z)$ is a Laurent polynomial if and only if $\deg f(z) = 0$. So the obstacle for row reduction to upper-triangular form arises when $\deg a(z) \geq 1$. We may assume that $\deg c(z) \leq \deg a(z)$, since multiplying $\mathbf{F}(z)$ on the left by the matrices in (4.68) replaces $a(z)$ by $-c(z)$ and $c(z)$ by $a(z)$. By the Euclidean division algorithm we can write $a(z) = c(z)g(z) + r(z)$, where $g(z)$ and $r(z)$ (the remainder) are Laurent polynomials with

$$\deg c(z) + \deg g(z) = \deg a(z) \quad \text{and} \quad \deg r(z) < \deg a(z).$$

We multiply by an update matrix to reduce the degree of the upper left-hand entry in $\mathbf{F}(z)$:

$$\begin{bmatrix} 1 & -g(z) \\ 0 & 1 \end{bmatrix} \begin{bmatrix} a(z) & b(z) \\ c(z) & d(z) \end{bmatrix} = \begin{bmatrix} (a(z) - c(z)g(z)) & (b(z) - b(z)g(z)) \\ c(z) & d(z) \end{bmatrix}$$
$$= \begin{bmatrix} r(z) & (b(z) - b(z)g(z)) \\ c(z) & d(z) \end{bmatrix}. \tag{4.70}$$

In the new matrix on the right the upper left-hand entry $a(z)$ of the original matrix $\mathbf{F}(z)$ has been replaced by $r(z)$ and we have reduced the degree of this entry: $\deg r(z) < \deg a(z)$.

We can always arrange that in the new matrix the degree of the upper left-hand entry is at least as big as the degree of the lower left-hand entry (use (4.68) to switch rows if necessary). Now we repeat this process a finite number of times to reduce the degree of the upper left-hand entry in the matrix to zero. Finally, we use (4.69) when $\deg a(z) = 0$ to obtain an upper-triangular matrix.[7] □

The proof of Theorem 4.10 involves many choices of prediction and update matrices to obtain the factorization. These choices are not unique; the only constraint is that the off-diagonal matrix entries must always be Laurent polynomials. Furthermore, it is not mandatory to factor a constant diagonal matrix of determinant 1 into prediction/update matrices as in (4.67); it can be moved to the left as a normalization matrix. We illustrate this with the following example.

Example 4.22. Consider the filter bank in Example 4.21 with polyphase analysis matrix

$$\mathbf{H}_p(z) = \begin{bmatrix} \left(\frac{1}{2}z + \frac{1}{2}\right) & 1 \\ \left(-\frac{3}{4} + \frac{1}{4}z^{-1}\right) & \frac{1}{2}z^{-1} \end{bmatrix}.$$

The analysis modulation matrix has determinant $H_0(z)H_1(-z) - H_0(-z)H_1(z) = -2z$, and $\det \mathbf{H}_p(z) = 1$.

We factor the polyphase analysis matrix following the method in the proof of Theorem 4.10. The first step is to find elementary prediction and update matrices to change $\mathbf{H}_p(z)$ into an upper-triangular matrix. The entries in column one have degree one, so we don't need to interchange the rows. We begin by dividing $H_{00}(z)$ by $H_{10}(z)$ with remainder:

$$H_{00}(z) = \frac{1}{2}z + \frac{1}{2} = \left(-\frac{3}{4} + \frac{1}{4}z^{-1}\right)\left(-\frac{2}{3}z\right) + \frac{2}{3} = H_{00}(z)g(z) + r(z),$$

---

[7]See [Daubechies and Sweldens (1998)] for a more detailed exposition.

where $g(z) = -\frac{2}{3}z$ has degree $1 - 1 = 0$ and the remainder $r(z) = \frac{2}{3}$ also has degree 0. We multiply the polyphase analysis matrix by the update matrix built with $-g(z)$ to reduce the degree of the upper left-hand entry:

$$\begin{bmatrix} 1 & \frac{2}{3}z \\ 0 & 1 \end{bmatrix} \begin{bmatrix} (\frac{1}{2}z + \frac{1}{2}) & 1 \\ (-\frac{3}{4} + \frac{1}{4}z^{-1}) & \frac{1}{2}z^{-1} \end{bmatrix} = \begin{bmatrix} \frac{2}{3} & \frac{4}{3} \\ (-\frac{3}{4} + \frac{1}{4}z^{-1}) & \frac{1}{2}z^{-1} \end{bmatrix}.$$

In the new matrix the element $2/3$ in the upper left position (the remainder in the division algorithm) has degree zero, so we can use a prediction matrix as in (4.69) to make the matrix upper triangular:

$$\begin{bmatrix} 1 & 0 \\ -\frac{3}{2}(-\frac{3}{4} + \frac{1}{4}z^{-1}) & 1 \end{bmatrix} \begin{bmatrix} \frac{2}{3} & \frac{4}{3} \\ (-\frac{3}{4} + \frac{1}{4}z^{-1}) & \frac{1}{2}z^{-1} \end{bmatrix} = \begin{bmatrix} \frac{2}{3} & \frac{4}{3} \\ 0 & \frac{3}{2} \end{bmatrix}.$$

Finally, we factor the upper triangular matrix on the right as a diagonal matrix times a unit upper triangular matrix:

$$\begin{bmatrix} \frac{2}{3} & \frac{4}{3} \\ 0 & \frac{3}{2} \end{bmatrix} = \begin{bmatrix} \frac{2}{3} & 0 \\ 0 & \frac{3}{2} \end{bmatrix} \begin{bmatrix} 1 & 2 \\ 0 & 1 \end{bmatrix}.$$

Going back to the original polyphase analysis matrix, we see that using left multiplications by prediction and update matrices we have transformed it into

$$\begin{bmatrix} 1 & 0 \\ (\frac{9}{8} - \frac{3}{8}z^{-1}) & 1 \end{bmatrix} \begin{bmatrix} 1 & \frac{2}{3}z \\ 0 & 1 \end{bmatrix} \mathbf{H}_p(z) = \begin{bmatrix} \frac{2}{3} & 0 \\ 0 & \frac{3}{2} \end{bmatrix} \begin{bmatrix} 1 & 2 \\ 0 & 1 \end{bmatrix}.$$

Thus the polyphase analysis matrix factors as

$$\mathbf{H}_p(z) = \begin{bmatrix} 1 & -\frac{2}{3}z \\ 0 & 1 \end{bmatrix} \begin{bmatrix} 1 & 0 \\ (-\frac{9}{8} + \frac{3}{8}z^{-1}) & 1 \end{bmatrix} \begin{bmatrix} \frac{2}{3} & 0 \\ 0 & \frac{3}{2} \end{bmatrix} \begin{bmatrix} 1 & 2 \\ 0 & 1 \end{bmatrix}. \qquad (4.71)$$

Here we have moved the unit upper/lower triangular matrices to the right side of (4.71) using the relations

$$\begin{bmatrix} 1 & f(z) \\ 0 & 1 \end{bmatrix}^{-1} = \begin{bmatrix} 1 & -f(z) \\ 0 & 1 \end{bmatrix} \quad \text{and} \quad \begin{bmatrix} 1 & 0 \\ f(z) & 1 \end{bmatrix}^{-1} = \begin{bmatrix} 1 & 0 \\ -f(z) & 1 \end{bmatrix}$$

(where $f(z)$ is any Laurent polynomial).

Let $D = \operatorname{diag}[\frac{2}{3}, \frac{3}{2}]$ be the diagonal matrix in the factorization (4.71). At this point we could use (4.67) to express D as a product of prediction and update matrices. Instead, we will treat D as a *normalization step* and move it to the left in the product using the relations

$$\begin{bmatrix} 1 & f(z) \\ 0 & 1 \end{bmatrix} D = D \begin{bmatrix} 1 & \frac{9}{4}f(z) \\ 0 & 1 \end{bmatrix} \quad \text{and} \quad \begin{bmatrix} 1 & 0 \\ f(z) & 1 \end{bmatrix} D = D \begin{bmatrix} 1 & 0 \\ \frac{4}{9}f(z) & 1 \end{bmatrix}.$$

This gives a lifting-step factorization of the polyphase analysis matrix:

$$\mathbf{H}_p(z) = \begin{bmatrix} \frac{2}{3} & 0 \\ 0 & \frac{3}{2} \end{bmatrix} \begin{bmatrix} 1 & -\frac{3}{2}z \\ 0 & 1 \end{bmatrix} \begin{bmatrix} 1 & 0 \\ (-\frac{1}{2} + \frac{1}{6}z^{-1}) & 1 \end{bmatrix} \begin{bmatrix} 1 & 2 \\ 0 & 1 \end{bmatrix} = D U_2 P U_1.$$

This factorization means that the filter bank can be implemented as follows. The signal $\mathbf{x}$ is split into $\mathbf{x}_{\text{even}}[n] = \mathbf{x}[2n]$ and $\mathbf{x}_{\text{odd}}[n] = \mathbf{x}[2n+1]$. Then the pair of signals $\mathbf{x}_{\text{even}}$ and $\mathbf{x}_{\text{odd}}$ are transformed into the trend $\mathbf{s}$ and detail $\mathbf{d}$ by the lifting steps

$$\text{(First Update)} \quad U_1: \quad \mathbf{s}^{(1)}[n] = \mathbf{x}_{\text{even}}[n] + 2\,\mathbf{x}_{\text{odd}}[n]\,,$$

$$\text{(Prediction)} \quad P: \quad \mathbf{d}^{(1)}[n] = \mathbf{x}_{\text{odd}}[n] - \tfrac{1}{2}\mathbf{s}^{(1)}[n] + \tfrac{1}{6}\mathbf{s}^{(1)}[n-1]\,,$$

$$\text{(Second Update)} \quad U_2: \quad \mathbf{s}^{(2)}[n] = \mathbf{s}^{(1)}[n] - \tfrac{3}{2}\mathbf{d}^{(1)}[n+1]\,,$$

$$\text{(Normalization)} \quad D: \quad \mathbf{s}[n] = \tfrac{2}{3}\mathbf{s}^{(2)}[n], \qquad \mathbf{d}[n] = \tfrac{3}{2}\mathbf{d}^{(1)}[n]\,.$$

Recall that multiplication by z^{-1} is the shift operator $(S\mathbf{y})[n] = \mathbf{y}[n-1]$ in the time domain. ∎

See [Jensen and la Cour-Harbo (2001), Ch. 12] for more examples and a discussion of the numerical analysis issues associated with the non-uniqueness of the factorization.

4.8 Biorthogonal Wavelet Bases

After all these constructions with z-transforms, we return to signal processing in the time domain using a two-channel PR filter bank. Let $\mathbf{h}_0$ (lowpass) and $\mathbf{h}_1$ (highpass) be the analysis filters, and let $\mathbf{g}_0$ and $\mathbf{g}_1$ be the corresponding synthesis filters. Recall that the PR property implies that the synthesis filters are uniquely determined by the analysis filters.

The analysis part of the filter bank takes an input signal $\mathbf{x}$ and passes it through the filters $\mathbf{h}_0$ and $\mathbf{h}_1$. Then the two filtered signals are downsampled to give the output

$$\begin{bmatrix} \mathbf{y}_0 \\ \mathbf{y}_1 \end{bmatrix} = \begin{bmatrix} \boxed{2\downarrow}(\mathbf{h}_0 \star \mathbf{x}) \\ \boxed{2\downarrow}(\mathbf{h}_1 \star \mathbf{x}) \end{bmatrix}.$$

The synthesis part of the filter bank takes the pair of signals $\mathbf{y}_0$ and $\mathbf{y}_1$, upsamples each of them, passes $\boxed{2\uparrow}\mathbf{y}_0$ and $\boxed{2\uparrow}\mathbf{y}_1$ through the synthesis filters, and then adds the result to give the output

$$\widetilde{\mathbf{x}} = \mathbf{g}_0 \star \left(\boxed{2\uparrow}\mathbf{y}_0 \right) + \mathbf{g}_1 \star \left(\boxed{2\uparrow}\mathbf{y}_1 \right).$$

The *signal processing*, which is usually a nonlinear operation (such as setting small values to zero), occurs between the analysis and synthesis stages, and an input $\mathbf{x}$ produces an output $\widetilde{\mathbf{x}}$:

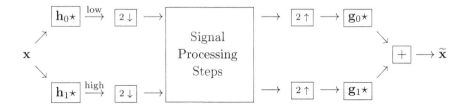

When the signal processing is absent, then $\tilde{\mathbf{x}} = \mathbf{x}$ and the PR property can be stated in the time domain as

$$\mathbf{x} = \underbrace{\mathbf{g}_0 \star \left(\boxed{2\uparrow} \; \boxed{2\downarrow} (\mathbf{h}_0 \star \mathbf{x}) \right)}_{\text{trend}} + \underbrace{\mathbf{g}_1 \star \left(\boxed{2\uparrow} \; \boxed{2\downarrow} (\mathbf{h}_1 \star \mathbf{x}) \right)}_{\text{detail}} = \mathbf{x}_s + \mathbf{x}_d \qquad (4.72)$$

for all finite signals $\mathbf{x}$. Here the *trend* part $\mathbf{x}_s$ contains the slow fluctuations in the signal, whereas the *detail* part $\mathbf{x}_d$ records the abrupt changes. The analogous formula for periodic signals is (3.43). Taking $\mathbf{x}$ as the unit impulse at 0, we see that the PR property for the filters can be expressed as

$$\mathbf{g}_0 \star \left(\boxed{2\uparrow} \; \boxed{2\downarrow} \mathbf{h}_0 \right) + \mathbf{g}_1 \star \left(\boxed{2\uparrow} \; \boxed{2\downarrow} \mathbf{h}_1 \right) = \delta_0 . \qquad (4.73)$$

Conversely, if equation (4.73) holds, then so does (4.72), as we see by writing $\mathbf{x} = \delta_0 \star \mathbf{x}$ and using the associativity of convolution.

Formula (4.72) expresses $\mathbf{x}$ as a linear combination of shifts of the synthesis filters $\mathbf{g}_0$ and $\mathbf{g}_1$. We will find a formula for the linear transformations $\mathbf{x} \to \mathbf{x}_s$ and $\mathbf{x} \to \mathbf{x}_d$ in terms of wavelet bases and dual bases for the trend and detail subspaces. Recall from Section 4.2 that given a FIR filter $\mathbf{h}$, we write $\mathbf{h}^{\mathrm{T}}$ for the time-reversed filter: $\mathbf{h}^{\mathrm{T}}[k] = \mathbf{h}[-k]$ for $k \in \mathbb{Z}$.

Lemma 4.3. *Suppose $\mathbf{h}$ and $\mathbf{g}$ are any FIR filters. Then for every finite signal $\mathbf{x}$*

$$\mathbf{g} \star \left(\boxed{2\uparrow} \; \boxed{2\downarrow} (\mathbf{h} \star \mathbf{x}) \right) = \sum_{m \in \mathbb{Z}} \langle S^{2m}\mathbf{h}^{\mathrm{T}}, \mathbf{x} \rangle \, S^{2m}\mathbf{g} . \qquad (4.74)$$

Furthermore, the coefficients $\langle S^{2m}\mathbf{h}^{\mathrm{T}}, \mathbf{x} \rangle$ are zero for $|m|$ sufficiently large.

Proof. The linear transformation $\boxed{2\uparrow} \; \boxed{2\downarrow}$ (downsampling followed by upsampling) projects a signal $\mathbf{y}$ onto its even part:

$$\boxed{2\uparrow} \; \boxed{2\downarrow} \mathbf{y}[k] = \begin{cases} \mathbf{y}[k] & \text{if } k \text{ is even,} \\ 0 & \text{if } k \text{ is odd.} \end{cases}$$

Hence

$$\mathbf{g} \star \left(\boxed{2\uparrow} \; \boxed{2\downarrow} (\mathbf{h} \star \mathbf{x}) \right)[n] = \sum_{m \in \mathbb{Z}} \mathbf{g}[n - 2m] \, (\mathbf{h} \star \mathbf{x})[2m]$$

$$= \sum_{m \in \mathbb{Z}} \left\{ \sum_{k \in \mathbb{Z}} \mathbf{h}[2m - k] \, \mathbf{x}[k] \right\} (S^{2m}\mathbf{g})[n]$$

$$= \sum_{m \in \mathbb{Z}} \langle S^{2m}\mathbf{h}^{\mathrm{T}}, \mathbf{x} \rangle \, (S^{2m}\mathbf{g})[n] ,$$

which proves (4.74). To prove the last statement we look at the overlap of the graphs of $\mathbf{x}$ and $S^{2m}\mathbf{h}^{\mathrm{T}}$. Since $\mathbf{x}$ is a finite signal and $\mathbf{h}$ is a FIR filter, there is an integer N such that $\mathbf{x}[k] = 0$ and $\mathbf{h}[k] = 0$ when $|k| > N$. Hence

$$\langle S^{2m}\mathbf{h}^{\mathrm{T}}, \mathbf{x} \rangle = \sum_{|k| \leq N} \mathbf{h}[2m - k] \, \mathbf{x}[k] .$$

But if $|m| > N$ and $|k| \leq N$, then $|2m - k| \geq 2|m| - |k| > 2N - N = N$, and so $\mathbf{h}[2m - k] = 0$. Thus $\mathbf{h}[2m - k] \, \mathbf{x}[k] = 0$ in this case, and hence $\langle S^{2m}\mathbf{h}^{\mathrm{T}}, \mathbf{x} \rangle = 0$. $\square$

Applying Lemma 4.3 to equation (4.72), we obtain the following generalization to nonperiodic signals of the *trend + detail* decomposition for periodic signals (equations (3.45) and (3.46)).

Theorem 4.11. *For the one-scale PR wavelet transform defined by the FIR analysis filters $\mathbf{h}_0$, $\mathbf{h}_1$ and synthesis filters $\mathbf{g}_0$, $\mathbf{g}_1$, the trend component of a finite signal $\mathbf{x}$ is*

$$\mathbf{x}_s = \sum_{m\in\mathbb{Z}} \langle S^{2m}\mathbf{h}_0^{\mathrm{T}}, \mathbf{x}\rangle S^{2m}\mathbf{g}_0 , \qquad (4.75)$$

and the detail component of the signal is

$$\mathbf{x}_d = \sum_{m\in\mathbb{Z}} \langle S^{2m}\mathbf{h}_1^{\mathrm{T}}, \mathbf{x}\rangle S^{2m}\mathbf{g}_1 . \qquad (4.76)$$

Every finite signal $\mathbf{x}$ has a decomposition $\mathbf{x} = \mathbf{x}_s + \mathbf{x}_d$.

Example 4.23. For the CDF$(2,2)$ transform, the filters are

$$\mathbf{h}_0 = c\left(-\delta_{-2} + 2\delta_{-1} + 6\delta_0 + 2\delta_1 - \delta_2\right) , \quad \mathbf{h}_1 = c\left(-2\delta_{-2} + 4\delta_{-1} - 2\delta_0\right) ,$$

$$\mathbf{g}_0 = c\left(2\delta_{-1} + 4\delta_0 + 2\delta_1\right) , \qquad\qquad \mathbf{g}_1 = c\left(-\delta_{-1} - 2\delta_0 + 6\delta_1 - 2\delta_2 - \delta_3\right) ,$$

where $c = \sqrt{2}/8$ (see Example 4.20). The normalization factors in the analysis and synthesis filters can be combined to give a single normalization of $c^2 = 1/32$ (binary shift) in the analysis filters, for example. Except for this normalization, all the filters have integer coefficients.

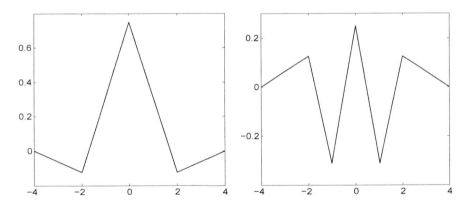

Fig. 4.10 Trend and detail for CDF$(2,2)$ decomposition of δ_0

To illustrate the expansion in Theorem 4.11, take $\mathbf{x} = \delta_0$. Then

$$\langle S^{2m}\mathbf{h}_0^{\mathrm{T}}, \delta_0\rangle = S^{2m}\mathbf{h}_0^{\mathrm{T}}[0] = \mathbf{h}^{\mathrm{T}}[-2m] = \mathbf{h}[2m] .$$

Thus the trend component of δ_0 is

$$\sum_{m\in\mathbb{Z}} \langle S^{2m}\mathbf{h}_0^{\mathrm{T}}, \delta_0\rangle S^{2m}\mathbf{g}_0 = \sum_{m\in\mathbb{Z}} \mathbf{h}_0[2m] S^{2m}\mathbf{g}_0 = \frac{\sqrt{2}}{8}\left\{-S^{-2}\mathbf{g}_0 + 6\mathbf{g}_0 - S^2\mathbf{g}_0\right\}$$

$$= \frac{1}{16}\left(-\delta_{-3} - 2\delta_{-2} + 5\delta_{-1} + 12\delta_0 + 5\delta_1 - 2\delta_2 - \delta_3\right) .$$

Likewise, the detail component of δ_0 is

$$\sum_{m\in\mathbb{Z}} \langle S^{2m}\mathbf{h}_1^{\mathrm{T}}, \delta_0\rangle S^{2m}\mathbf{g}_1 = \sum_{m\in\mathbb{Z}} \mathbf{h}_1[2m] S^{2m}\mathbf{g}_1 = \frac{\sqrt{2}}{4}\left\{-S^{-2}\mathbf{g}_1 - \mathbf{g}_1\right\}$$

$$= \frac{1}{16}\left(\delta_{-3} + 2\delta_{-2} - 5\delta_{-1} + 4\delta_0 - 5\delta_1 + 2\delta_2 + \delta_3\right).$$

It is clear that these two components add to δ_0. The sum of the trend entries is 1, while the detail component oscillates and the sum of its entries is 0. Figure 4.10 shows the piecewise linear graphs that interpolate the values of the trend and detail at integer times.   ∎

We now describe the trend-detail decomposition in Theorem 4.11 from the point of view of linear algebra, generalizing the case of periodic signals treated in Section 3.5.

Theorem 4.12. *For a PR filter bank, the set of all even-shifted filters $\{S^{2m}\mathbf{g}_0, S^{2n}\mathbf{g}_1 : m, n \in \mathbb{Z}\}$ is linearly independent. Hence the decomposition of a finite signal $\mathbf{x}$ into a trend $\mathbf{x}_s$ in equation (4.75) and a detail $\mathbf{x}_d$ in equation (4.76) is unique.*

Proof. Suppose some finite linear combination of the even-shifted filters adds up to zero:

$$\sum_{m\in\mathbb{Z}} c_m S^{2m}\mathbf{g}_0 + \sum_{n\in\mathbb{Z}} d_n S^{2n}\mathbf{g}_1 = 0.$$

Taking the z-transform of the left side of this equation, we obtain the relation

$$\varphi(z)G_0(z) + \psi(z)G_1(z) = 0,$$

where $\varphi(z) = \sum_{m\in\mathbb{Z}} c_m z^{-2m}$ and $\psi(z) = \sum_{n\in\mathbb{Z}} d_n z^{-2n}$. Since the Laurent polynomials φ and ψ only have even powers of z, they satisfy $\varphi(z) = \varphi(-z)$ and $\psi(z) = \psi(-z)$. Hence we get another linear relation

$$\varphi(z)G_0(-z) + \psi(z)G_1(-z) = 0.$$

These two relations can be written in matrix-vector form as

$$\mathbf{G}_m(z)\begin{bmatrix}\varphi(z)\\\psi(z)\end{bmatrix} = \begin{bmatrix}0\\0\end{bmatrix}, \tag{4.77}$$

where $\mathbf{G}_m(z)$ is the synthesis modulation matrix for the filter bank. From the PR property we know that $\mathbf{G}_m(z)$ is an invertible matrix. Hence the only solution to equation (4.77) is $\varphi(z) = 0$, $\psi(z) = 0$. This means that all the coefficients $c_m = 0$ and $d_n = 0$, proving linear independence.   □

Corollary 4.1. *The shifted synthesis filters satisfy the* biorthogonality relations

$$\langle S^{2m}\mathbf{h}_0^{\mathrm{T}}, S^{2n}\mathbf{g}_0\rangle = \delta_{m,n}, \quad \langle S^{2m}\mathbf{h}_1^{\mathrm{T}}, S^{2n}\mathbf{g}_0\rangle = 0,$$
$$\langle S^{2m}\mathbf{h}_0^{\mathrm{T}}, S^{2n}\mathbf{g}_1\rangle = 0, \quad \langle S^{2m}\mathbf{h}_1^{\mathrm{T}}, S^{2n}\mathbf{g}_1\rangle = \delta_{m,n} \tag{4.78}$$

for all integers m, n (where $\delta_{n,n} = 1$ and $\delta_{m,n} = 0$ for $m \neq n$).

Proof. By Theorem 4.11 the signal $\mathbf{x} = S^{2n}\mathbf{g}_0$ has a trend-detail decomposition

$$S^{2n}\mathbf{g}_0 = \sum_{m\in\mathbb{Z}} \langle S^{2m}\mathbf{h}_0^{\mathrm{T}}, S^{2n}\mathbf{g}_0 \rangle S^{2m}\mathbf{g}_0 + \sum_{m\in\mathbb{Z}} \langle S^{2m}\mathbf{h}_1^{\mathrm{T}}, S^{2n}\mathbf{g}_0 \rangle S^{2m}\mathbf{g}_1.$$

But by Theorem 4.12 we know that this decomposition is unique. Hence all the coefficients on the right side of this equation must be zero, except for the coefficient of $S^{2n}\mathbf{g}_0$, which must be one. Likewise, we take the signal $\mathbf{x} = S^{2n}\mathbf{g}_1$ and apply the same argument. □

Example 4.24. For the CDF$(2,2)$ transform in Example 3.13,

$$\mathbf{h}_0^{\mathrm{T}} = c\left(-\delta_{-2} + 2\delta_{-1} + 6\delta_0 + 2\delta_1 - \delta_2\right), \quad \mathbf{h}_1^{\mathrm{T}} = c\left(-2\delta_0 + 4\delta_1 - 2\delta_2\right),$$
$$\mathbf{g}_0 = c\left(2\delta_{-1} + 4\delta_0 + 2\delta_1\right), \qquad\qquad \mathbf{g}_1 = c\left(-\delta_{-1} - 2\delta_0 + 6\delta_1 - 2\delta_2 - \delta_3\right)$$

with $c = \sqrt{2}/8$. The biorthogonality relations (4.78) can be checked directly using the orthogonality of the unit impulses. When there is overlapping of supports, then cancellation produces biorthogonality:

$$\langle \mathbf{h}_0^{\mathrm{T}}, \mathbf{g}_0 \rangle = \frac{1}{32}(2\cdot 2 + 4\cdot 6 + 2\cdot 2) = 1,$$
$$\langle S^{\pm 2}\mathbf{h}_0^{\mathrm{T}}, \mathbf{g}_0 \rangle = \frac{1}{32}((-1)\cdot 4 + 2\cdot 2) = 0,$$
$$\langle \mathbf{h}_1^{\mathrm{T}}, \mathbf{g}_1 \rangle = \frac{1}{32}((-2)\cdot(-2) + 4\cdot 6 + (-2)\cdot(-2)) = 1,$$
$$\langle \mathbf{h}_0^{\mathrm{T}}, \mathbf{g}_1 \rangle = \frac{1}{32}((-2)\cdot(-1) + 2\cdot 6 + 6\cdot(-2) + 2\cdot(-1)) = 0,$$
$$\langle \mathbf{h}_1^{\mathrm{T}}, \mathbf{g}_0 \rangle = \frac{1}{32}((-2)\cdot 4 + 4\cdot 2) = 0.$$

But if $|m| \geq 2$ then $\langle S^{2m}\mathbf{h}_0^{\mathrm{T}}, \mathbf{g}_0 \rangle = 0$ since the supports of $S^{2m}\mathbf{h}_0^{\mathrm{T}}$ and $\mathbf{g}_0$ are disjoint in this case. Likewise for the other inner products. ■

4.9 Orthogonal Filter Banks

Assume we have a two-channel FIR filter bank with perfect reconstruction. Let $\mathbf{h}_0$ and $\mathbf{h}_1$ be the analysis filters and $\mathbf{g}_0$ and $\mathbf{g}_1$ the synthesis filters. As usual, we assume the filters are real-valued.

Definition 4.6. The filter bank is *orthogonal* if the synthesis filters are the *time-reversed* analysis filters: $\mathbf{g}_0 = \mathbf{h}_0^{\mathrm{T}}$ and $\mathbf{g}_1 = \mathbf{h}_1^{\mathrm{T}}$.

The use of the term *orthogonal* in this definition is explained by Corollary 4.1, since the biorthogonality relations now become orthogonality relations:

$$\langle S^{2m}\mathbf{g}_0, S^{2n}\mathbf{g}_0 \rangle = \delta_{m,n}, \quad \langle S^{2m}\mathbf{g}_1, S^{2n}\mathbf{g}_0 \rangle = 0,$$
$$\langle S^{2m}\mathbf{g}_0, S^{2n}\mathbf{g}_1 \rangle = 0, \qquad \langle S^{2m}\mathbf{g}_1, S^{2n}\mathbf{g}_1 \rangle = \delta_{m,n}. \tag{4.79}$$

In this case the one-scale wavelet decomposition is $\mathbf{x} = \mathbf{x}_s + \mathbf{x}_d$ with trend $\mathbf{x}_s$ and detail $\mathbf{x}_d$ given by

$$\mathbf{x}_s = \sum_{m \in \mathbb{Z}} \langle S^{2m}\mathbf{g}_0, \mathbf{x} \rangle S^{2m}\mathbf{g}_0 \,, \qquad \mathbf{x}_d = \sum_{m \in \mathbb{Z}} \langle S^{2m}\mathbf{g}_1, \mathbf{x} \rangle S^{2m}\mathbf{g}_1 \,.$$

Thus the set of vectors $\{S^{2m}\mathbf{g}_0, S^{2n}\mathbf{g}_1 : m, n \in \mathbb{Z}\}$ is an orthonormal basis for the vector space $\ell_0(\mathbb{Z})$ of finite signals. The trend component of the signal is orthogonal to the detail component, and the wavelet transform is energy-preserving:

$$\|\mathbf{x}\|^2 = \|\mathbf{x}_s\|^2 + \|\mathbf{x}_d\|^2 = \sum_{m \in \mathbb{Z}} |\langle S^{2m}\mathbf{g}_0, \mathbf{x} \rangle|^2 + \sum_{m \in \mathbb{Z}} |\langle S^{2m}\mathbf{g}_1, \mathbf{x} \rangle|^2$$

by Parseval's relation for an orthonormal basis.

Theorem 4.13. *A two-channel filter bank is orthogonal if and only if the analysis modulation matrix* $\mathbf{H}_m(z)$ *satisfies*

$$\mathbf{H}_m(z)\mathbf{H}_m(z^{-1})^{\mathrm{T}} = 2I \,. \tag{4.80}$$

Proof. The definition of orthogonality can be stated in terms of z-transforms of the filters as $G_0(z) = H_0(z^{-1})$ and $G_1(z) = H_1(z^{-1})$. This means that the synthesis and analysis modulation matrices satisfy $\mathbf{G}_m(z) = \mathbf{H}_m(z^{-1})^{\mathrm{T}}$. By Theorem 4.6 the PR property is equivalent to $\mathbf{H}_m(z)\mathbf{G}_m(z) = 2I$. Hence the filter bank is orthogonal if and only if the matrix equation (4.80) holds. □

Since the Laurent polynomials $H_0(z)$ and $H_1(z)$ have real coefficients, it follows that $\mathbf{H}_m(z^{-1})^{\mathrm{T}} = \overline{\mathbf{H}_m(z)}^{\mathrm{T}}$ when $z = \mathrm{e}^{\mathrm{i}\omega}$. So condition (4.80) for an orthogonal filter bank implies that the normalized matrix $(1/\sqrt{2})\mathbf{H}_m(\mathrm{e}^{\mathrm{i}\omega})$ is *unitary* for all real frequencies ω. The converse is also true and easy to prove.

Example 4.25. For the normalized Haar wavelet transform, the analysis modulation matrix is

$$\mathbf{H}_m(z) = \frac{1}{\sqrt{2}} \begin{bmatrix} (1+z) & (1-z) \\ (1-z) & (1+z) \end{bmatrix}$$

(see Example 4.19). In this case

$$\mathbf{H}_m(z^{-1})^{\mathrm{T}} = \frac{1}{\sqrt{2}} \begin{bmatrix} (1+z^{-1}) & (1-z^{-1}) \\ (1-z^{-1}) & (1+z^{-1}) \end{bmatrix} \,.$$

We calculate that $\mathbf{H}_m(z)\mathbf{H}_m(z^{-1})^{\mathrm{T}} = 2I$. Thus the normalized Haar transform is orthogonal. ∎

We now prove that a two-channel FIR orthogonal filter bank is determined by the lowpass analysis filter (which must satisfy a single quadratic relation) and the time shift between the lowpass channel and the highpass channel (such a shift is given by multiplying the z-transform of the lowpass filter by a power of z):

Theorem 4.14. *Let* $\mathbf{h}_0$ *be a FIR filter whose z-transform satisfies* $H_0(-1) = 0$.

(1) *If* $\mathbf{h}_0$ *is the lowpass analysis filter for an orthogonal filter bank, then* $H_0(z)$
satisfies the half-band condition

$$H_0(z)H_0(z^{-1}) + H_0(-z)H_0(-z^{-1}) = 2. \qquad (4.81)$$

(2) *Conversely, if* $H_0(z)$ *satisfies the half-band condition* (4.81), *let* K *be an integer*
and define $\mathbf{h}_1$ *to be the FIR filter with* z-*transform*

$$H_1(z) = z^{-2K+1}H_0(-z^{-1}). \qquad (4.82)$$

Then $\mathbf{h}_1$ *is a FIR highpass filter and the two-channel filter bank with analysis*
filters $\mathbf{h}_0$, $\mathbf{h}_1$ *is orthogonal.*

Proof. For any filter bank with the PR property, the synthesis modulation matrix
satisfies $\mathbf{G}_m(z) = 2\mathbf{H}_m(z)^{-1}$. If the filter bank is orthogonal, then (4.80) implies
that $\mathbf{G}_m(z) = \mathbf{H}_m(z^{-1})^{\mathrm{T}}$. Comparing matrix entries on each side of this equation,
we see that $G_0(z) = H_0(z^{-1})$. Thus equation (4.46) in Theorem 4.8 becomes
equation (4.81), proving statement (1).

To prove statement (2), let $\mathbf{h}_1$ be defined by (4.82). Note that $H_1(1) = H_0(-1) = 0$, so $\mathbf{h}_1$ is a highpass filter. The determinant of the analysis modulation matrix formed from $H_0(z)$ and $H_1(z)$ is

$$\begin{aligned}
d(z) &= H_0(z)H_1(-z) - H_0(-z)H_1(z) \\
&= -z^{-2K+1}\left\{H_0(z)H_0(z^{-1}) + H_0(-z)H_0(-z^{-1})\right\} \\
&= -2z^{-2K+1},
\end{aligned}$$

where we have used the half-band equation (4.81) in the last line. Thus the pair
of filters $\mathbf{h}_0$, $\mathbf{h}_1$ satisfy the conditions in Theorem 4.7 for a PR filter bank. The
corresponding synthesis filters $\mathbf{g}_0$ and $\mathbf{g}_1$ from equation (4.43) are

$$G_0(z) = \frac{2}{d(z)}H_1(-z) = -z^{2K-1}H_1(-z) = H_0(z^{-1}),$$

$$G_1(z) = -\frac{2}{d(z)}H_0(-z) = z^{2K-1}H_1(-z) = H_1(z^{-1}).$$

Hence $\mathbf{g}_0 = \mathbf{h}_0^{\mathrm{T}}$ and $\mathbf{g}_1 = \mathbf{h}_1^{\mathrm{T}}$, so we obtain an orthogonal filter bank. $\qquad\square$

Lemma 4.4. *Let* $\mathbf{h}_0$ *be a nonzero lowpass FIR filter. If* $H_0(z)$ *satisfies the half-band*
equation (4.81), *then the length of* $\mathbf{h}_0$ *is* $2K$, *where* $K \geq 1$ *is an integer.*

Proof. The z-transform is $H_0(z) = a_m z^{-m} + \cdots + a_n z^{-n}$, where $a_m \neq 0$, $a_n \neq 0$.
Since $H_0(-1) = 0$ by the lowpass assumption, we know that $m < n$, so the length
$n - m + 1$ of $\mathbf{h}_0$ is greater than one. Thus the products that occur in the half-band
condition (4.81) have the form

$$H_0(z)H_0(z^{-1}) = a_m a_n z^{m-n} + \cdots + a_m a_n z^{n-m}, \qquad (4.83)$$

$$H_0(-z)H_0(-z^{-1}) = (-1)^{m-n}\left\{a_m a_n z^{m-n} + \cdots + a_m a_n z^{n-m}\right\}, \qquad (4.84)$$

where the omitted terms on the right are linear combinations of monomials z^p
with $m - n < p < n - m$. If $m - n$ is even, then $(-1)^{m-n} = 1$ and the sum of

the right hand sides of equations (4.83) and (4.84) includes the non-constant terms $2a_m a_n \left(z^{m-n} + z^{n-m}\right)$. Since $m - n \neq 0$, this violates the half-band equation (4.81). Hence $m - n$ must be odd, and so the length $m - n + 1$ of $\mathbf{h}_0$ is even.   □

Suppose $\mathbf{h}_0$ is a lowpass filter such that $H_0(z)$ satisfies equation (4.81). Then $\mathbf{h}_0$ has even length $2K \geq 2$ by Lemma 4.4. For any integer q the Laurent polynomial $z^q H_0(z)$ also satisfies the half-band condition (4.81) and vanishes at $z = -1$; multiplying the z-transform by z^q corresponds to a delay or advance in the time domain. So in constructing lowpass filters for an orthogonal filter bank we may assume that $\mathbf{h}_0$ is a causal filter whose z-transform is

$$H_0(z) = \mathbf{h}_0[0] + \mathbf{h}_0[1]\, z^{-1} + \cdots + \mathbf{h}_0[2K - 1]\, z^{-2K+1} \tag{4.85}$$

with $\mathbf{h}_0[0] \neq 0$ and $\mathbf{h}_0[2K - 1] \neq 0$.

We now assume that $\mathbf{h}_0$ is a lowpass filter with z-transform (4.85). The corresponding highpass filter $\mathbf{h}_1$ in equation (4.82) has z-transform

$$H_1(z) = -\mathbf{h}_0[2K - 1] + \mathbf{h}_0[2K - 2]\, z^{-1} + \cdots - \mathbf{h}_0[1]\, z^{2K-2} + \mathbf{h}_0[0]\, z^{-2K+1}. \tag{4.86}$$

Thus we obtain $\mathbf{h}_1$ from $\mathbf{h}_0$ simply by reversing the coefficients and putting in alternating $\pm$ signs. In particular, $\mathbf{h}_1$ is a causal filter with the same length as $\mathbf{h}_0$. For example if

$$\mathbf{h}_0 = a\delta_0 + b\delta_1 + c\delta_2 + d\delta_3$$

has length 4 (for example, the lowpass filter for the Daub4 transform in Example 3.7), then

$$\mathbf{h}_1 = -d\delta_0 + c\delta_1 - b\delta_2 + a\delta_3 \,.$$

Definition 4.7.   The Laurent polynomial $P(z) = H_0(z)H_0(z^{-1})$ is the *power spectral response function* of $\mathbf{h}_0$.

Proposition 4.4. *If $\mathbf{h}_0$ is a causal lowpass FIR filter of the form (4.85) that satisfies the half-band condition (4.81), then the power spectral response function $P(z)$ satisfies the following conditions:*

(1) positivity: $P(e^{i\omega}) \geq 0$ *for all real values of ω,*
(2) symmetry: $P(z) = P(z^{-1})$ *and* $P(z) = cz^{-2K+1} + \cdots + cz^{2K-1}$ *with* $c \neq 0$,
(3) half-band: $P(z) + P(-z) = 2$ *(the only even term is the constant 2),*
(4) lowpass: $P(-1) = 0$.

Proof. Since the filter $\mathbf{h}_0$ is real, its z-transform satisfies $H_0(\bar{z}) = \overline{H_0(z)}$. Hence if ω is real then

$$P(e^{i\omega}) = H_0(e^{i\omega})H_0(e^{-i\omega}) = |H_0(e^{i\omega})|^2 \geq 0.$$

The other conditions are evident from the definition of $P(z)$ together with equations (4.81) and (4.85).   □

The converse of Proposition 4.4 is true: Given a Laurent polynomial $P(z)$ that satisfies conditions (1)–(4), we can find a causal lowpass filter $\mathbf{h}_0$ with power spectral response function $P(z)$. The construction of the filter uses the method of *spectral factorization*; we give the details for the family of Daubechies wavelet transforms in the next section.[8]

4.10   Daubechies Wavelet Transforms

We now construct the filters for the Daub$2K$ family of orthogonal wavelet transforms. The strategy is first to use the Bezout polynomials to obtain a Laurent polynomial $P(z)$ that satisfies the conditions in Proposition 4.4 for a power spectral response function, and then to show that $P(z)$ has a spectral factorization.

4.10.1   *Power spectral response function*

Take the Bezout polynomial $B_K(y)$ of degree $K - 1$ from (4.53) and make the quadratic change of variable

$$y = \frac{1}{4}\left(-z + 2 - z^{-1}\right),$$

as in equation (4.49). This gives the Laurent polynomial $P(z) = 2(1 - y)^K B_K(y)$, which we already used to construct the CDF(p, q) filters in Section 4.5. We can write this formula in terms of z as

$$P(z) = 2^{-2K+1}(1 + z)^K(1 + z^{-1})^K B_K\left(\frac{-z + 2 - z^{-1}}{4}\right). \qquad (4.87)$$

From equation (4.87) it is obvious that $P(z) = P(z^{-1})$, so the symmetry condition is satisfied. The half-band condition follows from the Bezout equation (4.52); note that we have multiplied $B_K(y)$ by 2 in defining $P(z)$. By definition $P(-1) = 0$, so the lowpass condition is satisfied. If $z = e^{\omega i}$ is on the unit circle, then $y = \sin^2(\omega/2)$ by (4.50), so $0 \le y \le 1$. Since the binomial coefficients in $B_K(y)$ are all positive and the constant term is 1, we have $B_K(y) \ge 1$ for $y \ge 0$. Also $(1 - y)^K \ge 0$ for $0 \le y \le 1$. Thus the positivity condition is satisfied. Furthermore, the only root of $P(z)$ on the unit circle is at $z = -1$, coming from the factor $(1 + z)^K(1 + z^{-1})^K$.

Now that we have a candidate for the power spectral response function, the next step is to find a factorization

$$P(z) = H_0(z)H_0(z^{-1}).$$

We start with the case $K = 2$ to obtain the Daub4 wavelet transform already constructed in Section 3.4 using lifting.

---

[8]See [Daubechies (1992), Lemma 6.1.3] for the general case.

4.10.2   *Construction of the Daub4 filters*

In the case of the Daub4 filter the spectral factorization is easy to obtain by solving two quadratic equations. The polynomial $B_2(y) = 1 + 2y$ has a single root at $y = -1/2$. The equation $y = -1/2$ is $z + z^{-1} = 4$, which we write as the quadratic equation $z^2 - 4z + 1 = 0$. This equation has two real roots $r = 2 - \sqrt{3}$ and $1/r = 2 + \sqrt{3}$. Here we have chosen r such that $0 < r < 1$. Thus we can factor

$$B_2(y) = 2 - \frac{1}{2}z - \frac{1}{2}z^{-1} = -\frac{1}{2}z^{-1}(z^2 - 4z + 1)$$

$$= -\frac{1}{2}z^{-1}(z - r)(z - r^{-1}) = \frac{1}{2r}(1 - rz^{-1})(1 - rz).$$

Using this factorization in equation (4.87) with $K = 2$, we can write

$$P(z) = \frac{1}{16r}(1 + z^{-1})^2(1 - rz^{-1})(1 + z)^2(1 - rz).$$

Thus if we define

$$H_0(z) = \frac{1}{4\sqrt{r}}(1 + z^{-1})^2(1 - rz^{-1}),$$

then $P(z) = H_0(z)H_0(z^{-1})$ is the spectral factorization. The corresponding filter $\mathbf{h}_0$ is lowpass and causal of length 4. Expanding the terms in the formula for $H_0(z)$ and using the numerical identity

$$\frac{1}{\sqrt{2 - \sqrt{3}}} = \frac{1 + \sqrt{3}}{\sqrt{2}}$$

(square both sides to verify this relation), we obtain the explicit filter coefficients

$$H_0(z) = \frac{1}{4\sqrt{2}}(a + bz^{-1} + cz^{-2} + dz^{-3}),$$

where $a = 1 + \sqrt{3}$, $b = 3 + \sqrt{3}$, $c = 3 - \sqrt{3}$, and $d = 1 - \sqrt{3}$. By Theorem 4.14 we can take

$$H_1(z) = z^{-3}H_0(-z^{-1}) = \frac{1}{4\sqrt{2}}\left(-d + cz^{-1} - bz^{-2} + az^{-3}\right)$$

for the z-transform of the highpass analysis filter. With these choices, the filters $\mathbf{h}_0$ and $\mathbf{h}_1$ are both causal.

The filter bank with these analysis filters gives the *inverse* of the Daub4 transform that was defined in Section 3.4, after we insert a time shift in the high frequency channel.[9] To see this, we recall the polyphase analysis matrix (3.28) for the Daub4 transform as defined in Section 3.4.2:

$$\mathbf{H}_p(z) = \frac{1}{4\sqrt{2}}\begin{bmatrix} (a + cz) & (b + dz) \\ (-b - dz^{-1}) & (a + cz^{-1}) \end{bmatrix}.$$

The filters $\widetilde{\mathbf{h}}_0$ and $\widetilde{\mathbf{h}}_1$ for this polyphase matrix have z-transforms

$$4\sqrt{2}\widetilde{H}_0(z) = a + cz^2 + z(b + dz^2) = a + bz + cz^2 + dz^3 = 4\sqrt{2}H_0(z^{-1}),$$

$$4\sqrt{2}\widetilde{H}_1(z) = -b - dz^2 + z(a + cz^{-2}) = a + bz + cz^2 + dz^3 = 4\sqrt{2}z^{-2}H_1(z^{-1}).$$

---

[9]The inversion and time shift are needed to obtain causal filters.

We conclude that $\widetilde{\mathbf{h}}_0 = \mathbf{h}_0^{\mathrm{T}}$ and $\widetilde{\mathbf{h}}_1 = \delta_2 \star \mathbf{h}_1^{\mathrm{T}}$, as claimed. In particular, this explains the choice of the constants a, b, c, d used to define the Daub4 transform in Section 3.4.

Remark 4.5. In any PR filter bank, the analysis filters can be interchanged with the synthesis filters to give a (different) PR filter bank. In the case of an orthogonal filter bank, this means that we can use the time-reversed analysis filters for the analysis bank. Also, arbitrary delays or advances can be inserted in the analysis filters without changing the PR or orthogonality properties. However, these shifts in alignment of the filters do effect the signal-processing effectiveness of the wavelet transform. See [Jensen and la Cour-Harbo (2001), §9.4.3] for further details and examples.

4.10.3 *Construction of the Daub2K filters*

To construct the Daub2K filters for an integer $K > 2$, we use the same root analysis strategy to factor the power spectral response function as for the case $K = 2$ just treated. Here are the steps.

(a) Define $Q_K(z) = z^{K-1}B_K\left(\dfrac{-z + 2 - z^{-1}}{4}\right)$. If m is a positive integer then
$$(z - 2 + z^{-1})^m = z^m + \cdots + z^{-m}.$$
Since $B_K(y)$ is a polynomial of degree $K - 1$ in y, it follows that
$$Q_K(z) = z^{K-1}\left\{cz^{K-1} + \cdots + cz^{-K+1}\right\} = cz^{2K-2} + \cdots + c,$$
where the constant $c \neq 0$. This shows that $Q_K(z)$ is a polynomial of degree $2K - 2$ in nonnegative powers of z and $Q_K(0) \neq 0$. Hence $Q_K(z)$ has $2K - 2$ nonzero complex roots (counted with multiplicity). Furthermore, none of these roots lie on the unit circle, by the argument given after equation (4.87).

(b) We can write
$$P(z) = 2^{-2K+1}(1 + z)^K(1 + z^{-1})^K z^{-K+1}Q_K(z). \tag{4.88}$$
The symmetry of $P(z)$ under $z \longleftrightarrow z^{-1}$ implies that each root r_j of $Q_K(z)$ with $0 < |r_j| < 1$ is paired with a root r_j^{-1} with $|r_j^{-1}| > 1$, for $j = 1, \ldots, K - 1$. Hence $Q_K(z)$ can be factored as
$$Q_K(z) = \alpha(z - r_1)\cdots(z - r_{K-1})(z - r_1^{-1})\cdots(z - r_{K-1}^{-1}),$$
where α is a nonzero constant. Multiplying $Q_K(z)$ by z^{-K+1} and using the algebraic identity $z^{-1}(z - r) = r(1 - rz^{-1})$, we obtain the factorization
$$z^{-K+1}Q_K(z) = \beta(1 - r_1 z^{-1})\cdots(1 - r_{K-1}z^{-1})(1 - r_1 z)\cdots(1 - r_{K-1}z),$$
where β is a nonzero constant. It follows from (4.88) that
$$
\begin{aligned}
P(z) = {}& \gamma(1 + z)^K(1 + z^{-1})^K \\
&\times (1 - r_1 z^{-1})\cdots(1 - r_{K-1}z^{-1})(1 - r_1 z)\cdots(1 - r_{K-1}z),
\end{aligned} \tag{4.89}
$$
where γ is a nonzero constant.

(c) Define

$$H_0(z) = \kappa(1 + z^{-1})^K (1 - r_1 z^{-1}) \cdots (1 - r_{K-1} z^{-1}), \qquad (4.90)$$

where $r_1, \ldots, r_{K-1}$ are the roots in (b) and κ is a normalizing constant. Since none of the roots r_j are 1, we can choose κ to make $H_0(1) = \sqrt{2}$. It now follows from (4.89) and (4.90) that $H_0(z)H_0(z^{-1}) = P(z)$, since we know that $P(1) = 2$. Let $\mathbf{h}_0$ be the causal lowpass filter with z-transform $H_0(z)$.

(d) We obtain the highpass filter from equation (4.82) by reversal of coefficients and alternating signs, as in (4.86).

The file `daub.m` in the UVI-WAVE directory `wfilter` constructs the Daub2K filters using the algorithm just described.[10]

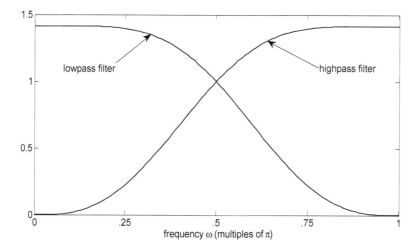

Fig. 4.11   Frequency response of Daub6 orthogonal filters

In an orthogonal filter bank there is mirror symmetry around the line $\omega = \pi/2$ between the frequency response of the lowpass and highpass filters (ignoring phase factors), unlike the case of the CDF(p, q) filters. Figure 4.11 shows this for the Daub6 orthogonal filters.

4.11   Computer Explorations

Some of the MATLAB m-files that are used in these explorations are from the public-domain (GNU license) MATLAB toolbox UVI-WAVE. See Section A.3.2 for details about downloading and using this toolbox. Set the `path` environment so MATLAB can find the UVI-WAVE directories and subdirectories. Then type `wtdemo` at the MATLAB prompt to see the complete set of demos in the UVI-WAVE `wdemo` directory.

---

[10]Finding all the complex roots of $Q_K(z)$ turns out to be a difficult numerical calculation when K is large since the roots occur in clusters; see [Strang and Nguyen (1997), Figure 5.6].

4.11.1 *Signal processing with the CDF(2,2) transform*

You will apply a three-scale CDF(2, 2) transform to analyze a signal, using the filter-bank approach (instead of matrix multiplication) to calculate the wavelet transforms (see Sections 4.5 and 4.8). For a test signal take the same sine wave with static noise and pops as in Section 3.7.6. Since the detail for a one-scale CDF(2, 2) transform is zero on linear functions (except near end points–see Figure 3.10), this transform should yield better results than the Haar transform on signals that are relatively smooth.

(a) Analysis of synthetic signals

Sample the signal $s(t) = \sin(4\pi t)$ at $2^9 = 512$ equidistant points in $0 \leq t \leq 1$ to obtain a finite signal $\mathbf{s}_9$:

```
s9 = sin([1:512]*4*pi/512);
```

Now add some random noise and random pops to the signal $\mathbf{s}_9$. Generate a two-component vector pop whose entries are integers between 100 and 400:

```
pop = round(100 + 300*rand(2,1))
```

The following code adds 1 to the two components of $\mathbf{s}_9$ whose indices are given by the numbers in pop, adds some random noise throughout, and plots the resulting signal:

```
s9(pop(1)) = s9(pop(1)) + 1.5;
s9(pop(2)) = s9(pop(2)) + 1.5;
s9 = s9 + 0.1*randn(1, 512);
figure, plot(s9), axis([0  550  -1.5  1.5])
```

The location of the two pops should be clear in the plot. Insert the title Signal with Pops and Static in the MATLAB figure.

To calculate a three-scale CDF(2, 2) wavelet transform of the signal, use the *filter bank* implementation in the UVI_WAVE toolbox (where the CDF family of transforms are named wspline):

```
[h,g,rh,rg] = wspline(2,2)
```

The four *filter vectors* $\mathbf{h}, \mathbf{g}, \mathbf{rh}, \mathbf{rg}$ are obtained from the one-stage CDF(2, 2) analysis and synthesis matrices; to see this use the m-files cdfamat.m and cdfsmat.m from Section 3.7.2 to generate these matrices:

```
Ta = cdfamat(6), Ts = cdfsmat(6)
```

Compare the filters and the matrices to answer the following questions (see Example 4.24):

(i) How are the rows of Ta obtained from the vectors $\mathbf{h}$ and $\mathbf{g}$?

(ii) How are the columns of Ts obtained from the transposes of the row vectors rh and rg?

Now obtain the three-scale CDF$(2,2)$ transform **st** of s_9 and plot the result by

```
st = wt(s9,h,g,3);
figure, isplit(st,3,'','r.')
```

The vector st from the UVI_WAVE function wt is the CDF$(2,2)$ transform of s9. It is a concatenation of the four vectors making up the transform. The UVI_WAVE function isplit separates st into its four parts (as you did by hand for the Haar transform in Section 3.7.6). It plots the 3-scale trend coefficients of s_6 and detail coefficients of d_6 in the first and second graphs (these are vectors of length $64 = 2^{9-3}$). The third graph gives the 2-scale detail coefficients of d_7 (a vector of length $128 = 2^{9-2}$). The fourth graph gives the 1-scale detail coefficients of d_8 (a vector of length $256 = 2^{9-1}$). Put title Three-Scale CDF$(2,2)$ Wavelet Coefficients at the top of the MATLAB figure, and then put titles s6, d6, d7, d8 to the left of each of the plot windows.

The multiresolution analysis of the signal is obtained by taking the inverse CDF$(2,2)$ transforms of the trend and detail coefficient vectors. Do this using the UVI_WAVE function multires:

```
y = multires(s9,h,rh,g,rg,3);
figure, split(y)
```

Here y is a 4×512 matrix whose rows contain d_8, d_7, d_6, and s_6. The split command displays a graph for each row. Put the title Multiresolution CDF$(2,2)$ Analysis of Signal at the top of the MATLAB figure, and put titles s6, d6, d7, d8 to the left of the corresponding plot windows.

(b) Filtering and compression

Just as in the case of the Haar transform used in Section 3.7.6, the pops are not evident in the trend but are obvious in the details. A crude way of filtering the signal to remove the pops and some of the static noise is to use only the trend s_6 and detail d_6 (rows 3 and 4 of the matrix y):

```
sfilter = y(3,:)  + y(4,:);
figure, subplot(311), plot(s9, 'r-'), hold on
subplot(312), plot(sfilter, 'g-')
```

To preserve the pops in the signal but remove the static noise you will need to use all the detail vectors. From the MATLAB figure in (a) you can see that the only large coefficients in these vectors come from the two pops. Compress and denoise the signal by setting to zero all the coefficients in the detail vectors that are less than the threshold level 0.2 in absolute value (do not change the transform coefficients of the trend vector $\mathbf{s}_6$). To do this efficiently, apply the `threshold` m-file from Section 3.7.6 to the detail entries (in positions 65–512 of the transform `st`):

```
cst = st;
cst(65:512) = threshold(st(65:512), 0.2);
```

Then calculate the 3-scale inverse $\mathrm{CDF}(2,2)$ transform of the compressed vector `cst` and plot it in the same window as the original signal and the filtered signal:

```
cs = iwt(cst,rh,rg,3);
subplot(313), plot(cs, 'b-')
```

Label the graphs as original signal, filtered signal, and filtered signal with pops. Put the title Filtering by Removing Wavelet Details and Noise at the top of the MATLAB figure.

4.11.2 *Two-dimensional discrete wavelet transforms*

You will explore the use of wavelet transforms for image processing (see Section 3.6). Set the `path` so that MATLAB can find the UVI_WAVE directories and subdirectories.

(a) Images as matrices

Type `gnimgdmo` at the MATLAB prompt to run the image-generating demo from the UVI_WAVE wdemo directory. Use the MATLAB text editor to open the m-file `genimg.m` in the `wdemo` directory. Notice how few lines of MATLAB code are needed to generate these complex images.

Now generate a random 8-bit integer matrix $\mathbf{X}$ of size 8×8 and display it as an image by the commands

```
X = floor(256*rand(8,8))
figure, show(X), colormap(gray)
```

Notice that the $(1,1)$ entry in $\mathbf{X}$ determines the gray scale of the upper left square in the figure, while the $(8,1)$ entry determines the gray scale of the lower left square (see Section 3.6.1). Now modify your matrix $\mathbf{X}$ to make row 4 black and column 2 white; then make the value of the $(4,2)$ square 128 (half-way between black and white) and leave all the other squares unchanged. Redisplay the matrix as a figure. Put the title Random Squares with Black Row and White Column on the MATLAB figure.

(b) One-scale wavelet transform of images

Let $\mathbf{W_a}$ be the $N \times N$ analysis matrix for a one-scale wavelet transform, where N is even. Then $\mathbf{W_a}$ can be written in 2×1 block form as

$$\mathbf{W_a} = \begin{bmatrix} \mathbf{A} \\ \mathbf{B} \end{bmatrix}$$

where the matrices $\mathbf{A}$ and $\mathbf{B}$ are of size $N/2 \times N$ (see Section 3.6.2). The rows of $\mathbf{A}$ are the *trend rows*, whereas the rows of $\mathbf{B}$ are the *detail rows*.

Generate and display the matrix $\mathbf{W_a}$ and the matrices $\mathbf{A}$ and $\mathbf{B}$ for the CDF(2, 2) analysis transform with $N = 8$ using the m-file `cdfamat.m` from Section 3.7.2 (use the colon operator to extract $\mathbf{A}$ and $\mathbf{B}$ from $\mathbf{W_a}$). Observe how each row of $\mathbf{A}$ and $\mathbf{B}$ is obtained from the row above by shifting to the right two places (with wraparound).

Now generate the matrix $\mathbf{W_a}$ and the matrices $\mathbf{A}$ and $\mathbf{B}$ for the CDF(2, 2) analysis transform with $N = 64$ by the same method. *Be sure to put ; at the end of each command when $N = 64$ so that the large matrices $\mathbf{W_a}$, $\mathbf{A}$, and $\mathbf{B}$ will not appear on screen.*

Generate and display a 64×64 test image $\mathbf{X}$ by

```
X = genimg(0,64,64);
figure, show(X, -1), colormap(gray), axis off
```

(The m-files `show.m` and `colormap.m` are from the UVI_WAVE toolbox.) Put the title 64 × 64 Synthetic Test Image on the MATLAB figure.

The one-scale wavelet transform of the image encoded by $\mathbf{X}$ is the matrix

$$\mathbf{Y} = \mathbf{W_a} \mathbf{X} \mathbf{W_a^T}.$$

The block decomposition of $\mathbf{W_a}$ gives a decomposition of $\mathbf{Y}$ into four $N/2 \times N/2$ blocks:

$$\mathbf{Y} = \begin{bmatrix} \mathbf{A} \\ \mathbf{B} \end{bmatrix} \mathbf{X} \begin{bmatrix} \mathbf{A^T} & \mathbf{B^T} \end{bmatrix} = \begin{bmatrix} \mathbf{AXA^T} & \mathbf{AXB^T} \\ \mathbf{BXA^T} & \mathbf{BXB^T} \end{bmatrix} = \begin{bmatrix} \mathbf{Y_{ss}} & \mathbf{Y_{sd}} \\ \mathbf{Y_{ds}} & \mathbf{Y_{dd}} \end{bmatrix}$$

(see Section 3.6.2 for an interpretation of each block in terms of trend/detail of rows and columns of $\mathbf{X}$). Calculate and display these matrices for the test image $\mathbf{X}$ generated above:

```
Yss = A*X*A'; Ysd = A*X*B'; Yds = B*X*A'; Ydd = B*X*B';
figure
subplot(2,2,1), show(Yss, -1), colormap(gray),
axis off, hold on
subplot(2,2,2), show(Ysd, -1), colormap(gray), axis off
subplot(2,2,3), show(Yds, -1), colormap(gray), axis off
subplot(2,2,4), show(Ydd, -1), colormap(gray), axis off
```

Here the UVI_WAVE m-file `show.m` automatically adjusts the gray scale to get maximum contrast. Put the title One-Scale CDF(2, 2) Transform of Synthetic Test Image on the MATLAB figure. Label the subplots as Trend, Vertical Detail, Horizontal Detail, and Diagonal Detail.

(c) Multiresolution representation of images

The synthesis matrix $\mathbf{W_s}$ is the inverse of the analysis matrix. It can be written in 1×2 block form as

$$\mathbf{W_s} = \begin{bmatrix} \mathbf{C} & \mathbf{D} \end{bmatrix}$$

where the matrices $\mathbf{C}$ and $\mathbf{D}$ are of size $N \times N/2$. The columns of $\mathbf{C}$ are the *trend vectors*, whereas the columns of $\mathbf{D}$ are the *detail vectors* (see Section 3.6.2).

Generate and display the matrix $\mathbf{W_s}$ and the matrices $\mathbf{C}$ and $\mathbf{D}$ for the CDF$(2,2)$ transform with $N = 8$ using the m-file cdfsmat.m from Section 3.7.2 (use the colon operator to extract $\mathbf{C}$ and $\mathbf{D}$ from $\mathbf{W_s}$). Observe how each column of $\mathbf{C}$ and $\mathbf{D}$ is obtained from the column to the left by shifting down two places (with wraparound).

Now generate the matrix $\mathbf{W_s}$ and the matrices $\mathbf{C}$ and $\mathbf{D}$ for the CDF$(2,2)$ synthesis transform with $N = 64$ by the same method.

The one-scale inverse wavelet transform of $\mathbf{Y}$ is the matrix

$$\mathbf{X} = \mathbf{W_s}\mathbf{Y}\mathbf{W_s^T}.$$

To obtain the *multiresolution representation* of the image $\mathbf{X}$, use the block decomposition of $\mathbf{Y}$ from part *(b)* and the trend and detail matrices $\mathbf{C}$ and $\mathbf{D}$:

$$\mathbf{X_{ss}} = \mathbf{C}\mathbf{Y_{ss}}\mathbf{C^T}, \quad \mathbf{X_{sd}} = \mathbf{C}\mathbf{Y_{sd}}\mathbf{D^T},$$
$$\mathbf{X_{ds}} = \mathbf{D}\mathbf{Y_{ds}}\mathbf{C^T}, \quad \mathbf{X_{dd}} = \mathbf{D}\mathbf{Y_{dd}}\mathbf{D^T}.$$

Calculate and display the multiresolution representation for the test image $\mathbf{X}$ generated above (see Section 3.6.3):

```
Xss = C*Yss*C'; Xsd = C*Ysd*D';
Xds = D*Yds*C'; Xdd = D*Ydd*D';
figure
subplot(2,2,1), show(Xss, -1), colormap(gray),
axis off, hold on
subplot(2,2,2), show(Xsd, -1), colormap(gray), axis off
subplot(2,2,3), show(Xds, -1), colormap(gray), axis off
subplot(2,2,4), show(Xdd, -1), colormap(gray), axis off
```

Put the title One-Scale CDF$(2,2)$ Multiresolution Representation of Synthetic Test Image on the MATLAB figure. Label the subplots as Trend, Vertical Detail, Horizontal Detail, and Diagonal Detail.

4.11.3 *Image compression and multiscale analysis*

Here you will perform a wavelet analysis of a famous photograph: the *Lena* image in the UVI_WAVE wdemo directory. This image has been used since the 1970's as a test image for comparing image-processing algorithms; go to *Wikipedia* for more information about its history. The image has a much more complex combination of textures and contours than the simple synthetic image in Section 4.11.2. At the MATLAB prompt, type wvt2ddmo. This is the two-dimensional wavelet transform demo in UVI_WAVE that includes the *Lena* image.

(a) One-scale analysis of image

To load and display the image type

```
load lena, X = lena;
size(X)
figure, show(X), colormap(gray), axis off
```

The size of the matrix $\mathbf{X}$ is 256×256. Perform a one-scale $CDF(2,2)$ wavelet transform of $\mathbf{X}$ and display the result:

```
Wa = cdfamat(256); Y = Wa*X*Wa' ;
Ybr = bandadj(Y,1);
figure, show(Ybr), colormap(gray)
```

Here, as explained in the demo, you use the UVI_WAVE m-file `bandadj.m` to adjust the brightness levels of the four parts of the wavelet transform (without this adjustment, only the $\mathbf{Y_{ss}}$ part would be visible). The 128×128 trend matrix $\mathbf{Y_{ss}}$ in the upper-left corner gives a rougher version of the original image, whereas the other 128×128 matrices in the picture give the horizontal, vertical and diagonal details. Insert the title One-Scale CDF(2, 2) Transform of Lena Image in the MATLAB figure.

To determine how much information about the image is contained in the three detail portions, set the trend block $\mathbf{Y_{ss}}$ in wavelet transform to zero and take the inverse transform:

```
Yd = Y; Yd(1:128, 1:128) = 0;
Ws = cdfsmat(256); Xd = Ws*Yd*Ws' ;
figure, show(Xd), colormap(gray), axis off
```

You should get a figure that you saw in the `wtdemo`; it shows the *edges* of the picture.

(b) Image compression

One of the most useful features of wavelet methods in image processing is the compression of images without loss of perceptible information. This is done by compressing the detail matrix `Yd`. Fix a threshold level of 30 and set to zero all entries in `Yd` whose absolute value is 30 or less (the value of 30 for this image gives good results; in general, experimentation is needed to find a good threshold for an image). Then create a *compressed* transformed matrix `Yc` by replacing the 128×128 upper left-hand block of zeros in `Yc` by the trend `Yss`, and take the inverse transform:

```
Ydc = threshold(Yd, 30) ;
Yc = Ydc; Yss = Y(1:128, 1:128);
Yc(1:128, 1:128) = Yss ;
Xc = Ws*Yc*Ws' ;
figure, show(Xc), colormap(gray), axis off
```

(this uses the m-file `threshold.m` from Section 3.7.6). Notice that the image displayed looks very much like the original. Insert the title Compressed Lena Image in the MATLAB figure and keep the figure window open.

To determine the number of nonzero entries in the compressed transform `Yc` and to find the compression ratio, calculate

```
numY = ones(256, 256).*(abs(Y)>0);
numYc = ones(256, 256).*(abs(Yc)>0);
sumY = sum(numY*ones(256, 1))
sumYc = sum(numYc*ones(256, 1))
compress = sumY/sumYc
```

Be careful to use the element-by-element multiplication .* (note the period mark before the asterisk) in the first two lines. This ensures that the matrices `numY` and `numYc` have entries that are either 1 or 0, depending on whether the corresponding entries of `Y` and `Yc` are nonzero or zero. Thus `sumY` and `sumYc` counts the number of nonzero entries in each matrix. Insert the label compression ratio = r (where r is the numerical value of `compress` that you have calculated) below the image in the MATLAB figure.

(c) Error measures in image processing

The difference between two eight-bit image matrices `X1` and `X2` (both assumed to be of size 256×256) can be measured by the *Mean Square Error* (MSE):

```
MSE = (norm(X1 - X2)^2)/2^16
```

Here the denominator $2^{16} = (256)^2$ is the total number of entries in the image matrices. With this normalization, the MSE between the 256×256 matrix with entries all 1 and the zero matrix is 1. A large MSE corresponds a big difference in the images. The largest possible value for MSE is $(255)^2$, since the differences between the individual entries of `X1` and `X2` are at most 255.

An equivalent measurement of image difference that corresponds more closely to how the human brain responds to light intensity is the *Peak Signal to Noise Ratio* (PSNR), which is defined on a logarithmic scale in *decibels* (dB) by

```
PSNR = 10*log10(255^2/MSE)
```

Here `log10` is the logarithm to base 10 in MATLAB. The ratio `255^2/MSE` is bigger than 1, so PSNR is a positive number. *Small* values of MSE correspond to *large* values of PSNR. As a *rule of thumb*, if the PSNR exceeds 40 dB, then the two images are perceived as the same.

Calculate the values of the MSE between the matrices `X` and `Xc` and the associated PSNR using these formulas with `X1 = X` and `X2 = Xc`. Insert the values of MSE and PSNR in the MATLAB figure below the image. Comment on the validity of the 40 dB rule of thumb criterion concerning the compressed image in this case.

(d) Two-scale analysis of image

To make a two-scale analysis of the *Lena* image, repeat the one-scale analysis on the trend matrix $\mathbf{Y_{ss}}$:

```
Yss = Y(1:128, 1:128);
W2a = cdfamat(128);
Y2 = Y; Y2(1:128, 1:128) = W2a*Yss*W2a';
```

Note that you are using the 128×128 CDF$(2,2)$ analysis matrix on $\mathbf{Y_{ss}}$ and then inserting the transformed matrix into the upper left-hand block of $\mathbf{Y}$. This gives the *two-scale* CDF$(2,2)$ transform $\mathbf{Y}^{(2)}$. To display the result, brighten the image using the bandadj.m utility:

```
Y2br = bandadj(Y2,2);
figure, show(Y2br), colormap(gray)
```

Insert the title CDF$(2,2)$ transform over two scales in the MATLAB figure.

4.11.4   *Fast two-dimensional wavelet transforms*

The matrix formulation of the two-dimensional wavelet transform and inverse transform in Sections 4.11.2 and 4.11.3 is helpful in understanding the mathematical theory of this transform. For numerical calculation, however, the matrix formulation is impractical, since an $N \times N$ image will require the order of N^3 arithmetic operations to calculate the product of three $N \times N$ matrices. Just as in the case of the one-dimensional wavelet transform, there is a fast implementation, either using lifting or using the two-channel filter bank (convolution) method. In both methods the one-dimensional wavelet transform is performed on the rows and columns of the image matrix.

You will use the *symlets* wavelet transform. This is a family of orthogonal transforms that has the most symmetry of the Daubechies family of transforms. Set the **path** so that MATLAB can find the UVI_WAVE directories and subdirectories.

(a) Multiscale analysis of image

To calculate the symlets(24) wavelet transform of an image, use the *filter bank* implementation in the UVI_WAVE toolbox.

```
[h,g,rh,rg] = symlets(24);
```

Here the vector $\mathbf{h}$ gives the 24 nonzero coefficients of the lowpass filter and the vector $\mathbf{g}$ gives the 24 nonzero coefficients of the highpass filter for the one-dimensional symlets(24) analysis transform. Plot these coefficients by

```
figure, subplot(2,2,1), plot([-11:12], h), hold on
subplot(2,2,2), plot([-11:12], g)
```

The vector **rh** gives the 24 nonzero coefficients of the lowpass filter and the vector **rg** gives the 24 nonzero coefficients of the highpass filter for the one-dimensional symlets(24) inverse wavelet transform. Plot these coefficients by

```
subplot(2,2,3), plot([-12:11], rh)
subplot(2,2,4), plot([-12:11], rg)
```

Answer the following questions:

(i) How are the graphs of **h** and **rh** related?
(ii) How are the graphs of **g** and **rg** related?

See Definition 4.6, where the notation is changed to $h_0 = h$ (lowpass) and $h_1 = g$ (highpass) for the analysis filters, and $g_0 = rh$ (lowpass) and $g_1 = rg$ (highpass) for the synthesis filters.

Label the four graphs lowpass analysis filter, highpass analysis filter, lowpass synthesis filter, and highpass synthesis filter. Insert the title Symlets(24) Wavelet Filters on the MATLAB figure.

Make a fast one-scale symlets(24) transform of the *Lena* image and then adjust the brightness levels for printing:

```
load lena, X = lena;
Y = wt2d(X,h,g,1);
Ybr = bandadj(Y,1);
figure, show(Ybr), colormap(gray), axis off
```

Calculate `norm(X)` and `norm(Y)`. Explain why your answers are predicted by orthogonal property of the symlets(24) transform.

Insert the title One-scale Symlets(24) Transform of Lena Image in the MATLAB figure. Notice that the four subregions in figure show more fine-scale edge detail than the CDF$(2, 2)$ transform in Section 4.11.3. This is due to the longer lengths of the symlets filters. The trade-off is that the more computational time is needed to obtain the transform.

(b) Multiresolution representation

Create a two-scale symlets(24) multiresolution representation of the *Lena* image using the fast transform/inverse transform:

```
Trend = mres2d(X,h,rh,g,rg,2,0);
Details = mres2d(X,h,rh,g,rg,2,4);
Image = Trend + Details;
figure, show(Trend), colormap(gray), axis off
figure, show(Details), colormap(gray), axis off
figure, show(Image), colormap(gray), axis off
```

Put titles Trend of Two-Scale Multiresolution Representation, Sum of Details of Two-Scale Multiresolution Representation, Sum of Trend and Details of Two-Scale Multiresolution Representation at the tops of the successive MATLAB figures. Calculate the values of the MSE between the matrices X and Image and the associated PSNR using the formulas in Section 4.11.3 *(c)* with X1 = X and X2 = Image. The Image plot is not quite as sharp as the original image; this is confirmed by the value less than 40 Db of PSNR.

4.11.5 *Denoising and compressing images*

Adapting audio terminology, the term *noise* is used to describe distortions of images due to numerical or physical imperfections. Often images are produced under less than idea conditions of lighting or signal transmission, and consequently are contaminated by significant amounts of noise. This is the case, for example, with many medical images or with digitized images of old photographs. One of the most important applications of wavelet techniques in image processing is to remove noise. You can do this by compression of the wavelet transform of the noisy image.

(a) Adding noise to an image

Generate a matrix of random normal integers (with mean zero, standard deviation 50) and add it to the *Lena* image:

```
Noise = floor(50*randn(256,256));
Xn = X + Noise;
figure, show(Xn), colormap(gray), axis off
```

You can see that the image is considerably degraded by the noise. Calculate the Mean Square Error (MSE) between the original image X and the noisy image Xn, and the associated Peak Signal to Noise Ratio (PSNR), as described in Section 4.11.3 *2(c)* with X1 = X and X2 = Xn. Put the title Noisy Lena Image above the image, and the values of the MSE and PSNR that you have calculated under the figure.

(b) Removing noise by wavelet techniques

Take a four-scale symlets(24) wavelet transform of the noisy image:

```
[h,g,rh,rg] = symlets(24);
Yn = wt2d(Xn,h,g,4);
```

The trend portion of the transformed matrix Yn is in the upper-left 16×16 block in the figure, whereas the first-scale detail is in the lower-right 128×128 block. Calculate the ratios of the largest element in the trend to the largest element in the detail block:

```
Yss = Yn(1:16,1:16); Ydd = Yn(129:256, 129:256);
a = max(max(abs(Yss)))
b = max(max(abs(Ydd)))
ratio = a/b
```

The ratio should be quite large. This indicates that a large proportion of the detail entries can be set to zero without changing the image substantially. Doing this will compress the image file and remove some of the noise. For this particular image and wavelet transform, some experimentation shows that setting 95% of the coefficients to zero (20:1 compression) is a good choice for maximizing the PSNR. The UVI_WAVE utility `elmin` will do this:

```
Ync95 = elmin(Yn,95);
Xnc95 = iwt2d(Ync95,rh,rg,4);
figure, show(Xnc95), colormap(gray), axis off
```

Put the title 20:1 Compressed and Denoised Lena Image above the MATLAB figure. Calculate the Mean Square Error and the Peak Signal to Noise Ratio between the original image matrix X and the matrix Xnc95 using the formulas in Section 4.11.3 (c). The MSE should be smaller and the PSNR larger than the MSE and PSNR for the noisy image (although the PSNR will still be below 40 Db).

Remark 4.6. More refined image-processing methods (such as are found in the MATLAB Wavelet toolbox) can be applied to the compressed image to improve its appearance.

4.12   Exercises

(1) Let $\mathbf{x}$ and $\mathbf{y}$ be the signals that are the following linear combinations of unit impulses: $\mathbf{x} = 3\delta_{-1} + 2\delta_0 - 5\delta_1 + 4\delta_2$, $\mathbf{y} = 7\delta_0 + 6\delta_1$.

   (a) Express $\mathbf{x}_0 = \boxed{2\downarrow}\mathbf{x}$, $\mathbf{x}_1 = \boxed{2\downarrow}(S^{-1}\mathbf{x})$, $\boxed{2\uparrow}\mathbf{x}_0$, and $S\boxed{2\uparrow}\mathbf{x}_1$ as linear combinations of unit impulses and draw stem graphs of all these signals. Then verify that $\mathbf{x} = \boxed{2\uparrow}\mathbf{x}_0 + S\boxed{2\uparrow}\mathbf{x}_1$.

   (b) Calculate the z-transforms $\mathbf{X}(z)$, $\mathbf{X}_0(z)$, $\mathbf{X}_1(z)$, and $\mathbf{Y}(z)$. Then verify that $\mathbf{X}_0(z^2) = 1/2\{\mathbf{X}(z) + \mathbf{X}(-z)\}$, $\mathbf{X}_1(z^2) = z/2\{\mathbf{X}(z) - \mathbf{X}(-z)\}$, and $\mathbf{X}(z) = \mathbf{X}_0(z^2) + z^{-1}\mathbf{X}_1(z^2)$.

   (c) Use the result of (b) to calculate the z-transform of the signal $\mathbf{w} = \mathbf{x} \star \mathbf{y}$.

   (d) Let $\mathbf{y}_{\mathrm{per}, 4}$ be the periodic extension of $\mathbf{y}$ of period 4. Write down the column vector in $\mathbb{R}^4$ corresponding to $\mathbf{y}_{\mathrm{per}, 4}$ and calculate its discrete Fourier transform as a vector in $\mathbb{C}^4$ using the Fourier matrix F_4.

   (e) Use your calculation in (b) to evaluate the discrete Fourier transform $\widehat{\mathbf{y}}_{\mathrm{per}, 4}[k]$ for $k = 0, 1, 2, 3$. The values should match up with those in (d).

   (f) What are the relations among $\|\mathbf{y}\|^2$, $\|\mathbf{y}_{\mathrm{per}, 4}\|^2$, and $\|\widehat{\mathbf{y}}_{\mathrm{per}, 4}\|^2$? Give the general result and then verify it numerically for this example.

(2) This exercise is about signals and FIR filters.

   (a) Suppose $\mathbf{x} = 2\delta_{-2} + 3\delta_{-1} + 4\delta_0 + 5\delta_1$. Express $\boxed{2\downarrow}\mathbf{x} =$ and $\boxed{2\uparrow}\boxed{2\downarrow}\mathbf{x}$ as linear combinations of unit impulses.

(b) Suppose $\mathbf{y} = \sum_{n \in \mathbb{Z}} \mathbf{y}[n]\delta_n$ is any finite signal. Express $\boxed{2\downarrow}\mathbf{y}$ as a linear combination of unit impulses. Then prove the formula $\boxed{2\uparrow}\,\boxed{2\downarrow}\mathbf{y} = \sum_{m \in \mathbb{Z}} \mathbf{y}[2m]\delta_{2m}$.

(c) Let $\mathbf{h}$ be a FIR filter and let $\mathbf{h}^{\mathrm{T}}$ be the *time-reversed* filter with $\mathbf{h}^{\mathrm{T}}[n] = \mathbf{h}[-n]$. Let S be the nonperiodic shift operator. Let $\mathbf{u}$ be any finite signal. Prove that $(\mathbf{h} \star \mathbf{u})[k] = \langle S^k \mathbf{h}^{\mathrm{T}}, \mathbf{u} \rangle$ for every integer k. (Note that the left side is the value of the convolution at time k, while the right side is an inner product with the shifted time-reversed filter.)

(d) Suppose that $\mathbf{h}$ is a FIR filter. Prove that $\langle \mathbf{h} \star \mathbf{v}, \mathbf{w} \rangle = \langle \mathbf{v}, \mathbf{h}^{\mathrm{T}} \star \mathbf{w} \rangle$ for every finite signal $\mathbf{v}$ and $\mathbf{w}$. (Note that both sides are inner products and the time-reversed filter appears on the right.)

(3) Consider the lazy filter bank in Section 4.3.1.

(a) Find the analysis filters, synthesis filters, the modulation matrices $\mathbf{H}_m(z)$ (analysis) and $\mathbf{G}_m(z)$ (synthesis), and the polyphase matrices $\mathbf{H}_p(z)$ (analysis) and $\mathbf{G}_p(z)$ (synthesis) for this filter bank. Verify that the modulation matrices satisfy $\mathbf{G}_m(z)\mathbf{H}_m(z) = 2I$ (2×2 identity matrix).

(b) Do the filters for this transform satisfy the lowpass/highpass conditions?

(4) Suppose the FIR filters $\mathbf{h}_0$ and $\mathbf{h}_1$ have z-transforms $H_0(z) = (1 + z)(1 + az)$ and $H_1(z) = (1 - z)(1 + bz)$, where a and b are real constants. Notice that $a = 0$ and $b = 0$ give the filters for the Haar transform, and that for all values of a and b these filters satisfy the lowpass/highpass conditions $H_0(-1) = 0$ and $H_1(1) = 0$.

(a) Find all values of a and b so that $\det \mathbf{H}_m(z)$ is a nonzero monomial, where $\mathbf{H}_m(z)$ is the modulation matrix for these filters.

(b) With $H_1(z)$ determined as in part (a), find the FIR synthesis filters $\mathbf{g}_0$ and $\mathbf{g}_1$ that go with $\mathbf{h}_0$ and $\mathbf{h}_1$ to give a two-channel PR filter bank.

(5) Suppose the FIR filters $\mathbf{h}_0$ and $\mathbf{h}_1$ have z-transforms $H_0(z) = (1 + z)^3$ and $H_1(z) = (1 - z)(1 + bz + cz^2)$, where b and c are real constants. Notice that for all values of b and c these filters satisfy the lowpass/highpass conditions $H_0(-1) = 0$ and $H_1(1) = 0$.

(a) Find all values of b and c so that $\det \mathbf{H}_m(z)$ is a nonzero monomial, where $\mathbf{H}_m(z)$ is the modulation matrix for these filters.

(b) With $H_1(z)$ determined as in part (a), find the synthesis filters $\mathbf{g}_0$ and $\mathbf{g}_1$ that go with $\mathbf{h}_0$ and $\mathbf{h}_1$ to give a two-channel PR filter bank.

(6) Consider a two-channel filter bank having FIR analysis filters $\mathbf{h}_0$ (lowpass) and $\mathbf{h}_1$ (highpass) with z-transforms $H_0(z)$ and $H_1(z)$. Suppose the polyphase analysis matrix is $\mathbf{H}_p(z) = \begin{bmatrix} (1 + z) & 2 \\ (1 - 3z) & 2 \end{bmatrix}$.

(a) Find $H_0(z)$ and $H_1(z)$ and show that the lowpass/highpass conditions $H_0(-1) = 0$ and $H_1(1) = 0$ are satisfied.

(b) Show that the condition for PR satisfied and find the synthesis filters $\mathbf{g}_0$ and $\mathbf{g}_1$.

(7) Consider a two-channel filter bank having FIR analysis filters $\mathbf{h}_0$ (lowpass) and $\mathbf{h}_1$ (highpass) with z-transforms $H_0(z)$ and $H_1(z)$. Suppose the polyphase analysis matrix is $\mathbf{H}_p(z) = \begin{bmatrix} 1 & (1-z) \\ (1+z) & (2-z^2) \end{bmatrix}$.

(a) Find $H_0(z)$ and $H_1(z)$ and determine whether the conditions $H_0(-1) = 0$ and $H_1(1) = 0$ for lowpass/highpass are satisfied.

(b) Is the condition for PR satisfied by these filters?

(8) Consider a two-channel filter bank having FIR analysis filter $\mathbf{h}_0$ (lowpass) and FIR synthesis filter $\mathbf{g}_0$ (lowpass).

(a) Suppose the high-pass filters $\mathbf{h}_1$ and $\mathbf{g}_1$ are obtained from the low-pass filters by the formulas $H_1(z) = z^k G_0(-z)$ and $G_1(z) = z^{-k} H_0(-z)$, where k is an odd integer. Use the modulation matrices to find the equation that the function $F(z) = H_0(z)G_0(z)$ must satisfy so that the filter bank has the PR property.

(b) Verify that your condition in part (a) is satisfied by $H_0(z) = z+2+z^{-1}$ and $G_0(z) = (3+2z^{-1}-z^{-2})/8$, and find the four Laurent polynomial entries in the polyphase matrix $\mathbf{H}_p(z)$ when $H_1(z) = z^3 G_0(-z)$ and $G_1(z) = z^{-3}H_0(-z)$. Be careful to distinguish between z and z^2 in the formula relating $\mathbf{H}_p(z^2)$ and $\mathbf{H}_m(z)$.

(9) This exercise is about formula (4.54) and the uniqueness of the Bezout polynomial $B_n(y)$.

(a) Write the Bezout equation (4.52) as $B_n(y) = (1-y)^{-n} - y^n D(y)$ where $D(y) = (1-y)^{-n}B_n(1-y)$. Then use this formula for $B_n(y)$ to show that

$$\left(\tfrac{d}{dy}\right)^k B_n(y)\big|_{y=0} = \left(\tfrac{d}{dy}\right)^k (1-y)^{-n}\big|_{y=0} = n(n+1)\cdots(n+k-1)$$

for $k = 0, 1, \ldots, n-1$. (HINT: The derivatives of $y^n D(y)$ of order less than n vanish at $y = 0$.)

(b) Use the result in (a) to obtain formula (4.54). (HINT: Since $B_n(y)$ is a polynomial of degree $n-1$, it is equal to its Taylor polynomial of degree $n-1$.)

(c) To prove the uniqueness of the Bezout polynomial, suppose $\widetilde{B}(y)$ is any polynomial of degree less than n that satisfies (4.52). Define $C(y) = B_n(y) - \widetilde{B}(y)$, so that $C(y)$ is also a polynomial of degree less than n. Show that $C(y) = y^n(1-y)^{-n}C(1-y)$. Use this to prove that $C(y) = 0$ for all y. (HINT: Show that the derivatives $C^{(k)}(0) = 0$ for $k = 0, 1, \ldots, n-1$ and use the fact that $C(y)$ is a polynomial of degree less than n.)

(10) Consider the CDF(p, q) filters $\mathbf{g}_0$ and $\mathbf{h}_0$ whose z-transforms are given by formulas (4.59) and (4.60). Here p and q are positive integers with $p + q = 2n$ even.

(a) Show that $\mathbf{g}_0$ has length $p + 1$.
(b) Show that $\mathbf{h}_0$ has length $p + 2q - 1$.
(c) Check that the results in (a) and (b) agree with the cases when (p, q) is $(2, 2)$, $(3, 1)$, or $(6, 2)$, as worked out in Section 4.5.

(11) In this exercise you will construct the Daubechies 9/7 filters described Example 4.18 by modifying the CDF$(6, 2)$ filters as follows.

(a) The Bezout polynomial $B_4(y) = 20y^3 + 10y^2 + 4y + 1$ has one real root $r_1 < 0$ and two complex conjugate roots r_2, r_3 (you can use the MATLAB `roots` command to find their approximate numerical values, but this is not needed for this exercise). We define a linear polynomial $p_1(y) = y - r_1$ and a quadratic polynomial $p_2(y) = (y - r_2)(y - r_3)$ in terms of these roots. Show that $B_4(y) = p_1(y)p_2(y)/\big(p_1(0)p_2(0)\big)$.
(b) Define the low-pass synthesis and analysis filters as

$$G_0(z) = \tfrac{\sqrt{2}}{16} z^{-2}(1 + z)^4 \, p_1\left(\tfrac{-z+2-z^{-1}}{4}\right)\Big/ p_1(0),$$

$$H_0(z) = \tfrac{\sqrt{2}}{16} z^{-2}(1 + z)^4 \, p_2\left(\tfrac{-z+2-z^{-1}}{4}\right)\Big/ p_2(0).$$

Use the Bezout equation (4.58) and (a) to verify that $G_0(z)H_0(z) + G_0(-z)H_0(-z) = 2$. Hence $G_0(z)$ and $H_0(z)$ are the low-pass filters of a PR filter bank.
(c) Let $\mathbf{g}_0$ be the FIR filter corresponding to $G_0(z)$ and $\mathbf{h}_0$ the FIR filter corresponding to $H_0(z)$. Show that $\mathbf{h}_0$ is symmetric around 0 and has length 9. Show that $\mathbf{g}_0$ is symmetric around 0 and has length 7.

(12) This exercise examines the CDF$(3, 1)$ wavelet transform.

(a) In the formulas in Example 4.16, replace $H_0(z)$ by $z^2 H_0(z)$ and $H_1(z)$ by $z^{-2}H_1(z)$. This modification corresponds to a time advance by two steps in the highpass channel and a time delay by two steps in the lowpass channel. Show that the modified filters still have the PR and lowpass/highpass properties, and involve the same powers of z.
(b) Calculate their analysis polyphase matrix $\mathbf{H}_p(z)$ for the modified filters from (a). Show that it is the same as the polyphase matrix in Example 3.11, where the CDF$(3, 1)$ transform was built up using lifting steps.

(13) Verify the matrix factorizations (4.67) and (4.68).

(14) Factor $\begin{bmatrix} 1 & -3z \\ 2z^{-1} & -5 \end{bmatrix} = \begin{bmatrix} 1 & 0 \\ f(z) & 1 \end{bmatrix}\begin{bmatrix} 1 & g(z) \\ 0 & 1 \end{bmatrix}$ with Laurent polynomials $f(z)$ and $g(z)$.

(15) Let $\mathbf{F}(z) = \begin{bmatrix} (z + 2) & (z + 4 + 2z^{-1}) \\ (z + 1) & (z + 3 + z^{-1}) \end{bmatrix}$.

(a) Verify that $\det \mathbf{F}(z) = 1$.

(b) The entries in the first column of $\mathbf{F}(z)$ both have degree 1. Find a constant c such that the upper left-hand entry in the matrix $\mathbf{G}(z) = \begin{bmatrix} 1 & c \\ 0 & 1 \end{bmatrix} \mathbf{F}(z)$ is a nonzero constant.

(c) Find a Laurent polynomial $f(z)$ such that $\begin{bmatrix} 1 & 0 \\ f(z) & 1 \end{bmatrix} \mathbf{G}(z)$ is upper triangular.

(d) Use (b) and (c) to obtain a factorization $\mathbf{F}(z) = U_2 P U_1$, where P is a prediction matrix (unit lower triangular) and U_1, U_2 are update matrices (unit upper triangular) with Laurent polynomial entries.

(e) Give the *lifting steps* corresponding to the factorization in (d), as in Example 4.22 (there is no normalization step in this case).

(16) Let $\mathbf{x} = \delta_1$. Follow the method of Example 4.23 to find the decomposition of $\mathbf{x}$ into a trend $\mathbf{s}$ and detail $\mathbf{d}$ for the CDF$(2, 2)$ transform. (Notice that $\mathbf{s}$ and $\mathbf{d}$ are *not* obtained by shifting the trend/detail vectors for δ_0.)

(17) Let $\mathbf{x} = \delta_0$. Follow the method of Example 4.23 to find the decomposition of $\mathbf{x}$ into a trend $\mathbf{s}$ and detail $\mathbf{d}$ for the CDF$(3, 1)$ transform (see Example 4.16 for the filters).

(18) Consider the PR filter bank with analysis filters $\mathbf{h}_0 = \delta_{-1} + 2\delta_0 + \delta_1$, $\mathbf{h}_1 = -3\delta_{-1} + 2\delta_0 + \delta_1$, and synthesis filters $\mathbf{g}_0 = (1/8)(3\delta_0 + 2\delta_1 - \delta_2)$, $\mathbf{g}_1 = (1/8)(\delta_0 - 2\delta_1 + \delta_2)$. Take the signal $\mathbf{x} = 3\delta_0 + 4\delta_1$.

(a) Sketch the stem graphs of $\mathbf{x}$, $\mathbf{h}_0^{\mathrm{T}}$, $S^{-2}\mathbf{h}_0^{\mathrm{T}}$, and $S^4\mathbf{h}_0^{\mathrm{T}}$, with each graph on a different axis. Then use these graphs to show that the inner products $\langle S^{2m}\mathbf{h}_0^{\mathrm{T}}, \mathbf{x} \rangle = 0$ for all $m > 1$ and all $m < 0$.

(b) Sketch the stem graphs of $\mathbf{x}$, $\mathbf{h}_1^{\mathrm{T}}$, $S^{-2}\mathbf{h}_1^{\mathrm{T}}$, and $S^4\mathbf{h}_1^{\mathrm{T}}$, with each graph on a different axis. Then use these graphs to verify $\langle S^{2m}\mathbf{h}_1^{\mathrm{T}}, \mathbf{x} \rangle = 0$ for all $m > 1$ and all $m < 0$.

(c) Calculate the following inner products of the signal with the shifted time-reversed analysis filters: $\langle \mathbf{h}_0^{\mathrm{T}}, \mathbf{x} \rangle$, $\langle \mathbf{h}_1^{\mathrm{T}}, \mathbf{x} \rangle$, $\langle S^2\mathbf{h}_0^{\mathrm{T}}, \mathbf{x} \rangle$, $\langle S^2\mathbf{h}_1^{\mathrm{T}}, \mathbf{x} \rangle$.

(d) Use the results of (a), (b), and (c) to calculate the coefficients a_m and b_m in the trend-detail decomposition

$$\mathbf{x} = \mathbf{x_s} + \mathbf{x_d} = \sum_{m \in \mathbb{Z}} a_m S^{2m} \mathbf{g}_0 + \sum_{m \in \mathbb{Z}} b_m S^{2m} \mathbf{g}_1 .$$

Use these coefficients to give explicit formulas for $\mathbf{x_s}$ and $\mathbf{x_d}$ as linear combinations of unit impulses. Finally, check that your formulas for $\mathbf{x_s}$ and $\mathbf{x_d}$ add up to $\mathbf{x}$.

(19) Consider the CDF(p, q) filters $\mathbf{g}_0$ and $\mathbf{h}_0$ whose z-transforms are given by formulas (4.59) and (4.60). Here p and q are positive integers with $p + q = 2n$ even.

(a) Let p be even. Show that $\widetilde{G}_0(z) = z^{-p/2} G_0(z)$ and $\widetilde{H}_0(z) = z^{p/2} H_0(z)$ are unchanged when z is replaced by z^{-1}. Hence the corresponding filters $\widetilde{\mathbf{g}}_0$

and $\widetilde{\mathbf{h}}_0$ are symmetric around 0.

(b) Check that the result in (a) agrees with the cases when (p, q) is $(2, 2)$ or $(6, 2)$, as worked out in Examples 4.20 and 4.17, respectively.

(20) In this exercise you will construct the Daub6 orthogonal filters following the general scheme in Section 4.10.

(a) Given the Bezout polynomial $B_3(y) = 1 + 3y + 6y^2$, calculate the associated polynomial $Q_3(z) = z^2 B_3\left(\frac{-z+2-z^{-1}}{4}\right)$ of degree 4. Check that $z^2 Q_3(z^{-1}) = z^{-2} Q_3(z)$.

(b) Use the MATLAB command `roots` to calculate the four roots of $Q_3(z)$. Verify that the two roots inside the unit circle are (to four decimal places) $r_1 = 0.2873 + 0.1529\mathrm{i}$ and $r_2 = \overline{r_1}$, whereas the two roots outside the unit circle are $1/r_1$ and $\overline{1/r_1}$. Explain why you know the other three roots once you have found r_1.

(c) Use equation (4.90) with $K = 3$ and the normalizing condition $H_0(1) = \sqrt{2}$ to obtain the lowpass Daub6 analysis filter $H_0(z)$. Show that $H_0(z) = \kappa(1 + z^{-1})^3(1 - \alpha z^{-1} + \beta z^{-2})$, where the constants are expressed in terms of the roots r_1 and r_2 as $\alpha = r_1 + r_2$, $\beta = r_1 r_2$, and $\kappa = \sqrt{2}/(8(1 - \alpha + \beta))$. Expand the formula for $H_0(z)$ and calculate the (approximate) numerical values of the filter coefficients.

(d) Use the MATLAB command `[h g rh rg] = daub(6)` with the *Uvi_Wave* package to generate the filters, where `h` is the lowpass analysis filter, `g` is the highpass analysis filter, `rh` is lowpass synthesis filter and `rg` is the highpass synthesis filter. Verify that `g` is obtained from `h` by time reversal and alternation of sign, following the scheme (4.86), and that the synthesis filters are the time reversals of the analysis filters. Observe that `rh` is the filter that you calculated in (c); see the remark at the end of Section 4.10.2.

Chapter 5

Wavelet Transforms for Analog Signals

5.1 Overview

This final chapter introduces wavelet transforms for analog signals, building on the wavelet analysis of digital signals from Chapters 3 and 4. These transforms employ shifts and dilations of a *scaling function* and *wavelet function*, which satisfy *two-scale* equations associated with downsampling. The coefficients in these equations come from the lowpass and highpass filters for PR filter banks studied in Chapter 4. We first examine in detail the simplest case, the Haar transform. Here the scaling and wavelet function are simple step functions. Then we consider more general orthogonal wavelet transforms and the construction of their scaling and wavelet functions by the iterative *cascade algorithm*. The decomposition of a signal into trend and detail subsignals and the associated pyramid algorithm that is the basic feature of the wavelet transform for digital signals also occurs for analog signals. To obtain this *multiresolution analysis* we will use the results about perfect reconstruction filter banks from Chapter 4.

To keep the exposition at a mathematical level consistent with the preceding chapters we only consider wavelet transforms constructed from orthogonal FIR filters such as the Daub2K family, even though wavelet transforms for analog signals constructed from biorthogonal filter pairs, such as the CDF(p, q) family, offer more flexibility in applications. We also make repeated citations of other books and articles for further details and some of the proofs.

5.2 Linear Transformations of Analog Signals

An *analog signal* is a real-valued function $f(t)$ defined for all real t. The term *analog* refers to the continuous range of the variable t (time). We assume that $f(t)$ is obtained as a limit of a sequence of piecewise constant functions. The integral of such a function is then defined as a limit of finite sums.[1] Such a function can have irregularities such as jumps and spikes, which are needed for a realistic mathematical

---

[1] This intuitive description is made precise via the Lebesgue theory of measurable functions and integration, which is beyond the scope of this book.

model of an electrical or acoustical signal.

5.2.1   *Finite-energy analog signals*

An analog signal $f(t)$ has *finite energy* if

$$\int_{-\infty}^{\infty} |f(t)|^2 \, dt < \infty. \tag{5.1}$$

For example, if $f(t)$ is piecewise continuous, bounded, and zero outside some interval $A \leq t \leq B$, then the integral in (5.1) has the usual definition from calculus and $f(t)$ is an analog signal with finite energy. We denote by L^2 the set of all finite-energy analog signals. If $f(t)$ and $g(t)$ are finite-energy signals then the function $h(t) = \alpha f(t) + \beta g(t)$ is also a finite-energy signal. Hence L^2 is a real vector space. Furthermore, the *inner product*

$$\langle f, g \rangle = \int_{-\infty}^{\infty} f(t)g(t) \, dx \tag{5.2}$$

is an absolutely convergent integral (since we are assuming that $f(t)$ is real-valued, no complex conjugate is needed in the inner product).[2] The *energy* of f is

$$\langle f, f \rangle = \int_{-\infty}^{\infty} |f(t)|^2 \, dt.$$

If $\langle f, f \rangle = 0$ then $f(t) = 0$ for all t.[3] This shows that L^2 is a real inner product space with norm $\|f\| = \sqrt{\langle f, f \rangle}$.

Definition 5.1. A sequence $\{g_n : n = 1, 2, \ldots\}$ of functions in L^2 is *Cauchy* if $\lim_{m \to \infty, n \to \infty} \|g_m - g_n\| = 0$.

A fundamental result from the Lebesgue theory of integration is that the normed vector space L^2 is *complete*:

Theorem 5.1. *If* $\{g_n : n = 1, 2, \ldots\}$ *is a Cauchy sequence in* L^2 *then there exists a function* $g \in L^2$ *such that* $\lim_{n \to \infty} \|g - g_n\| = 0$.

We assume this property without further discussion. See any book on Lebesgue measure and integration for details.

5.2.2   *Orthogonal projections*

Let $\ell^2(\mathbb{Z})$ be the inner-product space of all real-valued square-summable functions on $\mathbb{Z}$ (see Example 1.24). For $\mathbf{c} \in \ell^2(\mathbb{Z})$ we follow the digital signal notation in earlier chapters and write $\mathbf{c}[n]$ for the value of $\mathbf{c}$ at $n \in \mathbb{Z}$.

Theorem 5.2 (Riesz–Fischer). *Let* $\{\varphi_k : k \in \mathbb{Z}\}$ *be an orthonormal set of functions in* L^2.

---

[2] To prove these assertions use the same pointwise inequalities as for $\ell^2(\mathbb{Z})$ in Example 1.24.

[3] If f is not a continuous function, we can only conclude that $f(t) = 0$ except on a set of Lebesgue measure zero. In this case we treat f as the zero function.

(1) *Given* $\mathbf{c} \in \ell^2(\mathbb{Z})$, *define* $g_n = \sum_{|k|\leq n} \mathbf{c}[k]\,\varphi_k$. *Then there exists a function* $g \in L^2$ *such that* $\lim_{n\to\infty} \|g - g_n\| = 0$. *Furthermore*

$$\mathbf{c}[k] = \langle \varphi_k, g \rangle \quad and \quad \sum_{k\in\mathbb{Z}} |\mathbf{c}[k]|^2 = \int_{-\infty}^{\infty} |g(t)|^2\, dt. \tag{5.3}$$

(2) *Given* $f \in L^2$, *define a function* $\mathbf{c}$ *on* $\mathbb{Z}$ *by* $\mathbf{c}[k] = \langle \varphi_k, f \rangle$. *Then*

$$\sum_{k\in\mathbb{Z}} |\mathbf{c}[k]|^2 \leq \int_{-\infty}^{\infty} |f(t)|^2\, dt. \tag{5.4}$$

Hence $\mathbf{c} \in \ell^2(\mathbb{Z})$.

Proof. (1) Let $0 \leq m < n$. Then $g_n - g_m = \sum_{m<|k|\leq n} \mathbf{c}[k]\,\varphi_k$. Since $\{\varphi_k : k \in \mathbb{Z}\}$ is an orthonormal set of functions, Parseval's formula (1.26) gives

$$\|g_n - g_m\|^2 = \sum_{m<|k|\leq n} |\mathbf{c}[k]|^2 \leq \sum_{m<|k|} |\mathbf{c}[k]|^2.$$

On the right we have the tail of a convergent series since $\mathbf{c}$ is square summable. It follows that $\{g_n\}$ is a Cauchy sequence of functions in L^2. Let $g \in L^2$ be the function from Theorem 5.1 such that $\lim_{n\to\infty} \|g - g_n\| = 0$. Then

$$\langle \varphi_k, g \rangle = \lim_{n\to\infty} \langle \varphi_k, g_n \rangle \quad and \quad \|g\|^2 = \lim_{n\to\infty} \|g_n\|^2$$

(verification left as an exercise), proving (5.3).

(2) Define the functions g_n in terms of the coefficients $\mathbf{c}[k]$ as in (1). Then

$$\langle g_n, f \rangle = \sum_{|k|\leq n} \mathbf{c}[k]\langle \varphi_k, f \rangle = \sum_{|k|\leq n} |\mathbf{c}[k]|^2. \tag{5.5}$$

We know by Parseval's formula that $\|g_n\|^2 = \sum_{|k|\leq n} |\mathbf{c}[k]|^2$. Also, by the Cauchy–Schwarz inequality $|\langle g_n, f \rangle| \leq \|g_n\|\|f\|$. Using these facts in (5.5) we obtain the inequality

$$\sum_{|k|\leq n} |\mathbf{c}[k]|^2 \leq \|f\|^2$$

for all positive integers n. Now let $n \to \infty$ to obtain (5.4). $\qquad\square$

We denote the function g in Theorem 5.2(1) as[4]

$$g = \sum_{k\in\mathbb{Z}} \mathbf{c}[k]\,\varphi_k.$$

Let $V \subset L^2$ be the set of all functions g of this form, with $\mathbf{c}$ ranging over $\ell^2(\mathbb{Z})$. If $h \in V$ is similarly defined using $\mathbf{d} \in \ell^2(\mathbb{Z})$, then for any real scalars α, β

$$\alpha g + \beta h = \sum_{k\in\mathbb{Z}} \big(\alpha\mathbf{c}[k] + \beta\mathbf{d}[k]\big)\,\varphi_k,$$

$$\int_{-\infty}^{\infty} g(t)h(t)\, dt = \sum_{k\in\mathbb{Z}} \mathbf{c}[k]\mathbf{d}[k].$$

---

[4]The sign of equality must be understood as the L^2 norm limit of the finite partial sums, as in equation (1.37) for Fourier series. For a given value of t, the numerical series $\sum_{k\in\mathbb{Z}} \mathbf{c}[k]\,\varphi_k(t)$ does not necessarily converge.

Thus V is a subspace of L^2 that contains all the finite linear combinations of the functions φ_k, together with all limits (in the sense of the L^2 norm) of such linear combinations. In signal-processing terms, this means that linear operations on the analog signals in V and inner products of these signals can be carried out using the corresponding digital signals in $\ell^2(\mathbb{Z})$.

Definition 5.2. We write $V = \overline{\text{Span}}\{\varphi_k : k \in \mathbb{Z}\}$ and call V the *closed linear span* of the given orthonormal set.

Theorem 5.3. *For every $f \in L^2$ there are unique functions $g \in V$ and $h \perp V$ such that $f = g + h$. We write $g = P_V f$ and call g the orthogonal projection of f onto V. It is given by*

$$g = \lim_{n \to \infty} \sum_{|k| \leq n} \langle \varphi_k , f \rangle \, \varphi_k ,$$

where the limit exists in the L^2 norm sense. In particular, the Bessel inequality

$$\sum_{k \in \mathbb{Z}} |\langle \varphi_k , f \rangle|^2 \leq \int_{-\infty}^{\infty} |f(t)|^2 \, dt \tag{5.6}$$

is valid for all $f \in L^2$. Furthermore, $f \in V$ if and only if equality holds in (5.6).

Proof. Let $\mathbf{c}[k] = \langle \varphi_k , f \rangle$ as in Theorem 5.2(2). Then $\mathbf{c} \in \ell^2(\mathbb{Z})$ and we can define $g \in V$ from $\mathbf{c}$ as in Theorem 5.2(1). Set $h = f - g$. Then $\langle \varphi_k , h \rangle = 0$ for all $k \in \mathbb{Z}$ by (5.3), so $h \perp V$. The uniqueness of g (and hence h) is easy: if $g_1 + h_1 = g_2 + h_2$ with $g_j \in V$ and $h_j \perp V$, then the function $g_1 - g_2 = h_2 - h_1$ is perpendicular to V. Hence $\langle \varphi_k , g_1 \rangle = \langle \varphi_k , g_2 \rangle$ for all k, so $g_1 = g_2$. Finally, $\|f\|^2 = \|g\|^2 + \|h\|^2 \geq \|g\|^2$ by the Pythagorean formula. The left side of Bessel's inequality (5.6) is $\|g\|^2$, so equality holds if and only if $h = 0$, which means $f \in V$. $\square$

5.2.3   *Shift and dilation operators*

Define the *shift operator* S and the *dilation operator* (scaling operator) D on an analog signal $f(t)$ by

$$Sf(t) = f(t-1) \quad \text{and} \quad Df(t) = \sqrt{2}f(2t) .$$

The action of S shifts the graph of $f(t)$ one unit to the right. The action of D on the graph of $f(t)$ *compresses* the horizontal axis by a factor of 2 and *expands* the vertical axis by a factor of $\sqrt{2}$. These operators are invertible:

$$S^{-1}f(t) = f(t+1) \quad \text{and} \quad D^{-1}f(t) = \frac{1}{\sqrt{2}}f(t/2) .$$

Thus for every integer k (positive and negative)

$$S^k f(t) = f(t-k) \quad \text{and} \quad D^k f(t) = 2^{k/2} f(2^k t) .$$

Remark 5.1. With the factor $\sqrt{2}$ removed D becomes the downsampling operator on digital signals, which is *not* invertible, since $t/2$ is not an integer when t is odd. The invertibility of the dilation operator on analog signals is a fundamental difference between wavelet transforms for analog signals and for digital signals.

The operators S and D do not commute with each other. Applying first S^2 to a function $f(t)$ and then D to $S^2 f(t)$ yields

$$DS^2 f(t) = \sqrt{2}(S^2 f)(2t) = \sqrt{2} f(2t - 2).$$

In the opposite order, applying D to $f(t)$ and then S to $Df(t)$ gives the same result:

$$SDf(t) = Df(t - 1) = \sqrt{2} f(2t - 2).$$

This shows that the shift and dilation operators satisfy the *commutation relation*[5]

$$SD = DS^2. \tag{5.7}$$

If $f(t)$ and $g(t)$ are finite-energy signals, then

$$\langle Sf, Sg \rangle = \int_{-\infty}^{\infty} f(t-1)g(t-1)\, dt = \int_{-\infty}^{\infty} f(s)g(s)\, ds = \langle f, g \rangle,$$

$$\langle Df, Dg \rangle = \int_{-\infty}^{\infty} f(2t)g(2t)\, 2dt = \int_{-\infty}^{\infty} f(s)g(s)\, ds = \langle f, g \rangle.$$

To prove this, we make a change of variables in the integrals ($s = t - k$, $ds = dt$ in the first and $s = 2t$, $ds = 2dt$ in the second; the limits of integration don't change). Thus S and D are unitary operators on L^2 (they preserve the inner product and the norm). This means that the inverse operators satisfy

$$\langle S^{-1}f, g \rangle = \langle f, Sg \rangle, \qquad \langle D^{-1}f, g \rangle = \langle f, Dg \rangle \tag{5.8}$$

for all $f, g \in L^2$.

5.3 Haar Wavelet Transform for Analog Signals

The first wavelet transform for analog signals was developed by A. Haar in 1910 (however the term *wavelet* was only introduced in the 1980's). It is built from a simple *scaling function* and a corresponding *wavelet function*. Although this transform is too crude for most signal processing applications, it furnishes the best introduction to more refined wavelet transforms.

5.3.1 *Haar scaling function*

Define the *Haar scaling function* by

$$\phi_{\text{Haar}}(t) = \begin{cases} 1 & \text{for } 0 \le t < 1, \\ 0 & \text{otherwise}. \end{cases} \tag{5.9}$$

For every integer k the shifted function $S^k \phi_{\text{Haar}}(t)$ has a graph that is a box of height 1 above the interval $k \le t < k+1$. Hence if j, k, m are integers, then

$$\langle D^j S^k \phi_{\text{Haar}}, D^j S^m \phi_{\text{Haar}} \rangle = \langle \phi_{\text{Haar}}, S^{m-k} \phi_{\text{Haar}} \rangle = \begin{cases} 1 & \text{for } k = m, \\ 0 & \text{otherwise}. \end{cases}$$

---

[5]This is the analog-signal version of formulas (3.39) for periodic discrete wavelet transforms.

Here we have used the unitary property of D and (5.8) in the first step, and then the non-overlapping property of the graphs of ϕ_{Haar} and $S^n\phi_{\text{Haar}}$ when $n = m - k \neq 0$ to see that the integral giving the inner product is zero. Thus for each fixed integer j we have an orthonormal set of shifted and dilated functions

$$\phi_{j,k}(t) = D^j S^k \phi_{\text{Haar}}(t) = 2^{j/2}\phi_{\text{Haar}}(2^j t - k) \qquad \text{for } k \in \mathbb{Z}.$$

Suppose $|j|$ is large. When j is positive, the graph of $\phi_{j,k}$ is a narrow tall rectangle, whereas when j is negative the graph is a wide short rectangle. For all j the rectangle is centered at $k + (1/2)^{j+1}$, and the rectangles for different values of k are disjoint.

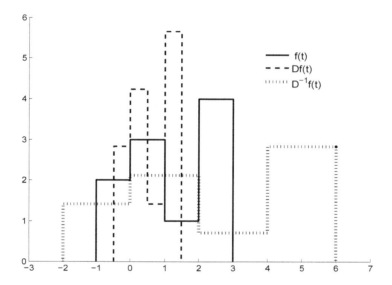

Fig. 5.1   Dilations of a piecewise constant function

Figure 5.1 compares the graph of the function $f = 2\phi_{0,-1} + 3\phi_{0,0} + \phi_{0,1} + 4\phi_{0,2}$ with the graphs of the functions $Df = 2\phi_{1,-1} + 3\phi_{1,0} + \phi_{1,1} + 4\phi_{1,2}$ and $D^{-1}f = 2\phi_{-1,-1} + 3\phi_{-1,0} + \phi_{-1,1} + 4\phi_{-1,2}$.

5.3.2   *Haar multiresolution analysis*

Given a finite-energy analog signal $f(t)$ and an integer j, we can define a digital signal $\mathbf{s}_j$ by

$$\mathbf{s}_j[k] = 2^{j/2}\langle \phi_{j,k}, f\rangle \quad \text{for } k \in \mathbb{Z}. \tag{5.10}$$

The values of this digital signal are called the level-j *Haar trend coefficients* of f:

$$\mathbf{s}_j[k] = 2^j \int_{k/2^j}^{(k+1)/2^j} f(t)\,dt = \text{ average value of } f(t) \text{ on } I_{j,k}, \tag{5.11}$$

where $I_{j,k} = \{t : k/2^j \leq t < (k+1)/2^j\}$. This formula holds because the function $2^{j/2}\phi_{j,k}(t)$ has the constant value 2^j on the dyadic interval $I_{j,k}$ and is zero outside this interval. Using these trend coefficients, we define a function $f_j(t)$ by

$$f_j(t) = 2^{-j/2} \sum_{k \in \mathbb{Z}} \mathbf{s}_j[k]\,\phi_{j,k}(t)\,. \tag{5.12}$$

This formula for f_j has two remarkable properties.[6]

(i) *Pointwise convergence*: For each t there is at most one term in the series (5.12) that is nonzero, since $\phi_{j,k}$ is nonzero only on $I_{j,k}$ and these intervals are mutually disjoint for different k.

(ii) *Locality*: If $f(t) = 0$ for $a \leq t \leq b$, then in the series (5.12) the trend coefficients $\mathbf{s}_j[k]$ are zero for $a2^j \leq k \leq b2^j - 1$.

Definition 5.3. The *Haar multiresolution space* V_j is the closed linear span of the orthonormal set $\{\phi_{j,k} : k \in \mathbb{Z}\}$ of shifted and dilated Haar scaling functions.

For $f \in L^2$ write $P_j f$ for the orthogonal projection of f onto V_j, as defined in Theorem 5.3.

Proposition 5.1. *The function $f_j(t)$ in equation (5.12) is equal[7] to $P_j f(t)$. Conversely, given any digital signal $\mathbf{s}_j \in \ell^2(\mathbb{Z})$, the analog signal $f_j(t)$ defined by equation (5.12) is in V_j. Thus V_j consists of all the finite-energy signals that are piecewise constant on the grid $2^{-j}\mathbb{Z}$.*

Proof. This follows directly from the definition of V_j and Theorem 5.3. $\qquad\square$

We can refine the grid $2^{-j}\mathbb{Z}$ by splitting each interval $I_{j,k}$ in half to get the grid $2^{-j-1}\mathbb{Z}$. Since $I_{j+1,k} \subset I_{j,k}$, a function that is constant on $I_{j,k}$ for all integers k is also constant on the smaller intervals $I_{j+1,k}$ for all integers k. Furthermore, the dilated interval $2I_{j+1,k} = I_{j,k}$, so $f(t)$ is in V_j if and only if $f(2t)$ is in V_{j+1}. These observations prove that

$$V_j \subset V_{j+1}\,, \quad DV_j = V_{j+1}\,, \quad \text{and} \quad D^{-1}V_j = V_{j-1} \quad \text{for all integers } j.$$

Figure 5.1 shows a function $f \in V_0$ together with the functions $Df \in V_1$ and $D^{-1}f \in V_{-1}$.

We call the increasing collection of subspaces

$$\cdots \subset V_{-2} \subset V_{-1} \subset V_0 \subset V_1 \subset V_2 \subset \cdots$$

the *Haar multiresolution* of the space of analog signals. It is the model for all the multiresolutions that we will construct using orthogonal filter banks, so we shall examine its properties in more detail.

---

[6]These properties do not hold for Fourier series or for the general orthogonal series in the Riesz–Fischer theorem.

[7]This means that $\|f_j - P_j f\| = 0$, so the functions are equal almost everywhere in the sense of Lebesgue measure on $\mathbb{R}$.

The Haar series expansion (5.12) has the following special property.[8]

Lemma 5.1. *If $f \in L^2$ and the function f_j is defined by equation (5.12), then*

$$|f_j(t)| \le 2^{j/2}\|f\| \quad \text{for all } t \in \mathbb{R}. \tag{5.13}$$

Proof. For a fixed t we can view the sum in (5.12) giving $f(t)$ as the inner product of two vectors in $\ell^2(\mathbb{Z})$, one with components $a_{j,k} = \langle \phi_{j,k}, f \rangle$ and the other with components $\phi_{j,k}(t)$ for $k \in \mathbb{Z}$. Applying the Cauchy–Schwarz inequality to this inner product, we conclude that

$$|f_j(t)| \le \left\{ \sum_{k \in \mathbb{Z}} |a_{j,k}|^2 \right\}^{1/2} \left\{ \sum_{k \in \mathbb{Z}} |\phi_{j,k}(t)|^2 \right\}^{1/2} \le \|f\| \left\{ \sum_{m < |k| \le n} |\phi_{j,k}(t)|^2 \right\}^{1/2},$$

where the last inequality comes from Bessel's inequality (5.6). But for each t there is only one value of k with $\phi_{j,k}(t) \ne 0$; for this k we have $\phi_{j,k}(t) = 2^{j/2}$. This proves inequality (5.13).   □

Let $f \in V_j$. From Lemma (5.1) we see that if j is negative and large in absolute value, then $|f(t)|$ is extremely small relative to $\|f\|$. In particular, if f is in V_j for *all* j then $f(t) = 0$ for all t, since $\|f\|$ is a fixed number. We write $\lim_{j \to -\infty} V_j = \bigcap_{j \in \mathbb{Z}} V_j$. The argument just given proves that

$$\lim_{j \to -\infty} V_j = \{\mathbf{0}\}, \tag{5.14}$$

the vector space whose only element is the zero vector.

Finally, we examine how the spaces V_j grow as $j \to +\infty$.

Theorem 5.4. *For every function $f(t)$ in L^2 the distance between $f(t)$ and $P_j f(t)$ (as measured by the L^2 norm) goes to zero as $j \to +\infty$. Thus $\lim_{j \to +\infty} \|f - P_j f\| = 0$.*

Proof. By definition of the space L^2 the function $f(t)$ can be approximated (within an arbitrarily small prescribed error in the L^2 norm) by a function $g(t)$ that is piecewise constant on a finite set of intervals and is zero for large $|t|$. By the triangle inequality

$$\|f - P_j f\| \le \|f - g\| + \|g - P_j g\| + \|P_j(g - f)\|$$
$$\le 2\|f - g\| + \|g - P_j g\|,$$

since $\|P_j(g - f)\| \le \|g - f\|$ by Theorem 5.3. Hence it is enough to prove the theorem for $g(t)$. But $g(t)$ is a finite linear combination of box functions

$$h(t) = \begin{cases} 1 & \text{if } a \le t < b, \\ 0 & \text{otherwise}, \end{cases}$$

where $a < b$ are real numbers. Since the projection P_j is a linear transformation, the same triangle inequality argument shows that it is enough to prove the theorem for the function $h(t)$.

---

[8]This property does not hold for most orthonormal sets of functions in the Riesz–Fischer theorem.

Let h be the function defined above. Since $(h - P_j h) \perp P_j h$, the Pythagorean theorem gives

$$\|h - P_j h\|^2 = \|h\|^2 - \|P_j h\|^2 = (b - a) - \|P_j h\|^2 \,.$$

So we need to prove that $\lim_{j \to \infty} \|P_j h\|^2 = b - a$. From Parseval's formula

$$\|P_j h\|^2 = \sum_{k \in \mathbb{Z}} |\langle \phi_{j,k}, h \rangle|^2 \,.$$

The inner products in the sum on the right are the integrals[9]

$$\langle \phi_{j,k}, h \rangle = 2^{j/2} \int_a^b \phi_{\text{Haar}}(2^j t - k)\, dt = 2^{-j/2} \int_{2^j a}^{2^j b} \phi_{\text{Haar}}(y - k)\, dy \,,$$

where we made a change of variable $y = 2^j t$ and $dy = 2^j dt$. We may assume $2^j(b - a) \geq 1$ since we are letting $j \to +\infty$. By looking at the graph of $\phi_{\text{Haar}}(y - k)$ (a unit square with base $k \leq y < k + 1$), we see that

$$\int_{2^j a}^{2^j b} \phi_{\text{Haar}}(y - k)\, dy = \begin{cases} 1 & \text{if } 2^j a \leq k \leq 2^j b - 1, \\ \kappa & \text{if } 2^j a - 1 < k < 2^j a \text{ or } 2^j b - 1 < k < 2^j b, \\ 0 & \text{otherwise}, \end{cases}$$

where $0 < \kappa < 1$. The first case occurs when the graph of $\phi_{\text{Haar}}(y-k)$ lies completely above the interval $2^j a \leq y \leq 2^j b$, which is true for all but at most two of the integers k in this interval. The second case occurs when only part of the graph lies above this interval. Hence when j is positive and large, then (with an error of order 2^{-j})

$$\sum_{k \in \mathbb{Z}} |\langle \phi_{j,k}, h \rangle|^2 \approx 2^{-j} \#\{\text{integers } k \text{ in } [2^j a, 2^j b]\} \approx b - a \,,$$

where $\#A$ denotes the number of elements in the set A. In the limit as $j \to +\infty$ these approximations become equalities. $\qquad \square$

We write[10] $\lim_{j \to +\infty} V_j = \overline{\bigcup_{j \in \mathbb{Z}} V_j}$ for the subspace of all functions that are limits (in the L^2 norm) of a sequence of functions $f_n \in V_{j_n}$. Then the conclusion of Theorem 5.4 can be stated as

$$\lim_{j \to +\infty} V_j = L^2 \,. \tag{5.15}$$

5.3.3 *Haar wavelet and wavelet transform*

The functions in V_1 can change values twice as often as those in V_0. The basic example is the *Haar wavelet function*

$$\psi_{\text{Haar}}(t) = \begin{cases} 1 & \text{for } 0 \leq t < 1/2, \\ -1 & \text{for } 1/2 \leq t < 1. \\ 0 & \text{otherwise}. \end{cases}$$

---

[9]We follow the argument in [Strichartz (1993), Lemma B1.2], which we also use later in Theorem 5.8.
[10]The bar denotes closure in the L^2 norm.

We can write this formula in terms of the shift and dilation operators as

$$\psi_{\text{Haar}} = D\phi_{\text{Haar}} - DS\phi_{\text{Haar}}.\tag{5.16}$$

Clearly ψ_{Haar} has integral 0, norm 1, and $\langle\psi_{\text{Haar}}, \phi_{\text{Haar}}\rangle = 0$.

For all integers j, k we define the shifted and dilated *Haar wavelets* by

$$\psi_{j,k}(t) = D^j S^k \psi_{\text{Haar}}(t) = 2^{j/2}\psi_{\text{Haar}}(2^j t - k).$$

Then $\psi_{j,k} \in V_{j+1}$ since $\psi_{\text{Haar}} \in V_1$. Since $D^j S^k$ is a unitary transformation, we have

$$\langle\psi_{j,k}, \phi_{j,m}\rangle = 0 \quad \text{for all } k, m \quad \text{and} \quad \langle\psi_{j,k}, \psi_{j,m}\rangle = \begin{cases} 1 & \text{if } k = m, \\ 0 & \text{otherwise}. \end{cases}$$

Figure 5.2 shows the basic Haar wavelet and two shifted and dilated Haar wavelets.

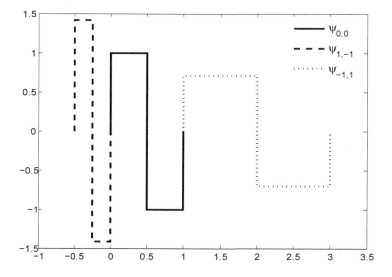

Fig. 5.2   Haar wavelet $\psi_{0,0} = \psi$ with $\psi_{1,-1} = DS^{-1}\psi$ and $\psi_{-1,1} = D^{-1}S\psi$

Given a finite-energy analog signal $f(t)$ and an integer j, we can define a digital signal $\mathbf{d}_j$ by

$$\mathbf{d}_j[k] = 2^{j/2}\langle\psi_{j,k}, f\rangle \quad \text{for } k \in \mathbb{Z}.\tag{5.17}$$

We call the values of $\mathbf{d}_j$ the *level-j Haar detail coefficients* of f. Using these coefficients, we define a function $g_j(t)$ by

$$g_j(t) = 2^{-j/2}\sum_{k\in\mathbb{Z}}\mathbf{d}_j[k]\,\psi_{j,k}(t) = \sum_{k\in\mathbb{Z}}\langle\psi_{j,k}, f\rangle\,\psi_{j,k}(t).\tag{5.18}$$

The series for g_j has the same pointwise convergence and locality properties as the series (5.12) for f_j.

Definition 5.4. The *Haar detail space* W_j is the closed linear span of the orthonormal set $\{\psi_{j,k} : k \in \mathbb{Z}\}$ of shifted and dilated Haar wavelet functions.

For $f \in L^2$ write $Q_j f$ for the orthogonal projection of f onto W_j (see Theorem 5.3).

Proposition 5.2. *The function $g_j(t)$ given by equation (5.18) is equal[11] to $Q_j f(t)$. Conversely, given any digital signal $\mathbf{d}_j \in \ell^2(\mathbb{Z})$, the analog signal $g_j(t)$ defined by the first sum in equation (5.18) is in W_j. Thus W_j consists of all the signals in V_j whose constant value on the left half of each interval $I_{j,k}$ is the negative of the constant value on the right half of this interval.*

Proof. This follows directly from the definition of W_j and Theorem 5.3. □

The space W_j is a subspace of V_{j+1}. Furthermore, every function $g(t)$ in W_j is orthogonal to every function $f(t)$ in V_m if $m \le j$, since $f(t)$ has a constant value on each interval $I_{j,k}$. Thus

$$W_j \perp V_m \quad \text{if } j \ge m. \tag{5.19}$$

Since $W_m \subset V_m$, relation (5.19) gives

$$W_j \perp W_m \quad \text{for all integers } j \ne m. \tag{5.20}$$

We write $V_j \oplus W_j$ for the subspace of all functions in V_{j+1} of the form $f(t)+g(t)$, where $f(t) \in V_j$ and $g(t) \in W_j$.

Theorem 5.5. *Every function $f_{j+1} \in V_{j+1}$ has a unique decomposition $f_{j+1} = f_j + g_j$ with $f_j \in V_j$ and $g_j \in W_j$. Thus V_{j+1} is the direct sum*

$$V_{j+1} = V_j \oplus W_j. \tag{5.21}$$

Proof. The key step is to go from level 1 to level 0 using the *two-scale equations*

$$D\phi_{\text{Haar}} = \frac{1}{\sqrt{2}}\big(\phi_{\text{Haar}} + \psi_{\text{Haar}}\big) \quad \text{and} \quad DS\phi_{\text{Haar}} = \frac{1}{\sqrt{2}}\big(\phi_{\text{Haar}} - \psi_{\text{Haar}}\big). \tag{5.22}$$

These equations hold because

$$\phi_{\text{Haar}}(t) + \psi_{\text{Haar}}(t) = \begin{cases} 2 & \text{when } 0 \le t < 1/2, \\ 0 & \text{otherwise}. \end{cases}$$

$$\phi_{\text{Haar}}(t) - \psi_{\text{Haar}}(t) = \begin{cases} 2 & \text{when } 1/2 \le t < 1, \\ 0 & \text{otherwise}. \end{cases}$$

From (5.22) and the commutation relation (5.7) we get similar two-scale equations from level $j+1$ to level j:

$$\phi_{j+1,2k} = D^j D S^{2k} \phi_{\text{Haar}} = D^j S^k D \phi_{\text{Haar}}$$

$$= \frac{1}{\sqrt{2}}\big(D^j S^k \phi_{\text{Haar}} + D^j S^k \psi_{\text{Haar}}\big) = \frac{1}{\sqrt{2}}\big(\phi_{j,k} + \psi_{j,k}\big) \tag{5.23}$$

---

[11]This means that $\|g_j - Q_j f\| = 0$, so the functions are equal almost everywhere in the sense of Lebesgue measure on $\mathbb{R}$.

and

$$\phi_{j+1,2k+1} = D^j D S^{2k} S \phi_{\text{Haar}} = D^j S^k D S \phi_{\text{Haar}}$$

$$= \frac{1}{\sqrt{2}} \left(D^j S^k \phi_{\text{Haar}} - D^j S^k \psi_{\text{Haar}} \right) = \frac{1}{\sqrt{2}} \left(\phi_{j,k} - \psi_{j,k} \right). \tag{5.24}$$

We now use these formulas to obtain decomposition (5.21).

Let f_{j+1} be a function in V_{j+1}. It has a Haar expansion (5.12) with j replaced by $j+1$. We write the expansion as a sum of even and odd shifts of the level-$(j+1)$ Haar scaling function:

$$f_{j+1} = 2^{-(j+1)/2} \sum_{k \in \mathbb{Z}} \mathbf{s}_{j+1}[2k] \, \phi_{j+1,2k} + 2^{-(j+1)/2} \sum_{k \in \mathbb{Z}} \mathbf{s}_{j+1}[2k+1] \, \phi_{j+1,2k+1} \,.$$

From equations (5.23) and (5.24) we can write this formula for f in terms of level-j Haar scaling functions and wavelets:

$$f_{j+1} = 2^{j/2} \sum_{k \in \mathbb{Z}} \mathbf{s}_{j+1}[2k] \frac{1}{2} \left(\phi_{j,k} + \psi_{j,k} \right) + 2^{j/2} \sum_{k \in \mathbb{Z}} \mathbf{s}_{j+1}[2k+1] \frac{1}{2} \left(\phi_{j,k} - \psi_{j,k} \right)$$

$$= 2^{j/2} \sum_{k \in \mathbb{Z}} \mathbf{s}_j[k] \, \phi_{j,k} + 2^{j/2} \sum_{k \in \mathbb{Z}} \mathbf{d}_j[k] \, \psi_{j,k} = f_j + g_j$$

with $f_j \in V_j$ and $g_j \in W_j$. This proves (5.21). □

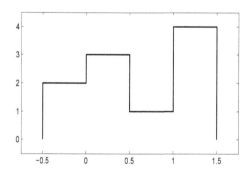

Fig. 5.3   Analog signal f_1 in V_1

Example 5.1. Let $f_1 = (1/\sqrt{2})(2\phi_{1,-1} + 3\phi_{1,0} + \phi_{1,1} + 4\phi_{1,2})$ be the function in V_1 shown in Figure 5.3. We use equations (5.23) and (5.24) to express $f_1 = f_0 + g_0$, where $f_0 = \phi_{0,-1} + 2\phi_{0,0} + 2\phi_{0,1} \in V_0$ and $g_0 = -\psi_{0,-1} + \psi_{0,0} + 2\psi_{0,1} \in W_0$. The graphs of f_0 and g_0 are shown in Figures 5.4 and 5.5. We can see from the graphs that f_0 gives the average value of f_1 over each of the three intervals $-1 \le t < 0$, $0 \le t < 1$, $1 \le t < 2$, while g_0 gives the correction to these averages needed to obtain f_1 on the six subintervals $-1 \le t < -0.5$, ..., $1.5 \le t < 2$. ∎

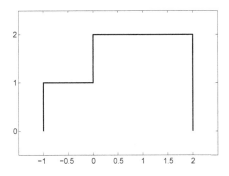

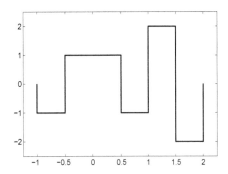

Fig. 5.4   Projection f_0 onto V_0 Fig. 5.5   Projection g_0 onto W_0

Remark 5.2. In the splitting $f_{j+1} = f_j + g_j$ from the proof of Theorem 5.5 the coefficients of f_j and g_j are

$$\mathbf{s}_j[k] = \frac{1}{2}\left(\mathbf{s}_{j+1}[2k] + \mathbf{s}_{j+1}[2k+1]\right),$$

$$\mathbf{d}_j[k] = \frac{1}{2}\left(\mathbf{s}_{j+1}[2k] - \mathbf{s}_{j+1}[2k+1]\right).$$

If we apply a one-scale Haar discrete wavelet analysis transform to the digital signal $\mathbf{s}_{j+1}$ as in Section 3.2, we obtain the same digital signals $\mathbf{s}_j$ and $\mathbf{d}_j$ as the *trend* and the *detail* at level j.

5.4   Scaling and Wavelet Functions from Orthogonal Filter Banks

The pioneering paper [Daubechies (1988)] constructs new families of scaling and wavelet functions that perform better than the Haar functions for analog signal processing, but still have similar localization properties. The starting point is to use the two-channel orthogonal filter banks with finite-impulse response (FIR) filters from Chapter 4.

Let $\mathbf{g}_0$ (lowpass) and $\mathbf{g}_1$ (highpass) be real-valued FIR synthesis filters for a two-channel orthogonal filter bank (see Section 4.9). Let $\mathbf{h}_0 = \mathbf{g}_0^{\mathrm{T}}$ and $\mathbf{h}_1 = \mathbf{g}_1^{\mathrm{T}}$ be the analysis filters. Recall from Theorem 4.14 that one of these four filters, together with an integer specifying the relative alignment of the lowpass and highpass filters, determines the other three filters. For simplicity in writing the formulas in this section we normalize the filters as follows.

From Lemma 4.4 the filter $\mathbf{g}_0$ must have even length $L+1$, where $L \geq 1$ is an odd integer. By making a time shift, we may arrange that

$$\begin{aligned}
\mathbf{g}_0[k] &= 0 \quad \text{if } k < 0 \text{ or } k > L, \\
\mathbf{g}_1[k] &= (-1)^k \mathbf{g}_0[L-k] \quad \text{for all integers } k.
\end{aligned} \tag{5.25}$$

The half-band equation (4.81) and the lowpass condition $G_1(1) = 0$ imply that

$G_0(1)^2 = 2$, so we can take $G_0(1) = \sqrt{2}$. Thus

$$\sum_{k=0}^{L} (-1)^k \mathbf{g}_0[k] = 0 \,, \quad \sum_{k=0}^{L} \mathbf{g}_0[k] = \sqrt{2} \,, \quad \sum_{k=0}^{L} \mathbf{g}_1[k] = 0 \,. \tag{5.26}$$

The first two equations in (5.26) give the *sum rule*

$$\sqrt{2} \sum_{k \text{ even}} \mathbf{g}_0[k] = \sqrt{2} \sum_{k \text{ odd}} \mathbf{g}_0[k] = 1 \,. \tag{5.27}$$

For the rest of this chapter we will assume that the filters $\mathbf{g}_0$, $\mathbf{g}_1$ satisfy (5.25) and (5.26).

5.4.1   *Cascade algorithm*

The Fourier transform for analog signals uses oscillating sinusoidal functions that are solutions to the differential equation for simple harmonic motion. A wavelet transform for analog signals uses a *scaling function* and its companion *wavelet function* that satisfy two-scale equations involving the lowpass and highpass filter coefficients.

Definition 5.5.   A nonzero function $\phi(t)$ in L^2 is a scaling function for the orthogonal FIR filter $\mathbf{g}_0$ if it satisfies the two-scale equation

$$\phi(t) = \sqrt{2} \sum_{k \in \mathbb{Z}} \mathbf{g}_0[k] \phi(2t - k) \quad \text{for all } t \in \mathbb{R} \,.$$

The wavelet function associated with a scaling function ϕ for $\mathbf{g}_0$ is the function ψ defined by

$$\psi(t) = \sqrt{2} \sum_{k=0}^{L} \mathbf{g}_1[k] \, \phi(2t - k) \quad \text{for all } t \in \mathbb{R} \,.$$

One way to understand these equations is to define linear transformations M_0 and M_1 on L^2 by

$$M_0 = \sum_{k=0}^{L} \mathbf{g}_0[k] \, DS^k \quad \text{and} \quad M_1 = \sum_{k=0}^{L} \mathbf{g}_1[k] \, DS^k \,. \tag{5.28}$$

Since $DS^k \phi(t) = \sqrt{2} \phi(2t - k)$, the two-scale equation for ϕ is the homogeneous equation

$$\phi = M_0 \phi \,, \tag{5.29}$$

and the wavelet ψ is obtained from ϕ by

$$\psi = M_1 \phi \,. \tag{5.30}$$

Thus ϕ is an *eigenfunction* of M_0 with eigenvalue $\lambda = 1$ (a *fixed-point* of the transformation M_0).

We call M_0 and M_1 the *cascade operators* associated with the filter $\mathbf{g}_0$. From equation (5.7) we obtain the same commutation relation satisfied by S and D:

$$SM_\alpha = \sum_{k\in\mathbb{Z}} \mathbf{g}_\alpha[k]\, SDS^k = \sum_{k\in\mathbb{Z}} \mathbf{g}_\alpha[k]\, DS^{k+2} = M_\alpha S^2 \qquad (5.31)$$

for $\alpha = 0, 1$. This is called a *two-scale relation*. Unlike filtering by convolution, the cascade operators are *not* shift-invariant. They combine shifts with a dilation that transfers a signal to the next finer scale.

We will obtain a scaling function and the associated wavelet by an iterative method. We start with $\phi^{(0)} = \phi_{\text{Haar}}$ and $\psi^{(0)} = \psi_{\text{Haar}}$. Then we use the so-called *cascade algorithm* to calculate

$$\phi^{(j+1)} = M_0\phi^{(j)} \quad \text{and} \quad \psi^{(j+1)} = M_1\phi^{(j)} \quad \text{for } j = 0, 1, 2, \ldots. \qquad (5.32)$$

At the first step of this algorithm the functions $\phi^{(1)}$, $\psi^{(1)}$ are simply the linear combinations, using the filter values as coefficients, of the dilated translates $DS^k\phi_{\text{Haar}}$.

Example 5.2 (Haar). If $\mathbf{g}_0 = (1/\sqrt{2})(\delta_0 + \delta_1)$ and $\mathbf{g}_1 = (1/\sqrt{2})(\delta_0 - \delta_1)$ are the normalized Haar filters, then

$$
\begin{aligned}
\phi^{(1)} &= \frac{1}{\sqrt{2}} D\big(\phi_{\text{Haar}} + S\phi_{\text{Haar}}\big) = \phi_{\text{Haar}}\,, \\
\psi^{(1)} &= \frac{1}{\sqrt{2}} D\big(\phi_{\text{Haar}} - S\phi_{\text{Haar}}\big) = \psi_{\text{Haar}}\,,
\end{aligned}
\qquad (5.33)
$$

by (5.16) and (5.22). Hence $\phi^{(j)} = \phi_{\text{Haar}}$ and $\psi^{(j)} = \psi_{\text{Haar}}$ for all j, so the cascade iteration converges at the first step. ∎

For any choice of filters $\mathbf{g}_0$ and $\mathbf{g}_1$, the functions $\phi^{(j)}$ and $\psi^{(j)}$ are in the Haar space V_j for every positive integer j. Figure 5.6 shows $\phi^{(j)}$ and $\psi^{(j)}$ for the Daub4 filters from Section 4.10.2 when $j = 2$ and $j = 10$. When $j = 2$, these functions are constant on dyadic intervals of length 2^{-2}. When $j = 10$, the functions are constant on dyadic intervals of length 2^{-10}; these intervals are visible after zooming in on the figure.

5.4.2 *Orthogonality relations*

Each iteration of the cascade algorithm gives functions that are piecewise constant on dyadic intervals and zero for large t. Calculating inner products in L^2 of shifts of these functions reduces to calculating a finite sum with a large number of terms. But these functions are complicated (see Figure 5.6), so we will rely on properties from Section 4.9 of the filters, which only have at most $L + 1$ nonzero coefficients, to compute the integrals.

Proposition 5.3. *The functions generated by the cascade algorithm satisfy the following orthogonality conditions for $j = 0, 1, 2, \ldots$:*

(a) $\{S^n\phi^{(j)} : n \in \mathbb{Z}\}$ is an orthonormal set of functions in L^2,

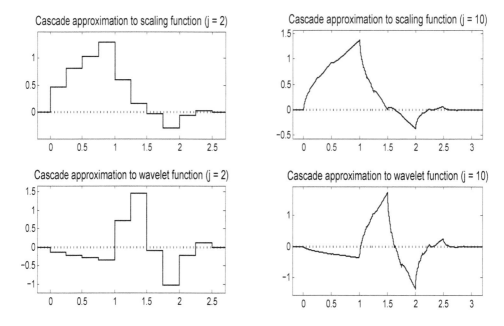

Fig. 5.6   Cascade iterations using Daub4 filters

(b) $\{S^n \psi^{(j)} : n \in \mathbb{Z}\}$ is an orthonormal set of functions in L^2,
(c) $S^k \phi^{(j)}$ is orthogonal to $S^n \psi^{(j)}$ in L^2 for all integers k, n.

Proof. Assertions (a), (b), and (c) are true for $j = 0$, since $\phi^{(0)} = \phi_{\text{Haar}}$ and $\psi^{(0)} = \psi_{\text{Haar}}$. Assume these assertions are true for j. We will use the induction hypothesis to replace inner products in L^2 with inner products in $\ell^2(\mathbb{Z})$. Then we use the orthogonality properties of the filters. The calculations for all three assertions are more transparent if we introduce the notation

$$\phi_\alpha^{(j)} = \begin{cases} \phi^{(j)} & \text{if } \alpha = 0, \\ \psi^{(j)} & \text{if } \alpha = 1. \end{cases}$$

Thus the cascade algorithm becomes $\phi_\alpha^{(j+1)} = M_\alpha \phi^{(j)}$ for $\alpha = 0, 1$.

Let α and β be 0 or 1. Then

$$\langle \phi_\alpha^{(j+1)}, S^n \phi_\beta^{(j+1)} \rangle = \sum_{k=0}^{L} \sum_{m=0}^{L} \mathbf{g}_\alpha[k] \mathbf{g}_\beta[m] \langle DS^k \phi^{(j)}, S^n DS^m \phi^{(j)} \rangle$$

$$= \sum_{k=0}^{L} \sum_{m=0}^{L} \mathbf{g}_\alpha[k] \mathbf{g}_\beta[m] \langle DS^k \phi^{(j)}, DS^{m+2n} \phi^{(j)} \rangle,$$

where we have used the commutation relation (5.7) in the second line. Since the operators D and S are unitary,

$$\langle DS^k \phi^{(j)}, DS^{m+2n} \phi^{(j)} \rangle = \langle S^k \phi^{(j)}, S^{m+2n} \phi^{(j)} \rangle = \langle \phi^{(j)}, S^{m-k+2n} \phi^{(j)} \rangle.$$

Using this simplification of the inner products in the formula above and making the substitution $\ell = m - k + 2n$ in the summation, we find that

$$\langle \phi_\alpha^{(j+1)}, S^n \phi_\beta^{(j+1)} \rangle = \sum_{\ell \in \mathbb{Z}} \left\{ \sum_{k=0}^{L} \mathbf{g}_\alpha[k] \mathbf{g}_\beta[k + \ell - 2n] \right\} \langle \phi^{(j)}, S^\ell \phi^{(j)} \rangle.$$

By the induction hypothesis

$$\langle \phi^{(j)}, S^\ell \phi^{(j)} \rangle = \begin{cases} 1 & \text{if } \ell = 0, \\ 0 & \text{if } \ell \neq 0. \end{cases}$$

Our calculation thus gives

$$\langle \phi_\alpha^{(j+1)}, S^n \phi_\beta^{(j+1)} \rangle = \sum_{k=0}^{L} \mathbf{g}_\alpha[k] \mathbf{g}_\beta[k - 2n] = \langle \mathbf{g}_\alpha, S^{2n} \mathbf{g}_\beta \rangle.$$

In this equation the inner product on the left is for analog signals in L^2, while the inner product on the right is for discrete signals in $\ell^2(\mathbb{Z})$. Since $\{\mathbf{g}_0, \mathbf{g}_1\}$ are the lowpass/highpass filters for an orthogonal filter bank,

$$\langle \mathbf{g}_\alpha, S^{2n} \mathbf{g}_\beta \rangle = \begin{cases} 1 & \text{if } \alpha = \beta \text{ and } n = 0, \\ 0 & \text{otherwise}, \end{cases}$$

by the orthogonality relations (4.79) for the filters. This proves that assertions (a), (b), (c) are true for $j + 1$ and completes the induction step. □

The approximate scaling and wavelet functions shown in Figure 5.6 are zero except in the range $0 \leq t \leq 3$ corresponding to the nonzero Daub4 filter coefficients $\mathbf{g}_\alpha[0], \dots, \mathbf{g}_\alpha[3]$. This property holds in general when the filter coefficients $\mathbf{g}_\alpha[n] = 0$ for $n < 0$ or $n > L$ and $\alpha = 0, 1$.

Lemma 5.2. *For every j the functions $\phi^{(j)}$ and $\psi^{(j)}$ are zero outside the interval $0 \leq t \leq L$.*

Proof. The assertion is true for $j = 0$. Assume it is true for j. With the same notation as in the proof of Proposition 5.3 we can write

$$\phi_\alpha^{(j+1)}(t) = \sum_{k=0}^{L} \mathbf{g}_\alpha[k] \sqrt{2} \phi^{(j)}(2t - k)$$

for $\alpha = 0, 1$. Let $0 \leq k \leq L$. If $t < 0$ then $2t - k < 0$. If $t > L$ then $2t - k > 2L - L = L$. In both cases $\phi^{(j)}(2t - k) = 0$, so the assertion is true for $j + 1$. □

The assertion of Lemma 5.2 is illustrated in Figure 5.7, where the nonzero filter coefficients are $\mathbf{g}_\alpha[0], \dots, \mathbf{g}_\alpha[5]$ on the left and $\mathbf{g}_\alpha[0], \dots, \mathbf{g}_\alpha[7]$ on the right.

Lemma 5.3. *For every $j = 0, 1, \dots$ the functions $\phi^{(j)}$ and $\psi^{(j)}$ satisfy*

$$\int_{-\infty}^{\infty} \phi^{(j)}(t) \, dt = 1, \qquad \int_{-\infty}^{\infty} \psi^{(j)}(t) \, dt = 0, \tag{5.34}$$

$$\sum_{n \in \mathbb{Z}} \phi^{(j)}(t + n) = 1 \quad \text{for all } t \in \mathbb{R}. \tag{5.35}$$

Proof. From the definitions (5.9) and (5.16) of ϕ_{Haar} and ψ_{Haar} equations (5.34) and (5.35) are true for $j = 0$. Assume they are true for j. With the same notation as in the proof of Proposition 5.3 we can write

$$\int_{-\infty}^{\infty} \phi_{\alpha}^{(j+1)}(t)\, dt = \sum_{k=0}^{L} \mathbf{g}_{\alpha}[k] \int_{-\infty}^{\infty} \sqrt{2}\phi^{(j)}(2t - k)\, dt$$

for $\alpha = 0, 1$. Note that the integrands on the right are piecewise constant functions that are zero outside the interval $0 \leq t \leq L$, by Lemma 5.2. Making the change of variable $s = 2t - k$ and $ds = 2dt$, we obtain by the induction hypothesis

$$\int_{-\infty}^{\infty} \sqrt{2}\phi^{(j)}(2t - k)\, dt = \frac{1}{\sqrt{2}} \int_{-\infty}^{\infty} \phi^{(j)}(s)\, ds = \frac{1}{\sqrt{2}}.$$

Hence

$$\int_{-\infty}^{\infty} \phi_{\alpha}^{(j+1)}(t)\, dt = \frac{1}{\sqrt{2}} \sum_{k=0}^{L} \mathbf{g}_{\alpha}[k] = \begin{cases} 1 & \text{if } \alpha = 0, \\ 0 & \text{if } \alpha = 1. \end{cases}$$

The last step follows from (5.26). This completes the induction step for (5.34).

To verify the induction step for (5.35), use the two-scale equation and the sum rule (5.27). The details are left as an exercise. $\qquad\square$

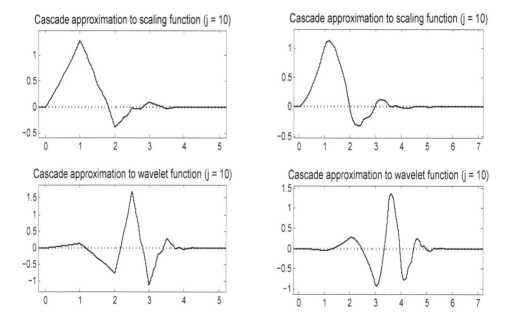

Fig. 5.7   Cascade iterations using Daub6 (left) and Daub8 (right) filters

The graphs in Figures 5.6 and 5.7 suggest that the iterations from the cascade algorithm converge. These graphs also clarify the term "wavelet" (little wave): a function that oscillates between positive and negative values, has average value zero, and has amplitude diminishing to zero outside a finite interval.[12]

Convergence of the cascade algorithm does indeed occur for the Daubechies family of filters (see Example 5.3). However, this does not always happen for other orthogonal FIR filters (see Example 5.4); we will give a sufficient condition for *strong convergence* (as measured by the L^2 norm) in Theorem 5.7. At this point we *assume* that the cascade iteration converges in this strong sense. From this assumption we can use the properties of the approximating functions $\phi^{(j)}$ and $\psi^{(j)}$ just proved to show that the limit functions satisfy the requirements for scaling and wavelet functions.

Theorem 5.6. *Suppose there is a function $\phi \in L^2$ such that $\|\phi - \phi^{(j)}\| \to 0$ as $j \to \infty$. Set $\psi = M_1\phi$. Then the following holds:*

(1) $M_0\phi = \phi$, so ϕ is a scaling function for the filter $\mathbf{g}_0$;
(2) $\|\psi - \psi^{(j)}\| \to 0$ as $j \to \infty$;
(3) The functions ϕ and ψ are zero outside the interval $0 \leq t \leq L$;
(4) $\int_0^L \phi(t)\,dt = 1$ and $\int_0^L \psi(t)\,dt = 0$.

Proof. We can write

$$M_0\phi - \phi = M_0(\phi - \phi^{(j)}) + \phi^{(j+1)} - \phi,$$

since $M_0\phi^{(j)} = \phi^{(j+1)}$ and M_0 is a linear transformation. Hence by the triangle inequality (1.23) for the norm on L^2 we have

$$\|M_0\phi - \phi\| \leq \|M_0(\phi - \phi^{(j)})\| + \|\phi^{(j+1)} - \phi\|. \tag{5.36}$$

Let $C_0 = \sum_{k=0}^L |\mathbf{g}_0[k]|$. For any function $f \in L^2$ the triangle inequality gives the estimate

$$\|M_0 f\| \leq \sum_{k=0}^L |\mathbf{g}_0[k]|\,\|DS^k f\| = C_0\|f\|. \tag{5.37}$$

Here we have used the property $\|DS^k f\| = \|f\|$, which is true because D and S are unitary operators. Taking $f = \phi - \phi^{(j)}$ in (5.37) and using (5.36), we obtain the inequality

$$\|M_0\phi - \phi\| \leq C_0\|\phi - \phi^{(j)}\| + \|\phi^{(j+1)} - \phi\|.$$

By assumption, both terms on the right go to zero as $j \to \infty$. This proves statement (1).

To prove statement (2), we observe that $\psi - \psi^{(j+1)} = M_1(\phi - \phi^{(j)})$. For any function $f \in L^2$ we can use the triangle inequality, as in the proof of statement (1), to estimate $\|M_1 f\| \leq C_1\|f\|$, where $C_1 = \sum_{k=0}^L |\mathbf{g}_1[k]|$. Hence

$$\|\psi - \psi^{(j+1)}\| \leq C_1\|\phi - \phi^{(j)}\|,$$

---

[12]A sine wave is *not* a wavelet, since its amplitude is constant.

so the limit of the left side is zero when $j \to \infty$.

Let $I \subset \mathbb{R}$ be any interval disjoint from $[0, L]$. The convergence in norm of $\phi^{(j)}$ to ϕ and $\psi^{(j)}$ to ψ, together with Lemma 5.2, give

$$\int_I \phi(t)\,dt = \lim_{j\to\infty} \int_I \phi^{(j)}(t)\,dt = 0 \quad \text{and} \quad \int_I \psi(t)\,dt = \lim_{j\to\infty} \int_I \psi^{(j)}(t)\,dt = 0\,.$$

This implies statement (3) if ϕ is continuous.[13]

To prove statement (4), use the fact that $\phi^{(j)}$ has integral 1 (equation (5.34)) to write

$$\int_0^L \phi(t)\,dt - 1 = \int_0^L \left(\phi(t) - \phi^{(j)}(t)\right)\,dt\,.$$

View the integral on the right as the inner product of the function $\phi(t) - \phi^{(j)}(t)$ with the function $f(t) = 1$ for $0 \le t \le L$ and $f(t) = 0$ otherwise. Since $\|f\| = \sqrt{L}$, the Cauchy–Schwarz inequality gives

$$\left| \int_0^L \phi(t)\,dt - 1 \right| \le \sqrt{L}\,\|\phi - \phi^{(j)}\|\,.$$

The right side of this inequality goes to zero as $j \to \infty$. The same argument using ψ and $\psi^{(j)}$ applies to the integral of ψ. □

Corollary 5.1. *(Assumptions and notation of Theorem 5.6) The shifted scaling and wavelet functions have the following orthogonality properties:*

(a) $\{S^n\phi : n \in \mathbb{Z}\}$ *is an orthonormal set of functions in* L^2 *;*
(b) $\{S^n\psi : n \in \mathbb{Z}\}$ *is an orthonormal set of functions in* L^2 *;*
(c) $S^k\phi$ *is orthogonal to* $S^n\psi$ *in* L^2 *for all integers* k, n *.*

Proof. We introduce the notation $\phi_0 = \phi$ and $\phi_1 = \psi$. We will first prove

$$\langle \phi_\alpha, S^n\phi_\beta \rangle = \lim_{j\to\infty} \langle \phi_\alpha^{(j)}, S^n\phi_\beta^{(j)} \rangle \tag{5.38}$$

for α and β taking values 0 or 1. Then we can apply Proposition 5.3 to obtain properties (a), (b), (c).

Write $\mathbf{u} = \phi_\alpha$, $\mathbf{u}_j = \phi_\alpha^{(j)}$, $\mathbf{v} = S^n\phi_\beta$, and $\mathbf{v}_j = S^n\phi_\beta^{(j)}$. Then

$$\langle \mathbf{u}, \mathbf{v} \rangle = \langle \mathbf{u}_j + (\mathbf{u} - \mathbf{u}_j), \mathbf{v}_j + (\mathbf{v} - \mathbf{v}_j) \rangle = \langle \mathbf{u}_j, \mathbf{v}_j \rangle + c_j\,,$$

where $c_j = \langle (\mathbf{u} - \mathbf{u}_j), (\mathbf{v} - \mathbf{v}_j) \rangle + \langle (\mathbf{u} - \mathbf{u}_j), \mathbf{v}_j \rangle + \langle \mathbf{u}_j, (\mathbf{v} - \mathbf{v}_j) \rangle$. Thus to prove (5.38) we need to show that $\lim_{j\to\infty} c_j = 0$. By Proposition 5.3 we know that $\|\mathbf{u}_j\| = \|\mathbf{v}_j\| = 1$. Hence by the Cauchy-Schwarz inequality

$$|c_j| \le \|\mathbf{u} - \mathbf{u}_j\|\|\mathbf{v} - \mathbf{v}_j\| + \|\mathbf{u} - \mathbf{u}_j\| + \|\mathbf{v} - \mathbf{v}_j\|\,.$$

Since S is unitary we have $\|\mathbf{v} - \mathbf{v}_j\| = \|\phi_\beta - \phi_\beta^{(j)}\|$. Thus

$$|c_j| \le \|\phi_\alpha - \phi_\alpha^{(j)}\|\,\|\phi_\beta - \phi_\beta^{(j)}\| + \|\phi_\alpha - \phi_\alpha^{(j)}\| + \|\phi_\beta - \phi_\beta^{(j)}\|\,.$$

From this inequality and Theorem 5.6 we conclude that $c_j \to 0$ as $j \to \infty$. □

---

[13] If ϕ is only in L^2 then the conclusion is that ϕ and ψ are zero *almost everywhere* (in the sense of Lebesgue measure) outside the interval $0 \le t \le L$.

Here is a sufficient condition for the convergence assumption of the cascade algorithm in Theorem 5.6 to hold.

Theorem 5.7. *Let $G_0(z)$ be the z-transform of the filter $\mathbf{g}_0$. Suppose that*

$$G_0(e^{\omega \mathrm{i}}) \neq 0 \quad \text{for } -\pi/2 \leq \omega \leq \pi/2 . \tag{5.39}$$

Then the cascade iterations $\phi^{(j)}$ converge strongly in L^2 to a function $\phi \in L^2$. Furthermore, ϕ is the unique function in L^2 such that $M_0\phi = \phi$ and $\int_{-\infty}^{\infty} \phi(t) \, dt = 1$.

Proof. The strategy is to transfer the cascade iteration from the time domain to the frequency domain by the Fourier transform on L^2, where it gives rise to an infinite product formula for the Fourier transform of the scaling function. This gives existence and uniqueness of ϕ as a so-called *generalized function*, but not necessarily as a function in L^2. Convergence of the functions $\phi^{(j)}$ in the L^2 norm as $j \to \infty$ can be proved using condition (5.39) in this infinite product formula and the fact that the Fourier transform preserves L^2 norm. See [Strichartz (1993), Theorem 5.2], [Walnut (2002), Theorem 8.36], or [Frazier (1999), §5.5] for complete details. $\square$

Example 5.3. Condition (5.39) is satisfied by the Daub2K filters constructed in Section 4.10. To see this, we use equation (4.87) for the power spectral response function:

$$|G_0(e^{\omega \mathrm{i}})|^2 = P(e^{\omega \mathrm{i}}) = 2(1 - y)^K B_K(y) ,$$

where $B_K(y)$ is the Bezout polynomial and $y = \sin^2(\omega/2)$. If $-\pi/2 \leq \omega \leq \pi/2$ then $1 - y = \cos^2(\omega/2) \geq 1/2$ and $y \geq 0$. From the formula (4.53) for the Bezout polynomial we see that $B(y) \geq 1$. Hence $|G_0(e^{\omega \mathrm{i}})|^2 \geq 1/2$ for $-\pi/2 \leq \omega \leq \pi/2$. $\blacksquare$

For the Daub4 filter the scaling function is continuous but it doesn't have a derivative.[14] This property is suggested (but not proved) by the jagged graph in Figure 5.6. For the Daub6 filter the scaling function has one derivative that is continuous but with big fluctuations–note the peaks near $t = 1$ and $t = 2$ in Figure 5.7. For the Daub8 filter the scaling function has one derivative that is continuous with moderate fluctuations–see Figure 5.7. This pattern continues for all the Daub2K filters: larger values of K give smoother scaling functions, with the scaling function having continuous derivatives up to order approximately $K/5$. See [Strichartz (1993), §7] for a brief explanation and [Daubechies (1992), Ch. 7] for complete details.

Remark 5.3. Assume the approximating functions $\phi^{(j)}(t)$ converge *pointwise* to $\phi(t)$ for all $t \in \mathbb{R}$, as is the case for the Daub2K filters, for example. Then taking the limit as $j \to \infty$ in (5.35), we obtain

$$\sum_{n \in \mathbb{Z}} \phi(t + n) = 1 \quad \text{for all } t \in \mathbb{R} . \tag{5.40}$$

This is called the *partition of unity* property. Note that for each t there are only a finite number of nonzero terms in the sum.

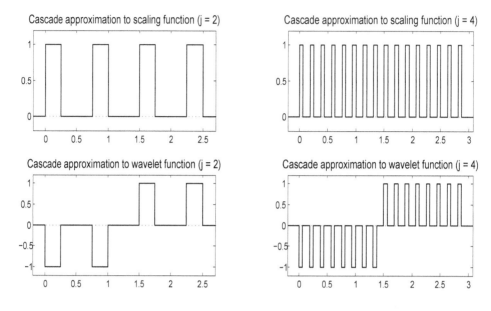

Fig. 5.8   Cascade approximations using filter $\mathbf{g}_0$ from Example 5.4

Example 5.4. Consider the lowpass filter $\mathbf{g}_0 = (1/\sqrt{2})(\delta_0 + \delta_3)$. This filter gives an orthogonal PR filter bank by Theorem 4.14, since

$$G_0(z)G_0(z^{-1}) + G_0(-z)G_0(-z^{-1})$$

$$= \frac{1}{2}\{(1 + z^{-3})(1 + z^3) + (1 - z^{-3})(1 - z^3)\} = 2\,.$$

But condition (5.39) is not satisfied by $\mathbf{g}_0$, since $|G_0(e^{\omega i})| = \sqrt{2}\,|\cos(3\omega/2)|$, which vanishes at $\omega = \pm\pi/3$. The cascade algorithm iterations do not converge in the L^2 norm as required in Theorem 5.6, although there is a *weaker* form of convergence in the sense of generalized functions. Namely, if $\phi(t) = (1/3)\phi_{\mathrm{Haar}}(t/3)$, then

$$\lim_{j\to\infty} \langle \phi^{(j)}, f \rangle = \langle \phi, f \rangle = \frac{1}{3}\int_0^3 f(t)\,dt \qquad (5.41)$$

for every continuous function f. This is suggested by Figure 5.8: the function $\phi^{(j)}(t)$ equals 1 on 2^j uniformly spaced subintervals of width 2^{-j} in the interval $0 \le t \le 3$, whereas $\phi^{(j)}(t)$ is 0 outside these subintervals. The spacing between the subintervals goes to zero as $j \to \infty$. So if $f(t)$ is continuous, then $\langle \phi^{(j)}, f \rangle$ is approximately $1/3$ of the integral of f over $0 \le t \le 3$. Although ϕ satisfies the two-scale equation (5.29), the shifted functions $S^n\phi$ are not mutually orthogonal. Thus we cannot use ϕ as a scaling function for a multiresolution synthesis. See [Strang and Nguyen (1997), §7.2] and [Daubechies (1992), §6.2] for more discussion of this example.   ∎

---

[14]It has slightly more than $1/2$ of a derivative in the sense of the mean-value theorem: if $\alpha < 0.55$ then $|\phi(x) - \phi(y)| \le C_\alpha |x - y|^\alpha$ for a suitable constant C_α and all x, y.

5.5 Multiresolution Analysis of Analog Signals

We continue with the setup of Section 5.4: $\mathbf{g}_0$ (lowpass) and $\mathbf{g}_1$ (highpass) are real-valued FIR synthesis filters for a two-channel orthogonal filter bank, and $\mathbf{h}_0 = \mathbf{g}_0^{\mathrm{T}}$, $\mathbf{h}_1 = \mathbf{g}_1^{\mathrm{T}}$ the analysis filters. These filters have even length $L+1$ and, as in Section 5.4, we may arrange so that they satisfy (5.25) and (5.26). We assume, as in Theorem 5.6, that the cascade iteration converges in the sense of the L^2 norm to a continuous[15] scaling function $\phi(t)$ that is normalized to have integral 1. This function is zero outside the interval $0 < t < L$. Let $\psi(t)$ be the corresponding wavelet function.

5.5.1 *Multiresolution spaces*

Following the same pattern as in Section 5.3.3 for the Haar scaling function and wavelet, we define the dilated and shifted scaling functions $\phi_{j,k} = D^j S^k \phi$ and the dilated and shifted wavelet functions $\psi_{j,k} = D^j S^k \psi$ for all integers j, k. Since D and S are unitary operators, we have $\langle D^j S^k f , D^j S^k g \rangle = \langle f , g \rangle$ for all finite-energy signals f and g. Hence at each fixed dilation scale j we know from Corollary 5.1 that

(a) $\{\, \phi_{j,k} \,:\, k \in \mathbb{Z} \}$ is an orthonormal set of functions in L^2,
(b) $\{\, \psi_{j,k} \,:\, k \in \mathbb{Z} \}$ is an orthonormal set of functions in L^2,
(c) $\phi_{j,k} \perp \psi_{j,n}$ in L^2 for all integers k, n.

Definition 5.6. The *trend space* at level j is $V_j = \overline{\mathrm{Span}}\{\, \phi_{j,k} \,:\, k \in \mathbb{Z} \}$ and the *detail space* at level j is $W_j = \overline{\mathrm{Span}}\{\, \psi_{j,k} \,:\, k \in \mathbb{Z} \}$.

We would like to show that the spaces V_j and W_j have the same properties as the trend and detail spaces for the Haar multiresolution analysis in Sections 5.3.2 and 5.3.3. From the definition of the basis functions and the commutation relation (5.7) we see immediately that

$$SV_j = V_j \quad \text{and} \quad SW_j = W_j ,$$
$$DV_j = V_{j+1} \quad \text{and} \quad DW_j = W_{j+1} .$$

The functions $\phi = M_0 \phi$ and $\psi = M_1 \phi$ are both in V_1, where M_0 and M_1 are the cascade operators on L^2 associated with the filters $\mathbf{h}_0$ and $\mathbf{h}_1$. Together with the orthogonality property (c) above, this implies that

$$V_j \subset V_{j+1} \quad \text{and} \quad W_j \subset V_{j+1} ,$$
$$V_j \perp W_j \quad \text{and} \quad W_j \perp W_m \text{ for all } j \neq m .$$

Since the dilation operator is invertible on L^2, we have the *two-scale* property for a function $f \in L^2$:

$$f(t) \in V_j \quad \text{if and only if} \quad f(2t) \in V_{j+1} . \tag{5.42}$$

---

[15]This assumption is satisfied, for example, by the Daub2K family of filters when $K \geq 2$—see [Daubechies (1992), Section 7.1].

Definition 5.7. The increasing collection of subspaces
$$\cdots \subset V_{-2} \subset V_{-1} \subset V_0 \subset V_1 \subset V_2 \subset \cdots$$
is the *multiresolution* generated by the scaling function ϕ.

Starting with the first example in Section 3.2.1, the basic theme in multiresolution analysis of digital or analog signals is that every signal in the trend space at level $j + 1$ is the sum of its projection onto the trend space at level j plus its projection onto the detail space at level j. For the Haar multiresolution analysis we proved this using two-scale equations (5.22) for $D\phi$ and $DS\phi$. We use the same strategy here to obtain the general *trend + detail* decomposition.

Proposition 5.4. *The functions $D\phi$ and $DS\phi$ satisfy*
$$D\phi = \sum_{n \in \mathbb{Z}} \mathbf{h}_0[2n]\, S^n \phi + \mathbf{h}_1[2n]\, S^n \psi\,, \tag{5.43}$$

$$DS\phi = \sum_{n \in \mathbb{Z}} \mathbf{h}_0[2n-1]\, S^n \phi + \mathbf{h}_1[2n-1]\, S^n \psi\,. \tag{5.44}$$

Hence $D^j S^k \phi \in V_j \oplus W_j$ for all $j, k \in \mathbb{Z}$, so $V_{j+1} = V_j \oplus W_j$.

Proof. Let f_0 be the orthogonal projection of $D\phi$ onto $V_0 \oplus W_0$ and f_1 the orthogonal projection of $DS\phi$ onto $V_0 \oplus W_0$. Then
$$f_0 = \sum_{n \in \mathbb{Z}} \langle S^n \phi,\, D\phi \rangle\, S^n \phi + \langle S^n \psi,\, D\phi \rangle\, S^n \psi\,,$$

$$f_1 = \sum_{n \in \mathbb{Z}} \langle S^n \phi,\, DS\phi \rangle\, S^n \phi + \langle S^n \psi,\, DS\phi \rangle\, S^n \psi\,.$$

Here the sums on the right only have a finite number of nonzero terms, since the functions in the inner products have disjoint supports for $|n|$ sufficiently large by Theorem 5.6(3). Denote $\phi = \phi_0$ and $\psi = \phi_1$. We can obtain more explicit formulas for f_0 and f_1 by combining the defining relations $\phi_\alpha = M_\alpha \phi$ for $\alpha = 0, 1$ with the two-scale relation (5.31) to write
$$S^n \phi_\alpha = S^n M_\alpha \phi = M_\alpha S^{2n} \phi\,. \tag{5.45}$$
From this equation we can express the inner products in the formulas for f_0 and f_1 as
$$\langle S^n \phi_\alpha,\, D\phi \rangle = \sum_{k \in \mathbb{Z}} \mathbf{g}_\alpha[k]\, \langle DS^{2n+k}\phi,\, D\phi \rangle = \mathbf{g}_\alpha[-2n] = \mathbf{h}_\alpha[2n]\,,$$

$$\langle S^n \phi_\alpha,\, DS\phi \rangle = \sum_{k \in \mathbb{Z}} \mathbf{g}_\alpha[k]\, \langle DS^{2n+k-1}\phi,\, D\phi \rangle = \mathbf{g}_\alpha[-2n+1] = \mathbf{h}_\alpha[2n-1]\,.$$

Here we have used the facts that the dilation D preserves inner products, together with the mutual orthogonality of different shifts of the scaling function ϕ (Corollary 5.1). Thus f_0 and f_1 are the linear combinations
$$f_0 = \sum_{n \in \mathbb{Z}} \mathbf{h}_0[2n] S^n \phi + \mathbf{h}_1[2n] S^n \psi\,,$$

$$f_1 = \sum_{n \in \mathbb{Z}} \mathbf{h}_0[2n-1] S^n \phi + \mathbf{h}_1[2n-1] S^n \psi\,.$$

Hence to prove (5.43) and (5.44) we need to show that $f_0 = D\phi$ and $f_1 = DS\phi$. For this we will use the perfect reconstruction property of the filter bank with filters $\mathbf{g}_0, \mathbf{g}_1, \mathbf{h}_0, \mathbf{h}_1$ as follows.

By (5.45) we can write the formulas just obtained as

$$f_0 = \sum_{n\in\mathbb{Z}} \{\mathbf{h}_0[2n]M_0 + \mathbf{h}_1[2n]M_1\} S^{2n}\phi,$$

$$f_1 = \sum_{n\in\mathbb{Z}} \{\mathbf{h}_0[2n-1]M_0 + \mathbf{h}_1[2n-1]M_1\} S^{2n}\phi.$$

Now we substitute the explicit forms of the cascade operators in these formulas to obtain the double summations

$$f_0 = \sum_{n\in\mathbb{Z}}\sum_{m\in\mathbb{Z}} \{\mathbf{h}_0[2n]\mathbf{g}_0[m] + \mathbf{h}_1[2n]\mathbf{g}_1[m]\} DS^{m+2n}\phi = \sum_{k\in\mathbb{Z}} \mathbf{c}_0[k]\, DS^k\phi,$$

$$f_1 = \sum_{n\in\mathbb{Z}}\sum_{m\in\mathbb{Z}} \{\mathbf{h}_0[2n-1]\mathbf{g}_0[m] + \mathbf{h}_1[2n-1]\mathbf{g}_1[m]\} DS^{m+2n}\phi$$

$$= \sum_{k\in\mathbb{Z}} \mathbf{c}_1[k]\, DS^k\phi,$$

where the coefficients in the second expressions for f_0 and f_1 are

$$\mathbf{c}_0[k] = \sum_{n\in\mathbb{Z}} \{\mathbf{g}_0[k-2n]\mathbf{h}_0[2n] + \mathbf{g}_1[k-2n]\mathbf{h}_1[2n]\},$$

$$\mathbf{c}_1[k] = \sum_{n\in\mathbb{Z}} \{\mathbf{g}_0[k-2n]\mathbf{h}_0[2n-1] + \mathbf{g}_1[k-2n]\mathbf{h}_1[2n-1]\}.$$

Viewing $\mathbf{c}_0$ and $\mathbf{c}_1$ as finite digital signals, we can write these formulas in terms of convolution, downsampling, and upsampling as

$$\mathbf{c}_0 = \mathbf{g}_0 \star \left(\boxed{2\uparrow}\ \boxed{2\downarrow}(\mathbf{h}_0 \star \delta_0)\right) + \mathbf{g}_0 \star \left(\boxed{2\uparrow}\ \boxed{2\downarrow}(\mathbf{h}_1 \star \delta_0)\right),$$

$$\mathbf{c}_1 = \mathbf{g}_0 \star \left(\boxed{2\uparrow}\ \boxed{2\downarrow}(\mathbf{h}_0 \star \delta_1)\right) + \mathbf{g}_0 \star \left(\boxed{2\uparrow}\ \boxed{2\downarrow}(\mathbf{h}_1 \star \delta_1)\right),$$

where δ_0 is the unit impulse at 0 and δ_1 is the unit impulse at 1. Now we use equations (4.73) and (4.72) for perfect reconstruction filter banks to conclude that $\mathbf{c}_0 = \delta_0$ and $\mathbf{c}_1 = \delta_1$. Thus the formulas for f_0 and f_1 in terms of $\mathbf{c}_0$ and $\mathbf{c}_1$ only have one nonzero term, namely $f_0 = D\phi$ and $f_1 = DS\phi$. This proves (5.43) and (5.44).

The subspaces V_0 and W_0 are invariant under the shift operator S. It follows that $DS^{2k}\phi = S^k D\phi = S^k f_0$ is in V_0 and $DS^{2k+1}\phi = S^k DS\phi = S^k f_1$ is in W_0. Hence $DS^m\phi \in V_0 \oplus W_0$ for all $m \in \mathbb{Z}$. Consequently $D^{j+1}S^m\phi \in V_j \oplus W_j$ for all $j, m \in \mathbb{Z}$. This proves that $V_{j+1} = V_j \oplus W_j$. $\square$

5.5.2 *Trend and detail projections*

Let f be a finite-energy analog signal. By Theorem 5.3 the projections of f onto the subspaces V_j and W_j are given by the orthogonal series

$$f_{\text{trend},\, j}(t) = 2^{-j/2}\sum_{k\in\mathbb{Z}} \mathbf{s}_j[k]\, \phi_{j,k}(t) \quad\text{with}\quad \mathbf{s}_j[k] = 2^{j/2}\langle\phi_{j,k}, f\rangle,$$

$$f_{\text{detail},\, j}(t) = 2^{-j/2}\sum_{k\in\mathbb{Z}} \mathbf{d}_j[k]\, \psi_{j,k}(t) \quad\text{with}\quad \mathbf{d}_j[k] = 2^{j/2}\langle\psi_{j,k}, g\rangle, \tag{5.46}$$

which we know *a priori* converge in the sense of the L^2 norm. Since f has finite energy, $\mathbf{s}_j$ and $\mathbf{d}_j$ are finite-energy digital signals by the Riesz–Fischer Theorem, but they don't have finite support, in general.

Definition 5.8. The function $f_{\text{trend}, j}$ is the *level-j trend* of f and the function $f_{\text{detail}, j}$ is the *level-j detail* of f. The coefficient sequence $\mathbf{s}_j$ is the *level-j discrete trend* of f and the coefficient sequence $\mathbf{d}_j$ is the *level-j discrete detail* of f.

The formulas (5.46) have two remarkable aspects that we have already seen in the case of the Haar multiresolution analysis and now prove in general.

Proposition 5.5. *The level-j trend and detail of f have the following properties.*

(1) Pointwise convergence: When $-T \leq t \leq T$, each of the series in (5.46) has only a finite number of nonzero terms (although the number of nonzero terms depends on j and T). Hence the series converge in the pointwise sense.
(2) Locality: The values at k of the level-j discrete trend and detail of f only depend on the values of f on the interval $2^{-j}k \leq t \leq 2^{-j}(k+L)$.

Proof. (1) By Theorem 5.6(3) we know that $\phi_{j,k}(t)$ and $\psi_{j,k}(t)$ are zero when t is outside the interval $[k, k + 2^{-j}L]$. Hence if $-T \leq t \leq T$, then the only nonzero terms in the series occur for k in the range $-T - 2^{-j}L \leq k \leq T$.

(2) The values at k of the level-j discrete trend and detail are the integrals[16]

$$\mathbf{s}_j[k] = \int_{-\infty}^{\infty} 2^j \phi(2^j t - k) f(t) \, dt = \int_0^L \phi(x) f\left(2^{-j}(k + x)\right) dx \,,$$

$$\mathbf{d}_j[k] = \int_{-\infty}^{\infty} 2^j \psi(2^j t - k) f(t) \, dt = \int_0^L \psi(x) f\left(2^{-j}(k + x)\right) dx \,. \tag{5.47}$$

Now note that when $0 \leq x \leq L$ then $2^{-j}k \leq 2^{-j}(k + x) \leq 2^{-j}(k + L)$. $\square$

Remark 5.4. The interval in Proposition 5.5(2) has length $2^{-j}L$. So when j is positive and large, this interval is very short. This *localization property* is one of the major advantages of wavelet series over Fourier series, since the Fourier coefficients of a periodic function are influenced by the values of the function over a whole period.

We can estimate the trend and detail coefficients as follows.

Lemma 5.4. *Suppose $f(t)$ has a continuous derivative $f'(t)$ for all t in some interval $I = [a, b]$. If j, k are such that the subinterval $[2^{-j}k, 2^{-j}(k + L)] \subset I$, then*

$$\left|\mathbf{s}_j[k] - f(2^{-j}k)\right| \leq CM2^{-j} \quad \text{and} \quad \left|\mathbf{d}_j[k]\right| \leq CM2^{-j}. \tag{5.48}$$

Here $M = \max_{t \in I} |f'(t)|$ and the constant C does not depend on f or j.

---

[16]To obtain the second integral in each case we make the change of variable $x = 2^j t - k$ and $dx = 2^j \, dt$ and use the fact that $\phi(t)$ and $\psi(t)$ are zero outside the interval $0 \leq t \leq L$.

Proof. By Theorem 5.6(4) we know that

$$\int_0^L \phi(x) f(2^{-j}k)\, dx = f(2^{-j}k) \quad \text{and} \quad \int_0^L \psi(x) f(2^{-j}k)\, dx = 0\,.$$

Hence from (5.47) we can write

$$\mathbf{s}_j[k] - f(2^{-j}k) = \int_0^L \phi(x) \left\{ f\left(2^{-j}(k+x)\right) - f(2^{-j}k) \right\} dx\,,$$

$$\mathbf{d}_j[k] = \int_0^L \psi(x) \left\{ f\left(2^{-j}(k+x)\right) - f(2^{-j}k) \right\} dx\,. \tag{5.49}$$

By the differentiability assumption on $f(t)$ and the mean-value theorem

$$\left| f\left(2^{-j}(k+x)\right) - f(2^{-j}k) \right| \le M |2^{-j}x| \le ML2^{-j} \quad \text{for } 0 \le x \le L\,. \tag{5.50}$$

Applying the Cauchy–Schwarz inequality to each of the integrals (5.49) gives

$$\left| \mathbf{s}_j[k] - f(2^{-j}k) \right| \le \|\phi\| \left\{ \int_0^L \left| f\left(2^{-j}(k+x)\right) - f(2^{-j}k) \right|^2 dx \right\}^{1/2} \le ML^{3/2}2^{-j}\,,$$

$$\left| \mathbf{d}_j[k] \right| \le \|\psi\| \left\{ \int_0^L \left| f\left(2^{-j}(k+x)\right) - f(2^{-j}k) \right|^2 dx \right\}^{1/2} \le ML^{3/2}2^{-j}\,,$$

where we used (5.50) and $\|\phi\| = \|\psi\| = 1$ in the second inequality in each case. This proves the lemma with $C = L^{3/2}$. $\qquad\square$

We have assumed the scaling function is continuous. Since it also vanishes outside $0 \le t \le L$, it is bounded: $|\phi(t)| \le M$ for some fixed constant M and all t.

Lemma 5.5. *Let $f \in L^2$. Then for all integers j and all $t \in \mathbb{R}$,*

$$|f_{\text{trend},\, j}(t)| \le C 2^{j/2} \|f\|\,, \tag{5.51}$$

where the constant C does not depend on f or j.

Proof. We use the same argument that gave the similar inequality (5.13) for the Haar series. Start with the expansion

$$f_{\text{trend},\, j}(t) = \sum_{k \in \mathbb{Z}} \langle \phi_{j,k},\, f \rangle\, \phi_{j,k}(t)\,.$$

Apply the Cauchy-Schwarz inequality to the right side to obtain the inequality

$$|f_{\text{trend},\, j}(t)| \le \left\{ \sum_{k \in \mathbb{Z}} |\langle \phi_{j,k},\, f \rangle|^2 \right\}^{1/2} \left\{ \sum_{k \in \mathbb{Z}} |\phi_{j,k}(t)|^2 \right\}^{1/2}$$

$$\le \|f\| \left\{ \sum_{k \in \mathbb{Z}} 2^j |\phi(2^j t - k)|^2 \right\}^{1/2}\,.$$

Here we used Parseval's formula for $\|f_{\text{trend},\, j}\|$, together with the triangle inequality $\|f_{\text{trend},\, j}\| \le \|f\|$.

There are at most $L+1$ integers k for which $\phi(2^j t - k) \ne 0$, since k must satisfy the inequalities $2^j t - L \le k \le 2^j t$, and each of these values of ϕ is bounded by M. Hence for every $t \in \mathbb{R}$,

$$\left\{ \sum_{k \in \mathbb{Z}} 2^j |\phi(2^j t - k)|^2 \right\}^{1/2} \le 2^{j/2} \sqrt{(L+1)M}\,.$$

This completes the proof of inequality (5.51). $\qquad\square$

Suppose $f \in V_j$ with j is negative and $|j|$ large. Then $f = f_{\text{trend}, j}$, so (5.51) shows that all the values $|f(t)|$ are extremely small compared to $\|f\|$. In particular, if f is in V_j for *all* j then inequality (5.51) is true of all j. Since $\|f\|$ is a fixed number, this implies that $f(t) = 0$ for all t. Following the same notation as for the Haar multiresolution analysis, we write $\lim_{j \to -\infty} V_j = \bigcap_{j \in \mathbb{Z}} V_j$. Then by the argument just given

$$\lim_{j \to -\infty} V_j = \{\mathbf{0}\}. \tag{5.52}$$

Theorem 5.8. *Every function $f(t)$ in L^2 satisfies $\lim_{j \to +\infty} \|f - P_j f\| = 0$. Thus the distance between $f(t)$ and $f_{\text{trend}, j}(t)$ (as measured by the L^2 norm) goes to zero as $j \to +\infty$.*

Proof. We use the same approximation argument as in the proof of Theorem 5.4. The only properties needed are that $\phi(t)$ is zero outside a finite interval $0 \leq t \leq L$, the integral of ϕ is 1, and $|\phi(t)| \leq M$ for all t and some constant M. If $h(t)$ is defined as in that proof, then

$$\int_{2^j a}^{2^j b} \phi(y - k)\, dy = \begin{cases} 1 & \text{if } 2^j a \leq k \leq 2^j b - L, \\ \kappa & \text{if } 2^j a - L < k < 2^j a \text{ or } 2^j b - L < k < 2^j b, \\ 0 & \text{otherwise}, \end{cases}$$

where $|\kappa| \leq ML$, since in the second case (which occurs for at most $L + 1$ integers k) the integration is over an interval of length L and the integrand is bounded by M. Just as in the proof of Theorem 5.4 we conclude that when j is positive and large,

$$\sum_{k \in \mathbb{Z}} |\langle \phi_{j,k}, h \rangle|^2 \approx 2^{-j} \#\{\text{integers } k \text{ in } [2^j a, 2^j b]\} \approx b - a.$$

In the limit as $j \to +\infty$ these approximations become equalities, and the approximation property stated in the theorem holds. $\qquad\square$

Following the same notation as for the Haar multiresolution analysis, we write $\lim_{j \to +\infty} V_j = \overline{\bigcup_{j \in \mathbb{Z}} V_j}$ for the subspace of all functions that are limits (in the L^2 norm) of a sequence of functions $f_n \in V_{j_n}$. Then the conclusion of Theorem 5.8 can be stated as

$$\lim_{j \to +\infty} V_j = L^2. \tag{5.53}$$

Remark 5.5. In the axiomatic *multiresolution analysis* approach to wavelet transforms for analog signals introduced in [Mallat (1989)], one starts with an increasing family of subspaces V_j in L^2 that satisfy the dilation condition (5.42) together with the limiting conditions (5.52) and (5.53) as $j \to \pm\infty$. One assumes the existence of a scaling function ϕ such that the set $\{S^k \phi : k \in \mathbb{Z}\}$ of shifted functions is an orthonormal basis for V_0. Since D and S are unitary transformations, it follows from the dilation condition that the set of functions $\{D^j S^k \phi : k \in \mathbb{Z}\}$ furnishes an orthonormal basis for V_j for every j. In particular, since $V_0 \subset V_1$ one obtains a two-scale equation (also called a *refinement equation*) by expanding $\phi \in V_0$ in terms of the orthonormal basis $\{DS^k\phi : k \in \mathbb{Z}\}$ for V_1. The lowpass filter coefficients in the two-scale equation are the inner products $\langle DS^k\phi, \phi \rangle$, which can be nonzero for infinitely many k in this general framework.

5.5.3 *Fast multiscale wavelet transform*

We obtain a fast multiscale wavelet transform from the multiresolution subspaces $\{V_j\}$ in several steps. As illustrated by the Haar multiresolution spaces, the spaces V_j give increasingly fine resolution as $j \to +\infty$. Starting with an analog signal $f \in L^2$ we replace f by $f_{\text{trend},J} \in V_J$ for some suitably large J. The choice of J will depend on the nature of f; given any error tolerance $\epsilon > 0$ we know from Theorem 5.8 that $\|f - f_{\text{trend},J}\| < \epsilon$ if we take J sufficiently large.

We now apply Proposition 5.4 to V_J, then to V_{J-1}, and so on to obtain a *K-stage multiresolution decomposition*

$$V_J = V_{J-1} \oplus W_{J-1} = V_{J-2} \oplus W_{J-2} \oplus W_{J-1} = \cdots$$
$$= V_{J-K} \oplus W_{J-K} \oplus \cdots \oplus W_{J-1}$$

for any integer $K \geq 1$. This gives an orthogonal expansion

$$f_{\text{trend},J} = f_{\text{trend},J-K} + f_{\text{detail},J-K} + \cdots + f_{\text{detail},J-1} \tag{5.54}$$

with $\|f_{\text{trend},J}\|^2 = \|f_{\text{trend},J-K}\|^2 + \|f_{\text{detail},J-K}\|^2 + \cdots + \|f_{\text{detail},J-1}\|^2$. For large K the functions in V_{J-K} have coarser and coarser resolution. Since $\|f_{\text{trend},J-K}\| \leq \|f_{\text{trend},J}\|$ for all $K \geq 0$, Lemma 5.5 implies $\|f_{\text{trend},J-K}\| \to 0$ exponentially fast as $K \to +\infty$. In practice, the choice of the number of stages K depends on the signal-processing requirements, such as compression or noise reduction.

Decomposition (5.54) of $f_{\text{trend},J}$ resembles the pyramid algorithm decomposition for periodic digital signals in Sections 3.3 and 3.5.4. This resemblance is not accidental. We can calculate this decomposition using the two-channel filter bank *trend-detail* decomposition from Section 4.8 applied to the discrete trend and detail of f, as follows.

Theorem 5.9. *The level j and level $j+1$ discrete trend and discrete detail of a finite-energy analog signal f are related by the two-channel filter bank algorithm for digital signals. The multiresolution analysis equations are*

$$\mathbf{s}_j = 2^{-1/2}\,\boxed{2\downarrow}(\mathbf{h}_0 \star \mathbf{s}_{j+1}) \quad \text{and} \quad \mathbf{d}_j = 2^{-1/2}\,\boxed{2\downarrow}(\mathbf{h}_1 \star \mathbf{s}_{j+1}). \tag{5.55}$$

The multiresolution synthesis equation is

$$\mathbf{s}_{j+1} = 2^{1/2}\big\{\mathbf{g}_0 \star (\,\boxed{2\uparrow}\mathbf{s}_j) + \mathbf{g}_1 \star (\,\boxed{2\uparrow}\mathbf{d}_j)\big\}. \tag{5.56}$$

Proof. By the two-scale relation (5.45) we can write

$$\phi_{j,k} = D^j S^k \phi = \sum_{n\in\mathbb{Z}} \mathbf{g}_0[n]\, D^{j+1} S^{n+2k}\phi = \sum_{n\in\mathbb{Z}} \mathbf{g}_0[n]\, \phi_{j+1,n+2k}\,, \tag{5.57}$$

$$\psi_{j,k} = D^j S^k \psi = \sum_{n\in\mathbb{Z}} \mathbf{g}_1[n]\, D^{j+1} S^{n+2k}\phi = \sum_{n\in\mathbb{Z}} \mathbf{g}_1[n]\, \phi_{j+1,n+2k}\,. \tag{5.58}$$

Hence

$$\mathbf{s}_j[k] = 2^{j/2}\langle \phi_{j,k}, f\rangle = 2^{j/2}\sum_{n\in\mathbb{Z}}\mathbf{g}_0[n]\,\langle \phi_{j+1,n+2k}, f\rangle$$

$$= 2^{-1/2}\sum_{n\in\mathbb{Z}}\mathbf{g}_0[n]\,\mathbf{s}_{j+1}[n+2k] = 2^{-1/2}\sum_{n\in\mathbb{Z}}\mathbf{h}_0[n]\,\mathbf{s}_{j+1}[2k-n],$$

since $\mathbf{h}_0[n] = \mathbf{g}_0[-n]$. This proves the first formula in (5.55). Likewise

$$\mathbf{d}_j[k] = 2^{j/2}\langle \psi_{j,k}, f\rangle = 2^{j/2}\sum_{n\in\mathbb{Z}}\mathbf{g}_1[n]\,\langle \phi_{j+1,n+2k}, f\rangle$$

$$= 2^{-1/2}\sum_{n\in\mathbb{Z}}\mathbf{g}_1[n]\,\mathbf{s}_{j+1}[n+2k] = 2^{-1/2}\sum_{n\in\mathbb{Z}}\mathbf{h}_1[n]\,\mathbf{s}_{j+1}[2k-n],$$

which is the second formula in (5.55).

To obtain the multiresolution synthesis equation, use equation (5.57) to express $\phi_{j,k}$ as a linear combination of the functions $\phi_{j+1,m}$. This gives

$$f_{\text{trend},j} = 2^{-j/2}\sum_{m\in\mathbb{Z}}\left\{\sum_{k\in\mathbb{Z}}\mathbf{g}_0[m-2k]\,\mathbf{s}_j[k]\right\}\phi_{j+1,m}.$$

Likewise, use equation (5.58) for $\psi_{j,k}$ to obtain

$$f_{\text{detail},j} = 2^{-j/2}\sum_{n\in\mathbb{Z}}\left\{\sum_{k\in\mathbb{Z}}\mathbf{g}_1[m-2k]\,\mathbf{d}_j[k]\right\}\phi_{j+1,m}.$$

By Proposition 5.4 we know that $f_{\text{trend},j+1} = f_{\text{trend},j} + f_{\text{detail},j}$. Substituting the level-j trend and detail formulas we just obtained and equating coefficients of $\phi_{j+1,m}$ on both sides of this equation, we find that

$$2^{-1/2}\mathbf{s}_{j+1}[m] = \sum_{k\in\mathbb{Z}}\mathbf{g}_0[m-2k]\,\mathbf{s}_j[k] + \sum_{k\in\mathbb{Z}}\mathbf{g}_1[m-2k]\,\mathbf{d}_j[k]$$

for all $m \in \mathbb{Z}$. This is equation (5.56). $\square$

The initial step in the fast multiscale wavelet transform algorithm for analog signals is to use sample values $f(k2^{-J})$ of f to approximate the values $\mathbf{s}_J[k]$ of the level-J discrete trend of f. When $f(t)$ is differentiable near the point $k2^{-k}$, Lemma 5.4 shows this to be a reasonable procedure provided J is taken sufficiently large.[17]

We use Theorem 5.9 to apply the pyramid algorithm for the (discrete) wavelet transform to the (approximate) discrete trend $\mathbf{s}_J$. For this we use the analysis filters $\mathbf{g}_0$ and $\mathbf{g}_1$ to calculate $\mathbf{s}_{J-1}, \mathbf{d}_{J-1}, \ldots, \mathbf{s}_{J-K}, \mathbf{d}_{J-K}$ as in Chapter 4. Notice that only the filter coefficients enter into this calculation. We never use the scaling function ϕ or the wavelet ψ.

Remark 5.6. There is a further important approximation needed to carry out the numerical calculation of the fast multiscale wavelet transform of a finite-energy analog signal f. Namely, we choose[18] a cutoff range $-T \leq t \leq T$ such that

---

[17]See Remark 5.7 for further discussion of this approximation.
[18]This is possible since f has finite energy.

$\int_{|t|>T} \|f(t)\|^2\, dt < \epsilon$, where ϵ is an acceptable error level, and replace f by the truncated function

$$\tilde{f}(t) = \begin{cases} f(t) & \text{for } |t| < T, \\ 0 & \text{for } |t| > T. \end{cases}$$

This truncation has two desirable properties:

(i) $\|f - \tilde{f}\|^2 < \epsilon$, so the multiscale wavelet transform of $\tilde{f}$ is a good approximation (in the energy norm) to the transform of f.
(ii) For every j the level-j discrete trend and detail for $\tilde{f}$ are finite digital signals, in the sense of chapter 4, since the scaling function ϕ and wavelet ψ are zero outside $0 \le t \le L$.

5.5.4 *Vanishing moments for wavelet functions*

The accuracy of the trend approximation to a signal and the estimate of the detail coefficients in Lemma 5.4 can be improved by better choice of the filter. Here we concentrate on the detail coefficients, since they can detect subtle changes in a signal that are not evident in the trend.

Definition 5.9. For $n = 0, 1, \ldots$ the nth *moment* of the wavelet function ψ is $\int_{-\infty}^{\infty} t^n\, \psi(t)dt$. The nth *moment* of the highpass filter $\mathbf{g}_1$ is $\sum_{k \in \mathbb{Z}} k^n\, \mathbf{g}_1[k]$.

In this definition the limits of integration can be restricted to $0 \le t \le L$ and the sum over k can be restricted to $0 \le k \le L$. This makes it evident that the moments of all orders are finite.

We already know that the moments of order zero for ψ and $\mathbf{g}_1$ both vanish. The nth moment of the filter can be calculated from derivatives of the z-transform evaluated at $z = 1$:

$$\sum_{k \in \mathbb{Z}} k^n\, \mathbf{g}_1[k] = (-1)^n \left(z\frac{d}{dz} \right)^n G_1(z)\Big|_{z=1}. \tag{5.59}$$

This follows because $\left(z\frac{d}{dz} \right) z^{-k} = -kz^{-k}$.

We say the filter $\mathbf{g}_1$ is *flat of order* K if all moments of order less than K vanish but the Kth moment is nonzero. By equation (5.59) this means that[19] the derivatives of $H_1(z)$ of order less than K vanish at $z = 1$ but the Kth derivative is nonzero at $z = 1$. In signal processing terms, the flatness of $\mathbf{g}_1$ means the impulse response function $G_1(e^{\omega i})$ of the filter vanishes to order K at frequency $\omega = 0$. The larger the value of K, the more effectively the filter removes the low frequency part of a signal.

Example 5.5. The Daub2K highpass filter is flat to order K for every integer $K = 1, 2, \ldots$ (where Daub2 is the Haar filter)—see Figure 4.11 for the Daub6 filter. Indeed, since $G_1(z)$ is a polynomial, flatness of order K means $z = 1$ is a root

---

[19]Since $\left(z\frac{d}{dz} \right)^n = z^n \left(\frac{d}{dz} \right)^n + R$, where R is a differential operator of order less than n.

of order K, so we need $G_1(z) = (1 - z)^K F(z)$, where $F(z)$ is a polynomial with $F(1) \neq 0$. Since $G_1(z) = H_0(-z^{-1})$, flatness follows from (4.90).   ■

Lemma 5.6. *The first and second moments of the wavelet function and the highpass filter are related as follows:*

(1) *The first moment of ψ equals the first moment of $\mathbf{g}_1$ multiplied by $1/(2\sqrt{2})$.*
(2) *If the first moment of $\mathbf{g}_1$ vanishes, then the moments of ψ of order less than 2 vanish, and the second moment of ψ equals the second moment of $\mathbf{g}_1$ multiplied by $1/(4\sqrt{2})$.*

Proof. We use the two-scale equation for ψ to write the moment integral as

$$\int_{-\infty}^{\infty} t\, \psi(t)\, dt = \sqrt{2} \sum_{k=0}^{L} \int_{-\infty}^{\infty} \mathbf{g}_1[k]\, t\, \phi(2t - k)\, dt$$

$$= 1/(2\sqrt{2}) \int_{-\infty}^{\infty} \left\{ s \sum_{k=0}^{L} \mathbf{g}_1[k] + \sum_{k=0}^{L} k\, \mathbf{g}_1[k] \right\} \phi(s)\, ds$$

$$= 1/(2\sqrt{2}) \sum_{k \in \mathbb{Z}} k\, \mathbf{g}_1[k]\,.$$

Here we made the change of variable $s = 2t - k$, $ds = 2dt$ in the integral and used the vanishing of the zeroth moment of $\mathbf{g}_0$. The last line follows from $\int_{-\infty}^{\infty} \phi(t)dt = 1$ in Theorem 5.6(4). This proves assertion (1). When the first moment of $\mathbf{g}_1$ vanishes, then a similar argument gives assertion (2).   □

Proposition 5.6. *Assume the highpass filter is flat to order $K \geq 2$. Let f be a finite-energy analog signal that has a continuous second derivative $f''(t)$ for all t in some interval $I = [a, b]$. If j, k are such that the subinterval $[2^{-j}k, 2^{-j}(k + L)] \subset I$, then the corresponding detail coefficient for $f(t)$ satisfies*

$$\left| \mathbf{d}_j[k] \right| \leq CM2^{-2j}\,. \tag{5.60}$$

Here $M = \max_{t \in I} |f''(t)|$ and the constant C does not depend on f or j.

Proof. Let $p(t) = f(2^{-j}k) + (t - 2^{-j}k)f'(2^{-j}k)$ be the Taylor polynomial of degree 1 for $f(t)$ centered at $2^{-j}k$. By Taylor's theorem with remainder we know that

$$|f(2^{-j}k + 2^{-j}x) - p(2^{-j}k + 2^{-j}x)| \leq \frac{M}{2} \left(2^{-j}x\right)^2 \tag{5.61}$$

if $0 \leq x \leq L$. But $\int_0^L \psi(x)\, p(2^{-j}k + 2^{-j}x)\, dx = 0$ since the moments of ψ of order less than 2 vanish by Lemma 5.6. Hence by equation (5.47) we can write

$$\mathbf{d}_j[k] = \int_0^L \psi(x) \left\{ f(2^{-j}k + 2^{-j}x) - p(2^{-j}k + 2^{-j}x) \right\} dx\,.$$

Now apply the Cauchy–Schwarz inequality to the integral and use inequality (5.61) to obtain the estimate

$$|\mathbf{d}_j[k]| \leq \|\psi\| \left\{ \int_0^L |f(2^{-j}k + 2^{-j}x) - p(2^{-j}k + 2^{-j}x)|^2 \, dx \right\}^{1/2}$$

$$\leq M2^{-2j-1} \left\{ \int_0^L x^4 \, dx \right\}^{1/2} = CM2^{-2j}.$$

Here $C = L^{5/2}/(2\sqrt{5})$. $\qquad\square$

Example 5.6. (Singularity Detection) We return to the one-scale Haar and Daub4 wavelet transforms of the signal in Example 3.8. Recall that this digital signal of length 64 is a sampled version of an analog signal that has two different types of singularities: a jump discontinuity at $n = 32$ and discontinuities in the first derivative (slope) at $n = 16$ and $n = 48$. The one-scale Haar and Daub4 transforms[20] for $j = 5$ are shown in Figure 5.9. As predicted by Lemma 5.5, the trend portions of both the Haar and Daub4 transforms (which are rescaled to $0 \leq n < 32$) are good approximations to the analog signal. The detail portions of the Haar and Daub4 transforms (which get rescaled to $32 \leq n < 64$) show the slope discontinuities at $n = 40$ (sharp peak for Daub4, spread out peak for Haar) and $n = 56$ (small peak for Daub4, small jump for Haar). The Haar detail does not detect the jump discontinuity at $n = 48$ that appears as a sharp peak in the Daub4 detail. This illustrates the contrast between the 2^{-j} bound in Lemma 5.4, valid for both the Haar and Daub4 detail signals on intervals where f is smooth, to the smaller 2^{-2j} bound in Proposition 5.6 which only applies to the Daub4 detail.[21] In particular, the Daub4 detail is zero on any interval where $f(t)$ is linear. $\qquad\blacksquare$

Remark 5.7. See [Strang and Nguyen (1997), §7.2] for comments on the "wavelet crime" of using single sample values of f to obtain the initial data $\mathbf{s}_J$ for the fast wavelet transform. Higher accuracy from single sample values can be obtained, for example, by modifying the Daub2K lowpass filters to obtain the *coiflet* orthogonal filters with vanishing moments—see [Daubechies (1992), §8.2], [Daubechies (1993), §4], and Exercises 5.7#10. An alternate approach with more computational expense is to use a *quadrature formula* that takes a weighted average of several sample values of f to approximate the integral for $\mathbf{s}_J[k]$.

Remark 5.8. In Chapters 3 and 4 we omitted several aspects of wavelet transforms for digital signals. Here are two of the most important topics that are also relevant for the fast multiscale wavelet transform of analog signals.

(i) *Boundary filters* that minimize undesirable boundary effects caused by periodic extension of finite non-periodic signals. See [Jensen and la Cour-Harbo (2001),

---

[20]The transforms are applied to the periodic extension of the signal. The non-periodic analog signal (extended by zero for $t > 1$ and $t < 0$) also has a slope discontinuity at $t = 1$ ($n = 64$).
[21]Of course, the constants C and M in these two estimates are different.

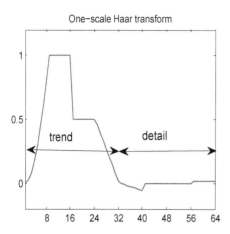
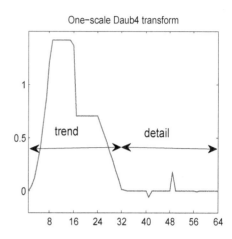

Fig. 5.9   Singularity detection using Haar and Daub4 transforms

Chapter 10], [Broughton and Bryan (2009), §6.5], [Van Fleet (2008), §11.3], and [Strang and Nguyen (1997), §8.5].

(ii) *Wavelet packet methods* that replace the basic pyramid algorithm by a tree structure adapted to the particular signal being processed. See [Jensen and la Cour-Harbo (2001), Chapter 8] and [Walnut (2002), Chapter 11].

5.5.5   *Guides to wavelet theory and applications*

Here we end this brief introduction to wavelet transforms for analog signals. Despite the technical details, the underlying message is quite simple and remarkable: once we have constructed a good FIR orthogonal lowpass filter, we can use it to do multiresolution signal processing on analog signals by a fast digital algorithm.

To explore the world of wavelets in more depth, the book [Hubbard (1998)] is an excellent starting place. It provides a highly readable account of the origins and basic ideas of wavelet theory in general mathematical terms and gives a broad sample of the scientific applications of wavelet methods. More detailed aspects of the construction of wavelet transforms for digital and analog signals with applications to signal and image processing are found in the introductory texts [Broughton and Bryan (2009)], [Frazier (1999)], [Jensen and la Cour-Harbo (2001)], [Van Fleet (2008)], [Walker (2008)] and the graduate-level texts [Boggess and Narcowich (2001)], [Burrus, Gopinath, and Guo (1998)], [Cohen and Ryan (1995)], [Daubechies (1992)], [Jaffard, Meyer, and Ryan (2001)], [Mallat (1999)], [Strang and Nguyen (1997)], [Walnut (2002)]. For links to wavelet information on the web, the *Wavelet Digest* (at `www.wavelet.org`) is a good resource.

5.6   Computer Explorations

The MATLAB m-files for these explorations are from the UVI_WAVE toolbox. See
Section A.3.2 for details about downloading and using them.

5.6.1   *Generating scaling and wavelet functions*

You will use cascade iteration from Section 5.4 to obtain approximations to some
of the Daub$2K$ scaling functions and wavelets.

(a) Daub4 scaling function and wavelet

Load the Daub4 filters using `daub.m` from the UVI_WAVE directory `wfilter`.

```
[h4, g4, rh4, rg4] = daub(4);
```

Here `h4` is the analysis low-pass filter with reversed filter `rh4` (low-pass synthesis
filter) of length 4, while `g4` is the analysis high-pass filter with reversed filter `rg4`
(the high-pass synthesis filter) of length 4.

Now use `wavelet.m` from the UVI_WAVE directory `wvt` to construct 10-scale approximations to the Daub4 scaling and wavelet functions from the cascade algorithm
and plot these functions in the time domain.[22]

```
figure, [s4, w4] = wavelet(rh4, rg4, 10, 0);
```

By Theorem 5.6 the scaling and wavelet functions for the Daub4 filters are zero
outside the continuous time interval $0 \le t \le 3$. Since you are carrying the cascade
algorithm to level 10, the approximations to these functions shown in the figure use
the discrete variable X varying from 1 to $3 \times 2^{10} = 3,072$ on the horizontal axis.
Edit the figure to obtain the labels as in Figure 5.6, as follows.

  (i) In the figure toolbar click on `tools` and `edit plot`. Then in the figure toolbar
      click on `edit` and `Axes properties`.
 (ii) Click on the plot of the scale function. In the boxes below the plots change
      the label to Daub4 Scaling Function and label the X axis as *continuous time*.
      Since $X = k$ is $t = k \times 2^{10}$, you must now change upper X limit for the axis to
      3072 $(= 3 \times 2^{10})$.
(iii) Click on `Ticks` and change table of X Ticks (locations, labels) to $(0,0)$,
      $(1024, 1.0)$, $(1024, 1.0)$, $(1548, 1.5)$, $(2,048, 2.0)$, $(2,560, 2.5)$, $(3,072, 3.0)$.
 (iv) Repeat steps (ii) and (iii) for the plot of the wavelet function.

Save the edited figure as `Daub4_sw`.

---

[22]This m-file uses the filter coefficients as the initial approximation instead of the Haar scaling and
wavelet functions as done in Section 5.4; as a result, the functions it generates must be multiplied
by $\sqrt{2}$ to obtain scaling and wavelet functions with norm 1.

(b) Smoothness of Daub4 scaling function and wavelet

The Daub4 scaling function $\phi(t)$ obtained as the limit of the cascade approximations is continuous. Despite the jagged appearance of the graphs of the approximating functions, $\phi(t)$ has a derivative at *almost every* point (in the sense of Lebesgue measure). There is a hierarchy of fractal sets where the graph is jagged to various degrees (like a mountain profile). At dyadic rational points the left-hand derivative exists, but not the right-hand derivative—see [Daubechies (1992), §7.2]. Explore this feature by clicking on the magnifying glass button, moving the cursor to the point on the scaling function graph where $t = 2.5$, and then clicking repeatedly to magnify the graph. To the left of $t = 2.5$ the graph is almost level, while to the right of $t = 2.5$ the slope is very steep.

The wavelet $\psi(t)$ is obtained as a linear combination of shifts and dilations of $\phi(t)$, so its graph has the same fractal nature.

(c) Daub6 and Daub8 scaling functions and wavelets

Load the Daub6 and Daub8 filters using `daub.m` from the Uvi_Wave directory `wfilter`.

```
[h6, g6, rh6, rg6] = daub(6);
[h8, g8, rh8, rg8] = daub(8);
```

Here `hn` is the analysis low-pass filter (for $n = 6, 8$) with reversed filter `rhn` (low-pass synthesis filter), while `gn` is the analysis high-pass filter with reversed filter `rgn` (the high-pass synthesis filter). These filters have length n.

Use `wavelet.m` from the Uvi_Wave directory `wvt` to construct 10-scale approximations to the Daub6 and Daub8 wavelet and scaling functions from the cascade algorithm and plot these functions in the time domain.

```
figure, [s6, w6] = wavelet(rh6, rg6, 10, 0);
figure, [s8, w8] = wavelet(rh8, rg8, 10, 0);
```

Edit the labels in the plots as in (a), with the appropriate modifications. Note that $0 \le t \le 5$ for Daub6 and $0 \le t \le 7$ for Daub8. It suffices to put tick marks at integer values of t. Save the edited figures as `Daub6_sw` and `Daub8_sw`.

(d) Smoothness of Daub6 and Daub8 scaling functions and wavelets

The Daub6 and Daub8 scaling functions have continuous first derivatives—see [Daubechies (1992), §7.2]. For Daub8 this is what your graph from (c) suggests, but for Daub6 the graph of the ten-scale approximation appears to have sharp peaks at $t = 1, 2, 3$, for example. Explore this feature by opening the `Daub6_sw` file, clicking on the magnifying glass button, moving the cursor to the point on the scaling function graph where $t = 1$, and then clicking repeatedly to magnify the graph. After many clicks, you will see the apparent "mountain peak" at this point in the graph flatten out. Use Tools, Reset View and repeat this exploration at the points on the graph where $t = 2$ and $t = 3$.

The Daub6 wavelet is obtained as a linear combination of shifts and dilations of the Daub6 scaling function, so it also has a continuous first derivative, despite the apparent peaks in the graph of the ten-scale approximation.

5.6.2   *Using wavelet transforms to find singularities*

Jumps in the slope of a curve (first-order singularities) are easily located. Jumps in the curvature (second-order singularities) are more subtle and not easily detected just by zooming in on the graph. You can use the detail part of a wavelet transform to detect these singularities.[23]

(a) Analog signal with slope and curvature singularities

Consider the piecewise-polynomial function

$$f(t) = \begin{cases} t^7 & \text{for } 0 < t \leq 1 \\ t & \text{for } 1 < t \leq 2 \\ -t^2 + 5t - 4 & \text{for } 2 < t \leq 4 \\ 0 & \text{for } t \leq 0 \text{ and } t > 4. \end{cases}$$

This function has the following properties that you should verify by calculating $\lim_{t \to \pm a} g(t)$, where $g(t)$ is $f(t)$, $f'(t)$, and $f''(t)$ and $a = 0, 1, 2, 4$.

(i) $f(t)$ is continuous for all t.
(ii) $f'(t)$ exists and is continuous except at $t = 1$ and $t = 4$ (slope singularities).
(iii) At the points where $f'(t)$ exists, then $f''(t)$ also exists and is continuous except at $t = 2$ (curvature singularity).

Convert $f(t)$ to a digital signal f by sampling at rate $N = 2^{10}$ on $0 \leq t \leq 4$, and make a line graph plot of f.

```
x = 2^(-10):2^(-10):4;   f = x;   n = 2^(10);
f(1:n-1) = x(1:n-1).^7;
f(2*n:4*n) = -x(2*n:4*n).^2 + 5*x(2*n:4*n) - 4;
figure, plot(x , f)
```

Label the figure Continuous function with slope and curvature singularities. The graph suggests that $f(t)$ is continuous and has slope singularities. Visual inspection of the graph doesn't reveal whether the transition from flat to curved at $t = 2$ is continuous.[24] Now you will examine these singularities using the detail part of the Daub2 ($=$ Haar), Daub4, and Daub6 wavelet transforms.

---

[23] Because of Newton's Law second derivatives play a major role in the equations of mathematical physics. If a massive particle (*e.g.* a railroad train) is traveling along a curve, then a curvature singularity causes an instantaneous force impulse perpendicular to the curve (*e.g.* a derailment).
[24] You know that it isn't by (iii), but if you had used a third-degree polynomial for $t > 2$, you could have achieved continuity.

(b) Wavelet transforms of signal

Generate the low-pass and high-pass analysis filters of lengths 2, 4, 6.

```
[h2, g2] = daub(2)
[h4, g4] = daub(4)
[h6, g6] = daub(6)
```

Take one-scale wavelet transform of the digital signal f using wt.m from the
UVI_WAVE wvt subdirectory. Then make separate plots of trend and detail using isplit.m from the UVI_WAVE wtutils subdirectory.

```
y2 = wt(f,h2,g2,1); figure, isplit(y2,1)
y4 = wt(f,h4,g4,1); figure, isplit(y4,1)
y6 = wt(f,h6,g6,1); figure, isplit(y6,1)
```

The trend plots for the three transforms are very similar to each other and to the
plot of f in (a). By looking at the vertical axes observe that the detail values are
several orders of magnitude smaller than the trend values in each figure except at
the singularities. This is predicted by Lemma 5.4 since the signal is smooth except
at the singular points. The detail plots for the three transforms differ significantly,
however. For Daub2 the detail is a rescaled discrete version of $f'(t)$, so it is rapidly
changing for $0 < n < 512$, constant for $512 < n < 1024$, and linear for $1024 <
n < 2048$. We will examine the details of the Daub4 and Daub6 transforms more
closely in (c) and (d). Put titles Daub2 trend, Daub2 detail, and so on the two plots
in each of the three figures by clicking on Tools, Edit Plot, Insert, Title. Save
the figures as Daub2_wt, Daub4_wt, Daub6_wt.

(c) Wavelet detection of slope singularities

Now examine the Daub4 and Daub6 detail plots closely around the slope singularity
at $t = 1$. Note that from the downsampling $t \to t/2$ in the wavelet transform this
singularity occurs at $n = 512$ in the detail part of the wavelet transform.

 Open the figure Daub4_wt generated in (b). Click on Tools, Edit Plot. Remove
the trend plot by clicking on it, and then Edit, Delete. Move the detail plot frame
to fill the figure space. Use Edit, Axes properties to set the X limits as 384 to
548 ($3/4 < t < 5/4$) and Y limits as -.5E-4 to .1E-3. Change the title to Daub4
detail at slope singularity and save the figure as Daub4_slope. Extract the Daub4
detail from the Daub4 wavelet transform y4 by

```
d4 = y4(2049:4096);
```

Notice that the detail is (essentially) zero to the right of $n = 512$ but is not zero
for $n < 512$. Enter d4(510:514) to examine the numerical values around the slope
singularity.

 Open the figure Daub6_wt generated in (b). Follow the same procedure as for
Daub4 to delete the trend plot and reset the X limits in the rescaled detail plot

as 384 to 548 and Y limits as $-.1E-2$ to $.5E-3$. Change the title to Daub6 detail at slope singularity and save the figure as `Daub6_slope`. Notice there are two sharp spikes at $n = 512, 513$; elsewhere in the plot the detail is many orders of magnitude smaller. Extract the Daub6 detail from the Daub6 wavelet transform y6 by

```
d6 = y6(2049:4096);
```

Enter `d6(510:514)` to examine the numerical values around the slope singularity.

(d) Wavelet detection of curvature singularities

Finally, examine the Daub4 and Daub6 detail plots closely around the curvature singularity at $t = 2$. Note that from the downsampling $t \to t/2$ in the wavelet transform this singularity occurs at $n = 1024$ in the detail part of the wavelet transform.

Open the `Daub4_slope` plot from (c). Click on `Tools, Edit Plot`. Use `Edit, Axes properties` to set the X limits as 992 to 1056 and the Y limits as $-.2E-5$ to $.2E-5$. Change the title to Daub4 detail at curvature singularity and save the file as `Daub4_curve`. Notice in the plot that the Daub4 detail is (essentially) zero to the left of $n = 1024$ and is constant to the right. Enter `d4(1022:1027)` to examine the numerical values near the curvature singularity.

Open the `Daub6_slope` plot from (c). Click on `Tools, Edit Plot`. Use `Edit, Axes properties` to set the X limits as 992 to 1056 and the Y limits as $-.1E-6$ to $.1E-5$. Change the title to Daub6 detail at curvature singularity and save the file as `Daub6_curve`. Notice in the plot that the Daub6 detail is (essentially) zero on both sides of the curvature singularity, with two spikes at the singularity. Enter `d6(1022:1027)` to examine the numerical values near the curvature singularity.

(e) Daub4 vs. Daub6 transform details

The contrasts between the Daub4 transform detail and Daub6 transform detail at the slope and curvature singularities are explained by Proposition 5.6 and Exercises 5.7 #10(b). For Daub4 the wavelet moments of order one or less vanish, but the second moment is nonzero, whereas for Daub6 the wavelet moments of order two or less vanish, but the third moment is nonzero (see Example 5.5 and Exercises 5.7 #9). Hence except at singularities the size of the Daub4 detail is controlled by the size of the second derivative of $f(t)$ and N^{-2}, where N is the sampling rate (2^{10} in this example), whereas the size of the Daub6 detail is controlled by the size of the third derivative of $f(t)$ and N^{-3}.

(i) (Slope singularity) Since $f''(t) = 7t^6 < 7$ and $f^{(3)}(t) = 42t^5 < 42$ for $0 < t < 1$, you expect the ratio of the Daub4 to the Daub6 detail for $n < 512$ to be on the order of $C(7N^{-2})/(42N^{-3}) = CN/6 = 171C$ (where C is a constant that doesn't depend on $f(t)$ or N). Check this estimate by calculating

```
d4(506:511)./d6(506:511)
```

Here the command ./ carries out element-by-element division of row vectors. Since $f''(t) = 0$ for $1 < t < 2$, both d4 and d6 will be (essentially) zero in the range $512 < n < 1024$. This explains the difference between the two plots in part (c).

(ii) (Curvature singularity) As already noted, the Daub4 and Daub6 details are (essentially) zero for $513 < n < 1024$. For $2 < t < 4$ one has $f''(t) = 4$, so for $1024 < n < 2048$ the Daub4 detail is on the order of $4N^{-2} \approx 4C \times 10^{-6}$ (where C is a constant that doesn't depend on $f(t)$ or N). Check this estimate by calculating d4(1025:1030). By contrast, since $f^{(3)}(t) = 0$ for $2 < t < 4$, the Daub6 detail is (essentially) zero on both sides of $n = 1024$, as observed in part (d).

This example shows the superior resolving power of the Daub6 wavelet transform over the Daub2 and Daub4 transforms as a mathematical microscope to detect the location of singularities.

5.7   Exercises

In Exercises 1–4 let V_j be the Haar multiresolution space (Definition 5.3), W_j the Haar detail space (Definition 5.4), $\phi_{j,k}$ the shifted and dilated Haar scaling functions, and $\psi_{j,k}$ the shifted and dilated Haar wavelet functions.

(1) Take the function f_1 in V_1 shown in Figure 5.10: $f_1(t) = 0$ when $t < -1$ or $t \geq 1$ and $f_1(t)$ has values 4, 2, 3, 1 in the four subintervals of $-1 \leq t < 1$.

   (a) Write f_1 as a linear combination of the functions $\phi_{1,k}$, as in Example 5.1.
   (b) Obtain the decomposition $f_1 = f_0 + g_0$ with $f_0 \in V_0$ and $g_0 \in W_0$.
   (c) Graph f_0 and g_0 on $-1 \leq t \leq 1$. Check that these graphs can be obtained from the graph of f_1 using the average value property (5.11).

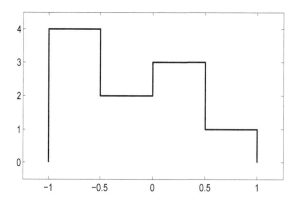

Fig. 5.10   Analog signal for Exercise #1

(2) Take the function f_2 in V_2 shown in Figure 5.11: $f_2(t) = 0$ when $t < 0$ or $t \geq 1$ and $f_2(t)$ has values 4, 2, 2, 1 in the four subintervals of $0 \leq t < 1$.

   (a) Write f_2 as a linear combination of the functions $\phi_{2,k}$.
   (b) Obtain the decomposition $f_2 = f_1 + g_1$ with $f_1 \in V_1$ and $g_1 \in W_1$.
   (c) Write f_1 as a linear combination of the functions $\phi_{1,k}$.
   (d) Obtain the decomposition $f_1 = f_0 + g_0$ with $f_0 \in V_0$ and $g_0 \in W_0$.
   (e) Graph f_0, g_0, and g_1 on $0 \leq t \leq 1$ and check from the graphs that the multiresolution decomposition $f_2 = f_0 + g_0 + g_1$ holds.

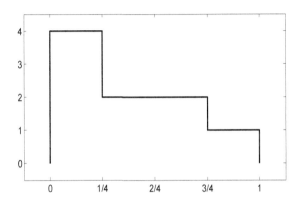

Fig. 5.11   Analog signal for Exercise #2

(3) Let $f(t) = t$ for $0 \leq t \leq 1$ and $f(t) = 0$ if $t < 0$ or $t > 1$. Assume $j \geq 0$.

   (a) Show that $\langle \phi_{j,k} , f \rangle = 0$ and $\langle \psi_{j,k} , f \rangle = 0$ when $k < 0$ or $k \geq 2^j$.
   (b) Calculate the level-j discrete trend $\mathbf{s}_j[k] = 2^{j/2} \langle \phi_{j,k} , f \rangle$ for $0 \leq k < 2^j$. Check that $\mathbf{s}_j[k]$ is the average value of $f(t)$ on $k2^{-j} \leq t \leq (k+1)2^{-j}$.
   (c) Calculate and graph the projection f_0 of f onto V_0 and the projection f_1 of f onto V_1.
   (d) Calculate the level-j discrete detail $\mathbf{d}_j[k] = 2^{j/2} \langle \psi_{j,k} , f \rangle$ for $0 \leq k < 2^j$.
   (e) Calculate and graph the projection g_0 of f onto W_0 and the projection g_1 of f onto W_1. Check that $f_1 = f_0 + g_0 + g_1$.

(4) Suppose the function $f(t)$ is a polynomial in t of degree n on some interval $a2^{-j} < t < b2^{-j}$, where $a < b$ are integers. Let $\mathbf{d}_j[k] = 2^{j/2} \langle \psi_{j,k} , f \rangle$ be the level-j discrete detail of $f(t)$. Prove that $\mathbf{d}_j[k]$ is a polynomial in k of degree at most $n-1$ for $a \leq k < b$. (This was already observed in Exercise 3(d) when $n = 1$.)

(5) Let V be an inner product space. Take f, g, and g_n $(n = 1, 2, \ldots)$ in V and assume that $\lim_{n \to \infty} \| g - g_n \| = 0$.

   (a) Prove that $\langle f, g \rangle = \lim_{n \to \infty} \langle f, g_n \rangle$.
   (b) Prove that $\| g \|^2 = \lim_{n \to \infty} \| g_n \|^2$.

(6) Consider the lowpass filter $g_0 = (1/\sqrt{2})(\delta_0 + \delta_3)$ in Example 5.4 which cannot be used to create a orthogonal multiresolution analysis of L^2.

    (a) Show that the function $\phi(t) = (1/3)\phi_{\text{Haar}}(t/3)$ satisfies the two-scale equation (5.29).

    (b) Find the associated highpass filter g_1 from (5.25) and use (5.30) to find the function $\psi = M_1\phi$.

    (c) Find the integers k such that $\langle S^k\phi, \phi \rangle$ is nonzero. Likewise for $\langle S^k\psi, \psi \rangle$ and $\langle S^k\phi, \psi \rangle$.

(7) Give a proof by induction of equation (5.35) in Lemma 5.3. (HINT: Use the two-scale equation and the sum rule (5.27).)

(8) Suppose that $\phi(t)$ is a normalized scaling function obtained, as in Theorem 5.6, from an orthogonal lowpass filter g_0 of length 4. Assume ϕ is continuous, so that $\phi(t) = 0$ for $t \le 0$ or $t \ge 3$ (Theorem 5.6). Write $\sqrt{2}\, g_0 = \begin{bmatrix} a\ b\ c\ d \end{bmatrix}$, so the two-scale equation is $\phi(t) = a\phi(2t) + b\phi(2t - 1) + c\phi(2t - 2) + d\phi(2t - 3)$.

    (a) Set $A = \begin{bmatrix} b\ a \\ d\ c \end{bmatrix}$ and $\mathbf{v} = \begin{bmatrix} \phi(1) \\ \phi(2) \end{bmatrix}$. Show that $\mathbf{v}$ is nonzero and is an eigenvector for A with eigenvalue 1. (HINT: Use the two-scale equation and the continuity of $\phi(t)$.)

    (b) Assume that the filter coefficients satisfy $b - a \ne 1$. Use the partition of unity property (5.40) and part (a) to find $\phi(1)$ and $\phi(2)$ in terms of a, b.

    (c) Find a 3×2 matrix B such that $\begin{bmatrix} \phi(1/2) \\ \phi(3/2) \\ \phi(5/2) \end{bmatrix} = B \begin{bmatrix} \phi(1) \\ \phi(2) \end{bmatrix}$.

    (d) Suppose g_0 is the Daub4 filter, with $a = (1 + \sqrt{3})/4$, $b = (3 + \sqrt{3})/4$, $c = (3 - \sqrt{3})/4$, and $d = (1 - \sqrt{3})/4$. Find $\phi(t)$ for $t = 1/2, 1, 3/2, 2, 5/2$. Give approximate decimal values and compare with the top right graph in Figure 5.6.

(9) Suppose that $\phi(t)$ and $\psi(t)$ are scaling and wavelet functions obtained as in Theorem 5.6 from an orthogonal lowpass filter g_0 and associated highpass filter g_1. Write $\mu_n = \int_{-\infty}^{\infty} t^n \phi(t)\, dt$ for the nth moment of ϕ and $\nu_n = \int_{-\infty}^{\infty} t^n \psi(t)\, dt$ for the nth moment of ψ. Write $\sigma_n = (1/\sqrt{2}) \sum_{k=0}^{L} k^n\, g_0[k]$ for the (normalized) nth moment of g_0 and $\tau_n = (1/\sqrt{2}) \sum_{k=0}^{L} k^n\, g_1[k]$ for the (normalized) nth moment of g_1. In particular, note that $\mu_0 = \sigma_0 = 1$ and $\nu_0 = \tau_0 = 0$.

    (a) Show that $(2^n - 1)\mu_n = \sum_{j=0}^{n-1} \binom{n}{j}\mu_j \sigma_{n-j}$ for $n \ge 1$, where $\binom{n}{j}$ is the binomial coefficient. Hence the moments of the scaling function are determined recursively by the moments of the lowpass filter (which can be calculated from the derivatives of its z-transform evaluated at $z = 1$). In particular, if $\sigma_n = 0$ for $1 \le n \le N - 1$, then $\mu_n = 0$ for $1 \le n \le N - 1$ and $(2^n - 1)\mu_N = \sigma_N$.

(b) Show that $2^n \nu_n = \sum_{j=0}^{n-1} \binom{n}{j} \mu_j \tau_{n-j}$ for $n \geq 1$. Hence by (a) the moments of the wavelet function are determined recursively by the moments of the lowpass and highpass filters. In particular, if $\tau_n = 0$ for $1 \leq n \leq N - 1$, then $\nu_n = 0$ for $1 \leq n \leq N - 1$ and $2^N \nu_N = \tau_N$.

(10) In the context of Proposition 5.6 assume $f(t)$ has a continuous Nth derivative for all t in some interval I and let $M = \max_{t \in I} |f^{(N)}(t)|$. Assume the scale j and shift k are such that the interval $[2^{-j}k, 2^{-j}(k + L)] \subset I$. Let σ_n and τ_n be the normalized nth moments of the filters $\mathbf{g}_0$ and $\mathbf{g}_1$ as in Exercise #9.

(a) Suppose $\sigma_n = 0$ for $1 \leq n \leq N - 1$ (the lowpass filter $\mathbf{g}_0$ is flat of order at least N at frequency $\omega = 0$). Show that $\left| s_j[k] - f(2^{-j}k) \right| \leq CM2^{-Nj}$, where the constant C does not depend on f, j, or k.

(b) Suppose $\tau_n = 0$ for $1 \leq n \leq N - 1$ (the highpass filter $\mathbf{g}_1$ is flat of order at least N at frequency $\omega = 0$). Show that $\left| d_j[k] \right| \leq CM2^{-Nj}$, where the constant C does not depend on f, j, or k.

Appendix A

Some Mathematical and Software Tools

A.1 Complex Numbers and Roots of Polynomials

A *complex number* z is a pair (x, y) of real numbers, denoted by $z = x + y\mathrm{i}$ (in engineering texts usually denoted by $z = x + y\mathrm{j}$). We call x the *real part* of z and y the *imaginary part* of z. Complex numbers are added by adding the real and imaginary parts separately, just as if they were vectors in $\mathbb{R}^2$. Multiplication is defined by

$$(a + b\mathrm{i})(x + y\mathrm{i}) = (ax - by) + (by + ax)\mathrm{i} \tag{A.1}$$

for $a, b, x, y \in \mathbb{R}$. In particular, $\mathrm{i}^2 = -1$ and $a(x + y\mathrm{i}) = ax + ay\mathrm{i}$ is the same as multiplying a vector in $\mathbb{R}^2$ by a real scalar.

We denote the set of all complex numbers by $\mathbb{C}$. It is clear from the definition that addition and multiplication of complex numbers is commutative:

$$(a + b\mathrm{i})(x + y\mathrm{i}) = (x + y\mathrm{i})(a + b\mathrm{i}) \,.$$

The distributive law $u(v + w) = uv + uw$ is also obvious from equation (A.1). A direct calculation shows that multiplication of complex numbers is associative:

$$u(vw) = (uv)w \quad \text{for } u, v, w \in \mathbb{C} \,. \tag{A.2}$$

For a complex number $c = a + \mathrm{i}b$ we define the *complex conjugate* $\bar{c} = a - b\mathrm{i}$ and the *modulus* $|c|$ by

$$|c| = \sqrt{\bar{c}c} = \sqrt{a^2 + b^2} \,.$$

Note that if $c \neq 0$ then $|c| > 0$ and $(|c|^{-2}\bar{c})c = 1$. Hence every nonzero complex number c has a *multiplicative inverse*

$$c^{-1} = |c|^{-2}\bar{c} = \left(\frac{a}{a^2 + b^2}\right) - \left(\frac{b}{a^2 + b^2}\right)\mathrm{i} \,.$$

It is easy to check that $\overline{zw} = \bar{z}\bar{w}$ for every $z, w \in \mathbb{C}$. Hence $|zw| = |z|\,|w|$ and $|z^n| = |z|^n$ for all integers n (if $n < 0$ then we must assume $z \neq 0$).

We view the set $\mathbb{R}$ of real numbers as a subset of $\mathbb{C}$ by identifying $x \in \mathbb{R}$ with the complex number $x + 0\mathrm{i}$. The algebraic operations in $\mathbb{C}$ just defined then reduce to the usual algebraic operations on real numbers under this identification.

As an algebraic system, the set of complex numbers satisfies the axioms of a *field* (relative to addition and multiplication), just like the real numbers: there are operations of addition, subtraction, multiplication, and division by nonzero elements. Addition and multiplication are commutative and distributive. However, there is no *order relation* $a < b$ for complex numbers, unlike the case of the real numbers.

Here is more geometric way to construct the field of complex numbers from the real numbers and matrix algebra. Given a complex number $z = x + iy$, define the real 2×2 matrix

$$T(z) = \begin{bmatrix} x & -y \\ y & x \end{bmatrix} = xI + yJ, \quad \text{where} \quad J = \begin{bmatrix} 0 & -1 \\ 1 & 0 \end{bmatrix}$$

and I is the 2×2 identity matrix. Note that J is the matrix for a $90°$ counter-clockwise rotation and $J^2 = -I$. Clearly $T(0) = \mathbf{0}$ (the 2×2 zero matrix) and $T(1) = I$. For any complex numbers z and w we calculate that $T(z + w) = T(z) + T(w)$ and $T(zw) = T(z)T(w)$ (matrix addition and multiplication on the right sides of the equalities). Obviously z is uniquely determined by the matrix $T(z)$. Thus addition and multiplication of complex numbers is the same as addition and multiplication of these special 2×2 real matrices. Since matrix multiplication is associative, this proves equation (A.2) for multiplication of complex numbers. Furthermore, $\det T(z) = |z|^2$ and $T(z^{-1}) = T(z)^{-1}$ (matrix inverse) when $z \neq 0$.

Let $f(z) = z^n + a_{n-1}z^{n-1} + \cdots + a_1 z + a_0$ be a nonzero polynomial with complex coefficients $a_{n-1}, \ldots, a_1, a_0$ (we can always make the coefficient of the leading term z^n equal 1 by division). A crucial property of the field of complex numbers[25] is that $f(z)$ has a *root* λ: a complex number such that $f(\lambda) = 0$. The example $f(z) = z^2 + 1$ shows that this property is not true for the field of real numbers.

If λ is a root of $f(z)$, then by the division algorithm for polynomials we can find a polynomial $g(z) = z^{n-1} + \cdots + b_0$ of degree $n - 1$ such that

$$f(z) = (z - \lambda)g(z) .$$

Continuing the division process, we can factor $f(z)$ as a product of n polynomials of degree one:

$$f(z) = (z - \lambda_1) \cdots (z - \lambda_n) .$$

Here some of the complex roots λ_k may be equal.

A.2 Exponential Function and Roots of Unity

The complex exponential function can be defined directly in terms of its Taylor series centered at zero:

$$e^t = 1 + t + \frac{t^2}{2!} + \cdots + \frac{t^n}{n!} + \cdots ,$$

---

[25]This property is sometimes called the *fundamental theorem of algebra*; the proof uses analysis.

where t can be any complex number. Since $|t^n/n!| = |t|^n/n!$, it follows from the ratio test that the partial sums of this series converge absolutely and uniformly in every disc $|t| \leq R$ for any value of R. Here *convergence* of a sequence of complex numbers means convergence of the real and imaginary parts of the sequence. The exponential function satisfies the *law of exponents*

$$e^{s+t} = e^s e^t \quad \text{for all } s, t \in \mathbb{C}. \tag{A.3}$$

This can be verified by using the binomial formula for $(s + t)^n$ in the exponential series for e^{s+t} and then rearranging the double series.

Every complex number $z = x + yi$ has a unique *polar decomposition*:

$$z = re^{i\theta} = r\cos(\theta) + r\sin(\theta)i, \quad \text{where} \quad r = |z| \quad \text{and} \quad \theta = \arctan(y/x). \tag{A.4}$$

This decomposition is just the formula for polar coordinates in $\mathbb{R}^2$ written in terms of complex numbers. We call θ the *argument* of z. Since $e^{(\theta+2\pi m)i} = e^{\theta i}$ for any integer m, the argument of z is only determined up to the addition of integer multiples of 2π. In particular, for any $\theta \in \mathbb{R}$, we have $|e^{\theta i}| = 1$ and the complex number $w = e^{\theta i}$ is a point on the unit circle in $\mathbb{R}^2$.

For any positive integer N the complex number

$$\omega = e^{2\pi i/N} = \cos(2\pi/N) + i\sin(2\pi/N)$$

has absolute value one and satisfies $\omega^N = 1$, since $(e^{2\pi i/N})^N = e^{2\pi i} = 1$ by equations (A.3) and (A.4). Every other complex number λ such that $\lambda^N = 1$ is of the form ω^k for some integer k. To prove this, note that

$$1 = |\lambda^N| = |\lambda|^N,$$

and hence $|\lambda| = 1$. Thus by the polar decomposition $\lambda = e^{i\theta}$. Since $\lambda^N = e^{iN\theta} = 1$, we see that $N\theta = 2k\pi$ for some integer k. Hence

$$\lambda = e^{2k\pi i/N} = \omega^k.$$

This shows that the polynomial $f(z) = z^N - 1$ has N distinct roots $1, \omega, \omega^2, \ldots,$ ω^{N-1}, which are located at N equidistant points around the unit circle $\{|z| = 1\}$. Thus it factors as

$$z^N - 1 = (z - 1)(z - \omega)(z - \omega^2) \cdots (z - \omega^{N-1}).$$

The number ω is called a *primitive Nth root of unity*. Complex multiplication of the set of numbers $\{\omega^k\}$ corresponds to addition of the exponents modulo N:

$$\omega^j \omega^k = \omega^p \quad \text{where } j + k \equiv p \mod (N).$$

For example, when $N = 8$ then $\omega^5 \omega^6 = \omega^3$ since $5 + 6 = 11 \equiv 3 \mod (8)$, and $\omega^{-5} = \omega^3$ since $-5 \equiv 3 \mod (8)$.

A.3   Computations in MATLAB and UVI_WAVE

A.3.1   *Introduction to* MATLAB

The basic mode of MATLAB is interactive. After you start the MATLAB program and obtain the prompt `>>`, you type commands that MATLAB then executes when you press the `Enter` key. If you have never used MATLAB before, you can type `demo` at the MATLAB prompt, click on `Desktop Environment`, and run the playback files. Then move on to the demo of `Matrices` and run `Basic matrix operations`. This gives a slide show that demonstrates how matrices are entered and displayed. You can also run the demonstration `Matrix manipulation`.

Creating matrices and vectors: The commands to do this are short and easy to remember because MATLAB specializes in matrix computations and uses standard linear algebra notation. The most direct way to create a matrix in MATLAB is to type the entries in the matrix between square brackets, one row at a time. To separate the entries in the same row, type a comma or press the space bar. To indicate the beginning of a new row, type a semicolon or press the `Enter` key. Try this by typing

    A = [1 2; 3 4; 5 6]

(followed by `Enter`). MATLAB should then display the 3×2 matrix

$$A =$$

$$1\ 2$$
$$3\ 4$$
$$5\ 6$$

(MATLAB displays matrices without braces). You could also generate this matrix by pressing the `Enter` key at the end of each row, instead of typing a semicolon.

Now use MATLAB to create the following matrix, row vector and column vector:

$$B = \begin{bmatrix} 1\ 2\ 3 \\ 4\ 5\ 6 \\ 7\ 8\ 9 \end{bmatrix} \qquad x = \begin{bmatrix} 4\ 3\ 2 \end{bmatrix} \qquad X = \begin{bmatrix} 1 \\ 2 \\ 3 \end{bmatrix}.$$

Type the names of each of these matrices and vectors that you have created at the MATLAB prompt. Note that x and X are different objects; MATLAB is *case sensitive*. Finally, type `whos` at the prompt to get a list of all the matrices and vectors that are in your current MATLAB workspace.

Size command: This command determines the number of rows and columns in a matrix. Every MATLAB command is documented in a `help` file, which you can access during a MATLAB session. Type `help size` now to get information about this command. Then use the `size` command to create a 4×2 matrix whose rows are the sizes of A, C, X, x (in that order), by typing `[size(A); size(B); size(X); size(x)]`

Give this matrix the name S by typing `S = ans`. Note that all matrices which occur in MATLAB must have names; if a matrix is unnamed, then it gets assigned the temporary name `ans` at the moment that it is created.

Modifying matrices: To access a given entry in a matrix, put the row and column number in parentheses following the matrix name. Type `a32 = A(3,2)` and check that $a32$ is the $(3,2)$ entry in A defined above. Observe that the equal sign $=$ in MATLAB (as in other programming languages) executes a *substitution*: the current value of the variable on the right side of the equal sign is placed into the location whose name is on the left side. Type `A(3,2) = 7` and check that the $(3,2)$ entry of A is now 7. Now change the $(3,2)$ entry of A back to 6. One way to do this is to type `A(3,2) = 6`. Another way that you should try for future use is the *up-arrow* key ↑. This lets you cycle through the commands that you have already typed. When you get to the command that generated A, press `Enter`. If you go too far with the up-arrow, you can use the down-arrow key ↓.

Colon operator: To access a whole row or column of a matrix, use the colon operator. For example, $A(:,2)$ is the second column of A, while $B(1,:)$ is the first row of B. Type

   `C(:,1) = B(:,1); C(:,2)= B(:,3)`

to create a 3×2 matrix C whose first column is the first column of B and whose second column is the third column of B. Then use the colon operator to create a 2×3 matrix D whose first row is the first row of B and whose second row is the third row of B. Use MATLAB to display the matrices C and D by typing

   `C, D   <Enter>` .

Block matrices: You can create *block matrices* by putting two matrices side by side (if they have the same number of rows), or one on top of the other (if they have the same number of columns). Use the matrices A, B, C, D, X created above to make the following block matrices (the semicolon means that the matrices are stacked one on top of the other). Not all of these are defined; you will get error messages when the matrix sizes are not compatible.

   `[A  X]` `[B  C]` `[C  D]` `[C;B]` `[B; D]`

Special matrices: Type each of the following commands that generate special matrices:

   `eye(4)` `zeros(3)` `zeros(3,5)` `ones(2,3)` `diag([4  5  6  7])`

Random matrices: The MATLAB command `rand(m, n)` creates an $m \times n$ matrix whose entries are random numbers uniformly distributed between 0 and 1. Try this by typing `R = rand(2, 3)`. Then use the up-arrow key to generate two more samples of the random matrix R. Notice how the entries in R change each time the command is executed.

Suppressing displays: When you place a semicolon at the end of a command, the command will be executed but the result will not be displayed on the screen. This is very useful when you are creating big matrices.

Matrix addition and multiplication: To obtain a linear combination $sA + tB$ of the matrices A and B (where s and t are real numbers) using MATLAB, you must type `s*A + t*B` (this is only defined when A and B are the same size). The conjugate transpose A^H of A is obtained by typing A'. To obtain matrix products AB or matrix-vector product Au using MATLAB, you must type `A*B` or `A*u` (remember that a vector is just a matrix with one column).

Function files: For more complicated MATLAB calculations you should use *function* files. A function file contains one or several MATLAB commands and is stored as a text file with a descriptive name such as `myfunction.m`, for example (the extension `*.m` is required). A *function file* requires the input of variables (which can be numbers or matrices), and gives as output the function evaluated at the input variables. When you type the name of a function file with the specific values for the input variables, then the function defined in your function file is evaluated at these variables.

Writing Function Files: Start MATLAB and click on *New* and then click on *Function*. This will open the MATLAB Editor/Debugger Window, and you can type the commands in this window. You can take any m-file, edit it (just as you would edit any text file), and then save it under a different name to obtain a new m-file.

Running Function Files: After you have created a function file and saved it to your directory, you must set the *Path* so that MATLAB can find this file. Click on *Set Path* and follow the directions to add your directory to the list of path names.

A.3.2 Uvi_Wave *software*

Download the Uvi_Wave collection of MATLAB files from the World Scientific supplementary page for this book:

http://www.worldscientific.com/worldscibooks/10.1142/9835#t=suppl

After you use a utility program to unzip the file and save it to your computer, you will obtain the main directory `Uvi_Wave.300` with several subdirectories `scal`, `wdemo`, `wfilter`, ... Start the MATLAB program on your computer and click `File` and `Set Path` in the toolbar. In the MATLAB search path find the `Uvi_Wave.300` directory on your computer and click on `Add with Subfolders`. Then click on `Save`. MATLAB will be now be able to access the Uvi_Wave files when you call them from the MATLAB prompt.

Appendix B

Solutions to Exercises

B.1 Solutions to Exercises 1.10

(1) (a) The vectors obtained by sampling are (with 4 decimal place accuracy)

$$\mathbf{x} = \begin{bmatrix} 0.0000 \\ 1.7500 \\ 3.5000 \\ 5.2500 \end{bmatrix}, \quad \mathbf{y} = \begin{bmatrix} 0.0000 \\ 3.5355 \\ -5.0000 \\ 3.5355 \end{bmatrix}, \quad \mathbf{z} = \begin{bmatrix} 0.0000 \\ 8.8211 \\ -6.5000 \\ 12.3211 \end{bmatrix} = \mathbf{x} + 2\mathbf{y}.$$

(b) Using the `floor` function we obtain the quantized vectors

$$q(\mathbf{x}) = \begin{bmatrix} 0 \\ 1 \\ 3 \\ 5 \end{bmatrix}, \quad q(\mathbf{y}) = \begin{bmatrix} 0 \\ 3 \\ -5 \\ 3 \end{bmatrix}, \quad q(\mathbf{z}) = \begin{bmatrix} 0 \\ 8 \\ -7 \\ 12 \end{bmatrix}, \quad q(\mathbf{x}) + 2q(\mathbf{y}) = \begin{bmatrix} 0 \\ 7 \\ -7 \\ 11 \end{bmatrix}.$$

Thus $q(\mathbf{z}) \neq q(\mathbf{x}) + 2q(\mathbf{y})$.

(2) Use the hint.

(3) Let $\mathbf{e}_{ij}$ be the 2×2 matrix with 1 in row i and column j, and 0 elsewhere.

    (a) M is a subspace with basis $\{\mathbf{e}_{11}, \mathbf{e}_{22}\}$.

    (b) M is a subspace with basis $\{\mathbf{e}_{11}, \mathbf{e}_{22}, \mathbf{e}_{12}\}$.

    (c) M is a subspace with basis $\{\mathbf{e}_{11}, \mathbf{e}_{22}, \mathbf{e}_{12} + \mathbf{e}_{21}\}$.

    (d) M is not a subspace since it is not closed under addition. For example $\mathbf{e}_{11}$ and $\mathbf{e}_{22}$ have determinant 0, but $\det(\mathbf{e}_{11} + \mathbf{e}_{22}) = 1$.

(4) Given $f, g \in U$ and $\alpha, \beta \in \mathbb{R}$, let $h(x) = \alpha f(x) + \beta g(x)$. Then $h(-1) = \alpha f(-1) + \beta g(-1) = 2\alpha f(1) + 2\beta g(1) = 2h(1)$. Hence $h \in U$. The shows that U is a subspace.

(5) (a) $\det(A) = 4$, so rank $A = 3$ and the columns of A are independent.

    (b) $A^{-1} = \dfrac{1}{4} \begin{bmatrix} 3 & 2 & 1 \\ 2 & 4 & 2 \\ 1 & 2 & 3 \end{bmatrix}$ (by MATLAB or Gauss-Jordan elimination). So the

dual basis is $\mathbf{u}_1 = \begin{bmatrix} 0.75 & 0.5 & 0.25 \end{bmatrix}$, $\mathbf{u}_2 = \begin{bmatrix} 0.5 & 1 & 0.5 \end{bmatrix}$, $\mathbf{u}_3 = \begin{bmatrix} 0.25 & 0.5 & 0.75 \end{bmatrix}$.

(c) Let $\mathbf{v} = \begin{bmatrix} 1 \\ 2 \\ 3 \end{bmatrix}$. Then $c_1 = \mathbf{u}_1\mathbf{v} = 2.5, \quad c_2 = \mathbf{u}_2\mathbf{v} = 4, \quad c_3 = \mathbf{u}_3\mathbf{v} = 3.5,$

giving the expansion $\begin{bmatrix} 1 \\ 2 \\ 3 \end{bmatrix} = 2.5 \begin{bmatrix} 2 \\ -1 \\ 0 \end{bmatrix} + 4 \begin{bmatrix} -1 \\ 2 \\ -1 \end{bmatrix} + 3.5 \begin{bmatrix} 0 \\ -1 \\ 2 \end{bmatrix}.$

(6) (a) The standard monomial basis in terms of these functions is
$$1 = -f_1(x) + f_2(x), \quad x = 3f_1(x) - 2f_2(x), \text{ and } x^2 = f_3(x) + 2f_2(x) - 3f_1(x).$$

(b) Since $\dim \mathcal{P}_3 = 3$ and the set $\{f_1, f_2, f_3\}$ spans $\mathcal{P}_3$, it must be independent.

(7) (a) $h(x) = g(x) - f(x)$, so $\dim V = 2$ and $\{f(x), g(x)\}$ is a basis for V.

(b) $\{f(x), g(x), h(x)\}$ is linearly independent: If $af(x) + bg(x) + ch(x) = 0$ (as a polynomial in x) then $(b+c)x^2 + ax + (c-b-a) = 0$ (as a polynomial in x). Hence $b + c = 0$, $a = 0$, and $c - b - a = 0$, implying $b = 0$ and $c = 0$. Thus $\dim V = 3$ and $\{f(x), g(x), h(x)\}$ is a basis for V. Note that $V = \mathcal{P}_3$ since $\dim \mathcal{P}_3 = 3$, so $\{1, x, x^2\}$ is also a basis for V. *A basis is never unique.*

(8) The polynomials in V are of the form $f(x) = a(x^2 + 4) + b(3x + 5)$ with a, b arbitrary scalars. Hence the polynomials $g(x) = x^2 + 4$ and $h(x) = 3x + 5$ span V and are linearly independent, since $g(x)$ is not a scalar multiple of $h(x)$. Thus $\{g(x), h(x)\}$ is a basis for V.

(9) (a) If $f(x) \in V$, then $f(0) = 0$ implies that $f(x) = xg(x)$ for some polynomial $g(x)$ of degree less than 3. But $f(1) = 0$, so $g(1) = 0$. This implies that $g(x) = (x-1)h(x)$ for some polynomial $h(x)$ of degree less than 2. Hence $h(x) = ax + b$ for some scalars a, b.

(b) Let $g(x) = x^2(x-1)$ and $h(x) = x(x-1)$. If $f(x) \in V$ then we know from (a) that $f(x) = ax^2(x-1) + bx(x-1) = ag(x) + bh(x)$ for some scalars a, b. Hence $g(x)$ and $h(x)$ span V and are linearly independent, since $g(x)$ is not a scalar multiple of $h(x)$. Thus $\{g(x), h(x)\}$ is a basis for V.

(10) (a) Since $2\mathbf{e}_1 = \begin{bmatrix} 1 \\ 1 \end{bmatrix} + \begin{bmatrix} 1 \\ -1 \end{bmatrix}$ and L is linear, $2L(\mathbf{e}_1) = \begin{bmatrix} 2 \\ 3 \end{bmatrix} + \begin{bmatrix} 4 \\ 5 \end{bmatrix} = \begin{bmatrix} 6 \\ 8 \end{bmatrix}.$

Likewise, $2\mathbf{e}_2 = \begin{bmatrix} 1 \\ 1 \end{bmatrix} - \begin{bmatrix} 1 \\ -1 \end{bmatrix}$, so $2L(\mathbf{e}_2) = \begin{bmatrix} 2 \\ 3 \end{bmatrix} - \begin{bmatrix} 4 \\ 5 \end{bmatrix} = \begin{bmatrix} -2 \\ -2 \end{bmatrix}.$

(b) The vectors in (a) give the columns of the matrix $\begin{bmatrix} 3 & -1 \\ 4 & -1 \end{bmatrix}$ of L.

(11) (a) $T\mathbf{v}_1 = \begin{bmatrix} 2 \\ 3 \\ 4 \end{bmatrix}, \quad T\mathbf{v}_2 = \begin{bmatrix} -2 \\ 0 \\ 4 \end{bmatrix}, \quad T\mathbf{v}_3 = \begin{bmatrix} 2 \\ 0 \\ 4 \end{bmatrix}.$

(b) If $f(x)$ and $g(x)$ are in V and $c \in \mathbb{R}$, then $T(f + cg) = \begin{bmatrix} 2(f(-1) + cg(-1)) \\ 3(f(0) + cg(0)) \\ 4(f(1) + cg(1)) \end{bmatrix} = \begin{bmatrix} 2f(-1) \\ 3f(0) \\ 4f(1) \end{bmatrix} + c \begin{bmatrix} 2g(-1) \\ 3g(0) \\ 4g(1) \end{bmatrix} = T(f) + cT(g).$

(c) The columns of A are the vectors $T\mathbf{v}_1$, $T\mathbf{v}_2$, $T\mathbf{v}_3$. Hence $A = \begin{bmatrix} 2 & -2 & 2 \\ 3 & 0 & 0 \\ 4 & 4 & 4 \end{bmatrix}$.

(12) (a) By definition $L(x) = 2 + (3x + 4)x = 3x^2 + 4x + 2$ and $L(1) = 3x + 4$.

(b) From (a) the matrix of L is $\begin{bmatrix} 3 & 0 \\ 4 & 3 \\ 2 & 4 \end{bmatrix}$.

(c) If $f(x) = 6x + 5$, then

$$Lf(x) = 2(6) + (3x + 4)(6x + 5) = 18x^2 + 39x + 32.$$

In terms of the ordered bases,

$$f(x) \longleftrightarrow \begin{bmatrix} 6 \\ 5 \end{bmatrix} \quad \text{and} \quad Lf(x) \longleftrightarrow \begin{bmatrix} 18 \\ 39 \\ 32 \end{bmatrix} = \begin{bmatrix} 3 & 0 \\ 4 & 3 \\ 2 & 4 \end{bmatrix} \begin{bmatrix} 6 \\ 5 \end{bmatrix}.$$

(13) (a) We calculate $L(\mathbf{v}_0)(x) = (3x + 4)0 + \int_0^1 12t \, dt = 6$. So in terms of the basis vectors, $L(\mathbf{v}_0) = 6\mathbf{v}_0$. Likewise, $L(\mathbf{v}_1)(x) = (3x + 4)1 + \int_0^1 12t^2 \, dt = 3x + 8$. So in terms of the basis vectors, $L(\mathbf{v}_1) = 8\mathbf{v}_0 + 3\mathbf{v}_1$. Finally, $L(\mathbf{v}_2)(x) = (3x + 4)(2x) + \int_0^1 12t^3 \, dt = 6x^2 + 8x + 3$. So in terms of the basis vectors, $L(\mathbf{v}_2) = 3\mathbf{v}_0 + 8\mathbf{v}_1 + 6\mathbf{v}_2$.

(b) From (a) the matrix for L is $A = \begin{bmatrix} 6 & 8 & 3 \\ 0 & 3 & 8 \\ 0 & 0 & 6 \end{bmatrix}$.

(c) When $f(x) = 6x^2 + 5x + 4$, then

$$Lf(x) = (3x + 4)(12x + 5) + \int_0^1 12t(6t^2 + 5t + 4) \, dt = 36x^2 + 63x + 82.$$

Thus $f = 4\mathbf{v}_0 + 5\mathbf{v}_1 + 6\mathbf{v}_2$ and $L(f) = 82\mathbf{v}_0 + 63\mathbf{v}_1 + 36\mathbf{v}_2$. Check by matrix multiplication: $\begin{bmatrix} 6 & 8 & 3 \\ 0 & 3 & 8 \\ 0 & 0 & 6 \end{bmatrix} \begin{bmatrix} 4 \\ 5 \\ 6 \end{bmatrix} = \begin{bmatrix} 82 \\ 63 \\ 36 \end{bmatrix}$.

(14) (a) The figure is a pentagon with vertices at $(0, 0)$, $(0, 1)$, $(1, 2)$, $(2, 1)$, $(1, 0)$.

(b) i. Scale by factor of 2 in horizontal direction and by factor of $1/2$ in vertical direction.

ii. Counterclockwise rotation by $\pi/3$ radians.

iii. Shear to left by -3.

iv. Translation to left by 2 units and up by 3 units.

(15)

(a) $R = \begin{bmatrix} -1/2 & -\sqrt{3}/2 & 0 \\ \sqrt{3}/2 & -1/2 & 0 \\ 0 & 0 & 1 \end{bmatrix}$

(c) $T = \begin{bmatrix} 1 & 0 & -3 \\ 0 & 1 & 5 \\ 0 & 0 & 1 \end{bmatrix}$

(b) $C = \begin{bmatrix} 1/3 & 0 & 0 \\ 0 & 1/3 & 0 \\ 0 & 0 & 1 \end{bmatrix}$

(d) $L = TC = \begin{bmatrix} 1/3 & 0 & -3 \\ 0 & 1/3 & 5 \\ 0 & 0 & 1 \end{bmatrix}$

(16) (a) Calculation.

(b) $\mathbf{x} = c_1\mathbf{u}_1 + c_2\mathbf{u}_2 + c_3\mathbf{u}_3$ with $c_1 = \mathbf{u}_1^T\mathbf{x} = \frac{-2}{3\sqrt{2}}$, $c_2 = \mathbf{u}_2^T\mathbf{x} = \frac{5}{3}$, and $c_3 = \mathbf{u}_3^T\mathbf{x} = 0$.

(c) $\|\mathbf{x}\|^2 = |c_1|^2 + |c_2|^2 + |c_3|^2 = \frac{4}{18} + \frac{25}{9} + 0 = \frac{27}{9} = 3$. This agrees with the calculation using the components for the standard basis: $1^2 + 1^2 + 1^2 = 3$.

(17) The calculation is the same as for the vectors $\mathbf{u} = \begin{bmatrix} 1\ 2\ 3 \end{bmatrix}^T$ and $\mathbf{v} = \begin{bmatrix} 4\ 5\ 0 \end{bmatrix}^T$ relative to the standard inner product on $\mathbb{R}^3$. Thus $\langle\mathbf{u},\mathbf{v}\rangle = 1\cdot 4 + 2\cdot 5 + 3\cdot 0 = 14$, $\|\mathbf{u}\| = \sqrt{1+4+9} = \sqrt{14}$, and $\|\mathbf{v}\| = \sqrt{16+25+0} = \sqrt{41}$. Note that $14 < \sqrt{14}\sqrt{41}$, in agreement with the Cauchy-Schwarz inequality.

(18) (a) $\|\mathbf{u}\| = \sqrt{2^2+3^2+4^2} = \sqrt{29}$, $\|\mathbf{v}\| = \sqrt{5^2+1^2+1^2} = \sqrt{27}$, $\langle\mathbf{u},\mathbf{v}\rangle = \mathbf{u}^H\mathbf{v} = 2\cdot 5i + (3-4i)\cdot(1+i) = 10i + 3 + 4 + 3i - 4i = 7 + 9i$.

(b) $\|\mathbf{u}+i\mathbf{v}\|^2 = \|\mathbf{u}\|^2 + \langle\mathbf{u},i\mathbf{v}\rangle + \langle i\mathbf{v},\mathbf{u}\rangle + \|i\mathbf{v}\|^2 = 9 + i(1+2i) - i(1-2i) + 16 = 21$. Hence $\|\mathbf{u}+i\mathbf{v}\| = \sqrt{21}$.

(19) (a) Since x is real, $U^H U = \frac{1}{25}\begin{bmatrix} x & -3i \\ -3i & x \end{bmatrix}\begin{bmatrix} x & 3i \\ 3i & x \end{bmatrix} = \frac{1}{25}\begin{bmatrix} x^2+9 & 0 \\ 0 & x^2+9 \end{bmatrix}$.

(b) U is unitary when $U^H U = I$. Hence we need $x^2 + 9 = 25$, so $x = \pm 4$.

(20) This is a generalization of the previous exercise.

(a) $U^H U = \begin{bmatrix} \bar{z} & \bar{w} \\ -w & z \end{bmatrix}\begin{bmatrix} z & -\bar{w} \\ w & \bar{z} \end{bmatrix} = \begin{bmatrix} |z|^2+|w|^2 & 0 \\ 0 & |z|^2+|w|^2 \end{bmatrix} = (|z|^2+|w|^2)\begin{bmatrix} 1 & 0 \\ 0 & 1 \end{bmatrix}$.

Hence U is unitary exactly when $|z|^2 + |w|^2 = 1$.

(b) Since $W^H W = I$ and $\det W^H = \det W^T = \overline{\det W}$, we get the equation $|\det W|^2 = \det I = 1$. Now take a real number θ such that $\det W = e^{2i\theta}$. Then $\det\left(e^{-i\theta}W\right) = \det(e^{-i\theta}I)\det W = e^{-2i\theta}\det W = 1$ (remember that $\det(cI) = c^2$ when I is the 2×2 identity matrix).

(c) If W is a 2×2 unitary matrix, choose θ as in (b) and define $U = e^{-i\theta}W$. Then U has determinant 1. Let U have columns $\mathbf{u}_1$ and $\mathbf{u}_2$. These vectors have length 1 and $\langle\mathbf{u}_1,\mathbf{u}_2\rangle = 0$. Write $\mathbf{u}_1 = \begin{bmatrix} z \\ w \end{bmatrix}$. Then $|z|^2 + |w|^2 = 1$.

Since $\mathbf{v} = \begin{bmatrix} -\bar{w} \\ \bar{z} \end{bmatrix}$ is orthogonal to $\mathbf{u}_1$ and the subspace of all vectors in $\mathbb{C}^2$ that are orthogonal to $\mathbf{u}_1$ has dimension one, we know that $\mathbf{u}_2$ must be a scalar multiple of $\mathbf{v}$. Thus $U = \begin{bmatrix} z & -c\bar{w} \\ w & c\bar{z} \end{bmatrix}$ for some complex number c. But the condition $\det U = 1$ gives $1 = c(|z|^2 + |w|^2) = c$. Hence U is given by a matrix as in (a) and $W = e^{i\theta}\begin{bmatrix} z & -\bar{w} \\ w & \bar{z} \end{bmatrix}$ with $|z|^2 + |w|^2 = 1$.

Geometric meaning: Write $z = x_1 + ix_2$ and $w = x_3 + ix_4$ with $x_j \in \mathbb{R}$. Then $x_1^2 + x_2^2 + x_3^2 + x_4^2 = 1$. Thus the 2×2 unitary matrices of determinant 1 correspond to the points on the 3-dimensional unit sphere in $\mathbb{R}^4$, just as the 2×2 real orthogonal matrices of determinant 1 (rotation matrices) correspond to the points on the unit circle in $\mathbb{R}^2$.

(21) (a) $\langle f,g\rangle = \int_0^1 x^3\,dx = 1/4$.

(b) $\|f\|^2 = \int_0^1 x^4\,dx = 1/5$ and $\|g\|^2 = \int_0^1 x^2\,dx = 1/3$. Hence $\|f\| = 1/\sqrt{5}$

and $\|g\| = 1/\sqrt{3}$.

(c) $\langle f, g \rangle = 1/4 = 1/\sqrt{16} < 1\sqrt{15} = \|f\|\,\|g\|$.

(d) $p(x) = \dfrac{\langle f, g \rangle}{\langle g, g \rangle} g(x) = (3/4)x$.

(e) $\langle f - p, g \rangle = \int_0^1 (x^2 - (3/4)x)x\, dx = \int_0^1 x^3\, dx - 3/4 \int_0^1 x^2\, dx = 1/4 - 1/4 = 0$.

(22) (a) $\langle u_1, u_1 \rangle = \int_0^1 1\, dx = 1$, $\langle u_1, u_2 \rangle = \sqrt{3} \int_0^1 (2x - 1)\, dx = 0$, and $\langle u_2, u_2 \rangle = 3 \int_0^1 (2x - 1)^2\, dx = 1$.

(b) $p(x) = c_1 u_1(x) + c_2 u_2(x)$ with coefficients $c_1 = \langle f, u_1 \rangle = \int_0^1 x^2\, dx = 1/3$ and $c_2 = \langle f, u_2 \rangle = \sqrt{3} \int_0^1 x^2(2x - 1)\, dx = \sqrt{3}/6$. Thus $p(x) = (1/3)u_1(x) + (\sqrt{3}/6)u_2(x) = x - 1/6$.

(c) From (b) we have $\|p\|^2 = c_1^2 + c_2^2 = (1/3)^2 + (\sqrt{3}/6)^2 = 7/36$.

(23) (a) $\langle f, g \rangle = \langle 2\varphi_{-1} + 3\varphi_1, 4\varphi_{-1} + (5+6\mathrm{i})\varphi_1 + 7\varphi_4 \rangle = 2 \cdot 4 + 3 \cdot (5 + 6\mathrm{i}) = 23 + 18\mathrm{i}$.
$\|g\|^2 = 4^2 + |5 + 6\mathrm{i}|^2 + 7^2 = 16 + 25 + 36 + 49 = 126$.

(b) $h(x) = 4\varphi_{-1}(x) + 0\varphi_0(x) + (5 + 6\mathrm{i})\varphi_1(x)$ and $\langle \varphi_4, h \rangle = 0$.

(24) (a) $c_0 = \frac{1}{2\pi} \int_0^{2\pi} x\, dx = \frac{1}{4\pi} x^2 \big|_0^{2\pi} = \pi$. For $k \neq 0$, integration by parts gives
$c_k = \frac{1}{2\pi} \int_0^{2\pi} x e^{-\mathrm{i}kx}\, dx = \frac{-1}{2k\pi\mathrm{i}} x e^{-\mathrm{i}kx} \big|_0^{2\pi} + \frac{-1}{2k\pi\mathrm{i}} \int_0^{2\pi} e^{-\mathrm{i}kx}\, dx = \frac{-1}{k\mathrm{i}}$. Note that $c_{-k} = \bar{c}_k$ since $f(x)$ is real valued.

(b) $\|f\|^2 = \frac{1}{2\pi} \int_0^{2\pi} x^2\, dx = \frac{1}{6\pi} x^3 \big|_0^{2\pi} = \frac{4\pi^2}{3}$.

(c) By the Parseval formula and (a), (b):
$$\tfrac{4\pi^2}{3} = \|f\|^2 = \sum_{k \in \mathbb{Z}} |c_k|^2 = |c_0|^2 + 2\sum_{k=1}^{\infty} |c_k|^2 = \pi^2 + \sum_{k=1}^{\infty} \tfrac{2}{k^2}.$$
Hence $\sum_{k=1}^{\infty} \frac{1}{k^2} = \frac{\pi^2}{6}$.

B.2   Solutions to Exercises 2.8

(1) (a) The first formula follows from $e^{2jk\pi\mathrm{i}/N} = \cos(2jk\pi/N) + \mathrm{i}\sin(2jk\pi/N)$ (Euler's formula). The second and third formulas then follow by taking real and imaginary parts.

(b) The aliasing relation for $\mathbf{E}_{N-k}$ follows from $e^{2j(N-k)\pi\mathrm{i}/N} = e^{-2jk\pi\mathrm{i}/N}$. The other aliasing relations then follow from (a).

(c) If $N = 4$ then $\mathbf{C}_0 = \begin{bmatrix} 1 \\ 1 \\ 1 \\ 1 \end{bmatrix}$, $\mathbf{C}_1 = \begin{bmatrix} 1 \\ 0 \\ -1 \\ 0 \end{bmatrix}$, $\mathbf{C}_2 = \begin{bmatrix} 1 \\ -1 \\ 1 \\ -1 \end{bmatrix}$, $\mathbf{S}_1 = \begin{bmatrix} 0 \\ 1 \\ 0 \\ -1 \end{bmatrix}$.

Also $\mathbf{C}_3 = \mathbf{C}_1$, $\mathbf{S}_3 = -\mathbf{S}_1$, and $\mathbf{S}_0 = \mathbf{S}_2 = \mathbf{0}$.

(2) (a) For $j = 0, \ldots, N - 1$ the entry in row $j + 1$ of $\mathbf{y}$ is $f(j/N) = \sum_{-m \leq k \leq m} c_k \omega^{kj}$. By definition of $\mathbf{E}_k$ this is the same as the entry in row $j + 1$ of $\sum_{-m \leq k \leq m} c_k \mathbf{E}_k$.

(b) Note that $\mathbf{E}_{-k} = \mathbf{E}_{N-k}$. If $N > 2m$ and $1 \leq k \leq m$ then $k < N - k \leq N - 1$. Hence the vectors $\mathbf{E}_k$ for $-m \leq k \leq m$ are all distinct and comprise an orthogonal set by Proposition 2.1. Now use the formula for $\mathbf{y}$ from (a) to evaluate the inner product $\langle \mathbf{E}_k, \mathbf{y} \rangle = N c_k$.

(c) Take $f(t) = f_m(t) - f_{m-N}(t)$, where $f_k(t) = e^{2\pi ikt}$. Then $f(t)$ is a nonzero function of t since $m \neq m - N$, and the functions $f_k(t)$ are mutually orthogonal for different values of k by Example 2.1. Also $-m \leq m - N \leq 0$ since $N \leq 2m$, so $f(t)$ is a trigonometric polynomial with frequencies at most m. But when we sample $f(t)$ at rate N we get $\mathbf{E}_m - \mathbf{E}_{m-N} = 0$.

(3) (a) Direct calculation gives $T = 4 \begin{bmatrix} \mathbf{e}_1 & \mathbf{e}_4 & \mathbf{e}_3 & \mathbf{e}_2 \end{bmatrix}$ and $T^2 = 16I_4$.

   (b) Let $\omega = e^{2\pi i/N}$ and index the rows and columns of T by $j, k = 0, \ldots, N-1$. Then $T_{jk} = \sum_{\ell=0}^{N-1} \omega^{-j\ell}\omega^{-\ell k} = \sum_{\ell=0}^{N-1} u^\ell$, where $u = \omega^{-(j+k)}$. Now $u = 1$ if $j + k = 0$ or $j + k = N$. Otherwise, $u \neq 1$ but $u^N = 1$. So by (2.5) we have $T_{jk} = N$ if $j = k = 0$ or if $j + k = N$; otherwise $T_{jk} = 0$. Hence $(1/N)T = \begin{bmatrix} \mathbf{e}_0 & \mathbf{e}_{N-1} & \cdots & \mathbf{e}_2 & \mathbf{e}_1 \end{bmatrix}$.

   (c) By definition of the DFT we have $\mathbf{g}[\ell] = \sum_{j=0}^{N-1} \omega^{-j\ell}\mathbf{y}[j]$. Hence

$$\widehat{\mathbf{g}}[k] = \sum_{\ell=0}^{N-1} \sum_{j=0}^{N-1} \omega^{-\ell k}\omega^{-j\ell}\mathbf{y}[j] = \sum_{j=0}^{N-1} \left\{ \sum_{\ell=0}^{N-1} \omega^{-\ell(j+k)} \right\}\mathbf{y}[j].$$

By equation (2.5) the sum over ℓ of $\omega^{-\ell(j+k)}$ equals N if $j + k = mN$ for some integer m, and otherwise equals 0. Hence the sum over j only has one term, and we get $\widehat{\mathbf{g}}[k] = N\mathbf{y}[-k + mN] = N\mathbf{y}[-k]$, since $\mathbf{y}$ is periodic with period N.

   (d) If $(1/\sqrt{N})F_N\mathbf{v} = \lambda\mathbf{v}$ for some vector $\mathbf{v} \neq \mathbf{0}$, then $\mathbf{v} = \lambda^4\mathbf{v}$, since $F_N^4 = N^2 I_N$. Hence $\lambda^4 = 1$, so that λ must be ± 1 or $\pm i$.

(4) (a) We can write $C = -2I + S + S^{N-1}$ in terms of the N-periodic shift operator S. Hence $\mathbf{f}[0] = -2$, $\mathbf{f}[1] = 1$, $\mathbf{f}[j] = 0$ for $j = 2, \ldots, N-2$, and $\mathbf{f}[N-1] = 1$.

   (b) Let $\omega = e^{2\pi i/N}$. Then $\widehat{\mathbf{f}}[k] = \sum_{j=0}^{N-1} \omega^{-kj}\mathbf{f}[j] = -2 + \omega^{-k} + \omega^k$. Theorem 2.29 gives $\widehat{(\mathbf{f} \star \mathbf{y})}[k] = \widehat{\mathbf{f}}[k]\widehat{\mathbf{y}}[k] = (\omega^{-k} - 2 + \omega^k)\widehat{\mathbf{y}}[k]$.

   (c) If $N = 4$, then $T = -2I + S + S^3$. So $C = \begin{bmatrix} -2 & 1 & 0 & 1 \\ 1 & -2 & 1 & 0 \\ 0 & 1 & -2 & 1 \\ 1 & 0 & 1 & -2 \end{bmatrix}$ is the circulant

matrix for T. The polynomial $p(z) = -2 + z + z^3$ has the values $p(1) = 0$, $p(-i) = -2$, $p(-1) = -4$, and $p(i) = -2$. These are the eigenvalues of C and also the values of $\widehat{\mathbf{f}}[k]$ at $k = 0, 1, 2, 3$.

   (d) $\mathbf{f} \star \mathbf{y} \longleftrightarrow C \begin{bmatrix} 2 \\ 3 \\ 1 \\ 5 \end{bmatrix} = \begin{bmatrix} 4 \\ -3 \\ 6 \\ -7 \end{bmatrix}$ and $\widehat{\mathbf{y}} \longleftrightarrow F_4 \begin{bmatrix} 2 \\ 3 \\ 1 \\ 5 \end{bmatrix} = \begin{bmatrix} 11 \\ 1 + 2i \\ -5 \\ 1 - 2i \end{bmatrix}$, so we have

$$\widehat{\mathbf{f} \star \mathbf{y}} \longleftrightarrow F_4 \begin{bmatrix} 4 \\ -3 \\ 6 \\ -7 \end{bmatrix} = \begin{bmatrix} 0 \\ -2 - 4i \\ 20 \\ -2 + 4i \end{bmatrix} = \begin{bmatrix} 0 & 0 & 0 & 0 \\ 0 & -2 & 0 & 0 \\ 0 & 0 & -4 & 0 \\ 0 & 0 & 0 & -2 \end{bmatrix}\begin{bmatrix} 11 \\ 1 + 2i \\ -5 \\ 1 - 2i \end{bmatrix}.$$ The en-

tries on the diagonal are the eigenvalues of C, and the last equality follows from Theorem 2.4.

(5) (a) The first column of C shows that $C = 4I + 7S + 5S^2$. Hence $C = \begin{bmatrix} 4 & 5 & 7 \\ 7 & 4 & 5 \\ 5 & 7 & 4 \end{bmatrix}$.

(b) The shift operator acts on periodic functions by $S\mathbf{y}[j] = \mathbf{y}[j-1]$. Hence $T\mathbf{y}[j] = 4\mathbf{y}[j] + 7\mathbf{y}[j-1] + 5\mathbf{y}[j-2]$. Using the periodicity of $\mathbf{y}[j]$, we get

$$T\mathbf{y}[0] = 4\mathbf{y}[0] + 7\mathbf{y}[-1] + 5\mathbf{y}[-2] = 4\mathbf{y}[0] + 5\mathbf{y}[1] + 7\mathbf{y}[2],$$
$$T\mathbf{y}[1] = 4\mathbf{y}[1] + 7\mathbf{y}[0] + 5\mathbf{y}[-1] = 7\mathbf{y}[0] + 4\mathbf{y}[1] + 5\mathbf{y}[2],$$
$$T\mathbf{y}[2] = 4\mathbf{y}[2] + 7\mathbf{y}[1] + 5\mathbf{y}[0] = 5\mathbf{y}[0] + 7\mathbf{y}[1] + 4\mathbf{y}[2].$$

Notice that the second coefficient array gives the entries in the matrix C.

(c) Let $p(z) = 4 + 7z + 5z^2$. The eigenvalues of C are obtained by evaluating $p(z)$ at the points $\omega^0 = 1$, ω^{-1} and ω^{-2} on the unit circle:

$$\lambda_0 = p(1) = 16, \quad \lambda_1 = p(\omega^{-1}) = 4 + 7\omega^{-1} + 5\omega^{-2},$$
$$\lambda_2 = p(\omega^{-2}) = 4 + 7\omega^{-2} + 5\omega^{-4} = 4 + 5\omega^{-2} + 7\omega^{-1}.$$

Note: Since $\omega^{-1} = (-1 - i\sqrt{3})/2$ and $\omega^{-2} = (-1 + i\sqrt{3})/2$, the eigenvalues are $\lambda_1 = -2 - \sqrt{3}i$ and $\lambda_2 = -2 + \sqrt{3}i$. As a check, the sum of the eigenvalues is $16 - 2 - 2 = 12$, which we know must always equal the trace of C (the sum of the diagonal elements).

(6) (a) The circular convolution formula when $N = 3$ is

$$\mathbf{f} \star \mathbf{g}[k] = \mathbf{f}[0]\mathbf{g}[k] + \mathbf{f}[1]\mathbf{g}[k-1] + \mathbf{f}[2]\mathbf{g}[k-2] = 4\mathbf{g}[k] + 5\mathbf{g}[k-1] + 6\mathbf{g}[k-2].$$

Using the periodicity of $\mathbf{g}$ this gives

$$\mathbf{f} \star \mathbf{g}[0] = 4\mathbf{g}[0] + 5\mathbf{g}[-1] + 6\mathbf{g}[-2] = 4\mathbf{g}[0] + 6\mathbf{g}[1] + 5\mathbf{g}[2],$$
$$\mathbf{f} \star \mathbf{g}[1] = 4\mathbf{g}[1] + 5\mathbf{g}[0] + 6\mathbf{g}[-1] = 5\mathbf{g}[0] + 4\mathbf{g}[1] + 6\mathbf{g}[2],$$
$$\mathbf{f} \star \mathbf{g}[2] = 4\mathbf{g}[2] + 5\mathbf{g}[1] + 6\mathbf{g}[0] = 6\mathbf{g}[0] + 5\mathbf{g}[1] + 4\mathbf{g}[2].$$

(b) The circulant matrix is the coefficient array from (a): $C = \begin{bmatrix} 4 & 6 & 5 \\ 5 & 4 & 6 \\ 6 & 5 & 4 \end{bmatrix}$.

(7) (a) It suffices to check that C^{H} commutes with the shift matrix S. By assumption $SC = CS$. Now $S^{\mathrm{H}} = S^{-1}$ since S is a unitary matrix. Also $S^{-1}CS = S^{-1}SC = C$, so multiplying on the right by S^{-1} we see that $S^{-1}C = CS^{-1}$. Putting these two facts together and remembering that taking Hermitian transpose reverses the order of the matrix product, we obtain $SC^{\mathrm{H}} = (CS^{\mathrm{H}})^{\mathrm{H}} = (CS^{-1})^{\mathrm{H}} = (S^{-1})^{\mathrm{H}}C^{\mathrm{H}} = SC^{\mathrm{H}}$.

(b) The information about the first column of C together with Theorem 2.2 imply that $C = c_0 I + c_1 S + \cdots + c_{N-1} S^{N-1}$. Since $S^{\mathrm{H}} = S^{-1} = S^{N-1}$, we get $C^{\mathrm{H}} = \overline{c_0} I + \overline{c_1} S^{N-1} + \cdots + \overline{c_{N-2}} S^2 + \overline{c_{N-1}} S$. By (a) we know that C^{H} is a circulant matrix, and hence by Theorem 2.2 its first column is $\begin{bmatrix} \overline{c_0} & \overline{c_{N-1}} & \cdots & \overline{c_2} & \overline{c_1} \end{bmatrix}^{\mathrm{T}}$.

(8) (a) The product of two circulant matrices is a circulant matrix, since it commutes with the shift operator. The first column of $C_{\mathbf{f}}C_{\mathbf{g}}$ is obtained by multiplying the first column of $C_{\mathbf{g}}$ on the left by $C_{\mathbf{f}}$. Now use equation (2.26) with $T = C_{\mathbf{f}}$.

(b) This follows from (a) by the associativity of matrix multiplication. Alternate proof of associativity by an index shift (modulo N) and interchange of order of summation:

$$((\mathbf{f} \star \mathbf{g}) \star \mathbf{h})[m] = \sum_{k=0}^{N-1} \left(\sum_{j=0}^{N-1} \mathbf{f}[j]\,\mathbf{g}[k-j] \right) \mathbf{h}[m-k]$$

$$= \sum_{j=0}^{N-1} \mathbf{f}[j] \left(\sum_{p=0}^{N-1} \mathbf{g}[p]\,\mathbf{h}[m-j-p] \right) = (\mathbf{f} \star (\mathbf{g} \star \mathbf{h}))[m] \,.$$

(c) This follows from (a) since circulant matrices of the same size all commute because they are polynomials in S. An alternate proof of commutativity of circular convolution uses index shifting (all indices read modulo N):

$$(\mathbf{f} \star \mathbf{g})[k] = \sum_{j=0}^{N-1} \mathbf{f}[j]\,\mathbf{g}[k-j] = \sum_{p=0}^{N-1} \mathbf{f}[k-p]\,\mathbf{g}[p] = (\mathbf{g} \star \mathbf{f})[k] \,.$$

B.3   Solutions to Exercises 3.8

(1) (a) Using $\mathbf{T}_a^{(3)}$ we get $\mathbf{s}_2 = \begin{bmatrix} 2 \\ 5 \\ 8 \\ 11 \end{bmatrix}$ and $\mathbf{d}_2 = \begin{bmatrix} 0 \\ -1 \\ 0 \\ 1 \end{bmatrix}$. Then using $\mathbf{T}_a^{(2)}$ we

get $\mathbf{s}_1 = \begin{bmatrix} 7/2 \\ 19/2 \end{bmatrix}$ and $\mathbf{d}_2 = \begin{bmatrix} -3/2 \\ -3/2 \end{bmatrix}$. Finally using using $\mathbf{T}_a^{(1)}$ we get $\mathbf{s}_0 = \begin{bmatrix} 13/2 \end{bmatrix}$ and $\mathbf{d}_0 = \begin{bmatrix} -3 \end{bmatrix}$.

(b) By direct calculation $\mathbf{W}_a^{(3)}\mathbf{x} = \begin{bmatrix} 13/2 & -3 & -3/2 & -3/2 & 0 & -1 & 0 & 1 \end{bmatrix}^{\mathrm{T}}$, which agrees with (a).

(c) $\mathbf{x} = \begin{bmatrix} 2 \\ 2 \\ 4 \\ 6 \\ 8 \\ 8 \\ 12 \\ 10 \end{bmatrix} = \dfrac{13}{2} \begin{bmatrix} 1 \\ 1 \\ 1 \\ 1 \\ 1 \\ 1 \\ 1 \\ 1 \end{bmatrix} - 3 \begin{bmatrix} 1 \\ 1 \\ 1 \\ 1 \\ -1 \\ -1 \\ -1 \\ -1 \end{bmatrix} - \dfrac{3}{2} \begin{bmatrix} 1 \\ 1 \\ -1 \\ -1 \\ 0 \\ 0 \\ 0 \\ 0 \end{bmatrix} - \dfrac{3}{2} \begin{bmatrix} 0 \\ 0 \\ 0 \\ 0 \\ 1 \\ 1 \\ -1 \\ -1 \end{bmatrix} - \begin{bmatrix} 0 \\ 0 \\ 1 \\ -1 \\ 0 \\ 0 \\ 0 \\ 0 \end{bmatrix} + \begin{bmatrix} 0 \\ 0 \\ 0 \\ 0 \\ 0 \\ 0 \\ 1 \\ -1 \end{bmatrix}.$

(d) $\widetilde{\mathbf{x}} = \begin{bmatrix} 2 & 2 & 5 & 5 & 8 & 8 & 11 & 11 \end{bmatrix}^{\mathrm{T}}$ (the sum of the first four vectors in (c)).

$\mathbf{x} - \widetilde{\mathbf{x}} = \begin{bmatrix} 0 & 0 & -1 & 1 & 0 & 0 & 1 & -1 \end{bmatrix}^{\mathrm{T}}$ (the sum of the last two vectors in (c)).

(e) Relative compression error $\|\mathbf{x} - \widetilde{\mathbf{x}}\|^2 / \|\mathbf{x}\|^2 = 4/432$ (less than 1%).

(2) (a) $P = \begin{bmatrix} 1 & 0 & 0 & 0 & 0 & 0 \\ 0 & 0 & 1 & 0 & 0 & 0 \\ 0 & 0 & 0 & 0 & 1 & 0 \\ 0 & 1 & 0 & 0 & 0 & 0 \\ 0 & 0 & 0 & 1 & 0 & 0 \\ 0 & 0 & 0 & 0 & 0 & 1 \end{bmatrix}$ and $\mathbf{T_a} = \frac{1}{2} \begin{bmatrix} 1 & 1 & 0 & 0 & 0 & 0 \\ 0 & 0 & 1 & 1 & 0 & 0 \\ 0 & 0 & 0 & 0 & 1 & 1 \\ 1 & -1 & 0 & 0 & 0 & 0 \\ 0 & 0 & 1 & -1 & 0 & 0 \\ 0 & 0 & 0 & 0 & 1 & -1 \end{bmatrix}.$

(b) $\mathbf{T_s} = 2(\mathbf{T_a})^{\mathrm{T}} = \begin{bmatrix} 1 & 0 & 0 & 1 & 0 & 0 \\ 1 & 0 & 0 & -1 & 0 & 0 \\ 0 & 1 & 0 & 0 & 1 & 0 \\ 0 & 1 & 0 & 0 & -1 & 0 \\ 0 & 0 & 1 & 0 & 0 & 1 \\ 0 & 0 & 1 & 0 & 0 & -1 \end{bmatrix}.$

(c) $\mathbf{s} = \begin{bmatrix} 1.5 & 2.5 & 3 \end{bmatrix}^{\mathrm{T}}$ and $\mathbf{d} = \begin{bmatrix} -0.5 & -0.5 & 0 \end{bmatrix}^{\mathrm{T}}$. Hence

$$\mathbf{x} = \begin{bmatrix} 1 \\ 2 \\ 2 \\ 3 \\ 3 \\ 3 \end{bmatrix} = 1.5 \begin{bmatrix} 1 \\ 1 \\ 0 \\ 0 \\ 0 \\ 0 \end{bmatrix} + 2.5 \begin{bmatrix} 0 \\ 0 \\ 1 \\ 1 \\ 0 \\ 0 \end{bmatrix} + 3.0 \begin{bmatrix} 0 \\ 0 \\ 0 \\ 0 \\ 1 \\ 1 \end{bmatrix} - 0.5 \begin{bmatrix} 1 \\ -1 \\ 0 \\ 0 \\ 0 \\ 0 \end{bmatrix} - 0.5 \begin{bmatrix} 0 \\ 0 \\ 1 \\ -1 \\ 0 \\ 0 \end{bmatrix}.$$

(d) $\tilde{\mathbf{x}} = \begin{bmatrix} 1.5 & 1.5 & 2.5 & 2.5 & 3.0 & 3.0 \end{bmatrix}^{\mathrm{T}}$ and $\mathbf{x} - \tilde{\mathbf{x}} = \begin{bmatrix} -0.5 & 0.5 & -0.5 & -0.5 & 0 & 0 \end{bmatrix}^{\mathrm{T}}$.

(e) $\|\mathbf{x} - \tilde{\mathbf{x}}\|^2 = 4 \cdot (0.5)^2 = 1$, $\|\mathbf{x}\|^2 = 36$, relative compression error $= 1/36$.

(3) (a) $\mathbf{W}_a^{(2)} = \begin{bmatrix} 1/4 & 1/4 & 1/4 & 1/4 \\ 1/4 & 1/4 & -1/4 & -1/4 \\ 1/2 & -1/2 & 0 & 0 \\ 0 & 0 & 1/2 & -1/2 \end{bmatrix}.$

(b) $\mathbf{s}_0 = (a + b + c + d)/4$, $\mathbf{d}_0 = (a + b - c - d)/4$, $\mathbf{d}_1 = \begin{bmatrix} (a - b)/2 \\ (c - d)/2 \end{bmatrix}.$

(4) Follow the hints.

(5) (a) $\mathbf{d}[0] = 7 - 4 - 2 \cdot 0 = 3$, $\mathbf{d}[1] = 3 - 0 - 2 \cdot 4 = -5$,
$\mathbf{s}[0] = 4 + 3 - 15 = -8$, $\mathbf{s}[1] = 0 - 5 + 3 \cdot 3 = 4$.

(b) $P = \begin{bmatrix} I & 0 \\ -I - 2S^{-1} & I \end{bmatrix} = \begin{bmatrix} 1 & 0 & 0 & 0 \\ 0 & 1 & 0 & 0 \\ -1 & -2 & 1 & 0 \\ -2 & -1 & 0 & 1 \end{bmatrix}.$

(c) $U = \begin{bmatrix} I & I + 3S \\ 0 & I \end{bmatrix} = \begin{bmatrix} 1 & 0 & 1 & 3 \\ 0 & 1 & 3 & 1 \\ 0 & 0 & 1 & 0 \\ 0 & 0 & 0 & 1 \end{bmatrix}.$

(6) (a) For the Daub4 transform, $a + c = b + d = (1 + \sqrt{3}) + (3 - \sqrt{3}) = 4$ and
$a + b + c + d = (1 + \sqrt{3}) + (3 + \sqrt{3}) + (3 - \sqrt{3}) + (1 - \sqrt{3}) = 8$.

(b) $\mathbf{T_a x} = \dfrac{1}{\sqrt{32}} \begin{bmatrix} a & b & c & d \\ c & d & a & b \\ -b & a & -d & c \\ -d & c & -b & a \end{bmatrix} \begin{bmatrix} 1 \\ 1 \\ 1 \\ 1 \end{bmatrix} = \dfrac{1}{\sqrt{32}} \begin{bmatrix} a+b+c+d \\ a+b+c+d \\ -b+a-d+c \\ -d+c-b+a \end{bmatrix} = \begin{bmatrix} \sqrt{2} \\ \sqrt{2} \\ 0 \\ 0 \end{bmatrix}.$

(c) The trend and detail vectors are $\mathbf{s} = \begin{bmatrix} \sqrt{2} \\ \sqrt{2} \end{bmatrix}$ and $\mathbf{d} = 0$. The squares of the norms are $\|\mathbf{x}\|^2 = 4 = \|\mathbf{s}\|^2$ and $\|\mathbf{d}\|^2 = 0$.

(7) (a) $\mathbf{s}^{(1)} = \mathbf{x}_{\text{even}} - \frac{1}{64} \left(3S^2 - 19S - 19I + 3S^{-1}\right) \mathbf{d}^{(1)}$.

   (b) $U = \begin{bmatrix} I & -\frac{1}{64}\left(3S^2 - 19S - 19I + 3S^{-1}\right) \\ 0 & I \end{bmatrix}$.

   (c) $DUP = \frac{\sqrt{2}}{128} \begin{bmatrix} A & B \\ -(32S^{-1}+32I) & 64I \end{bmatrix}$, with $A = 3S^{-2} - 16S^{-1} + 90I - 16S + 3S^2$ and $B = -6S^{-1} + 38I + 38S - 6S^2$.

(8) (a) $\mathbf{s}^{(2)} = \mathbf{s}^{(1)} + \frac{1}{36}\left(3S + 16I - 3S^{-1}\right) \mathbf{d}^{(1)}$.

   (b) $U_2 = \begin{bmatrix} I & \frac{1}{36}\left(3S + 16I - 3S^{-1}\right) \\ 0 & I \end{bmatrix}$.

   (c) $DU_2 P U_1 = \frac{\sqrt{2}}{64} \begin{bmatrix} (3S^{-2} - 7S^{-1} + 45I - 9S) & (-9S^{-1} + 45I - 7S + 3S^2) \\ -(8S^{-1} + 24I) & (24I + 8S) \end{bmatrix}$.

(9) To verify the first equation in (3.36), use Theorem 3.2 to write $\mathbf{u}_j = \mathbf{u}_0(S_N)^{-2j}$ and $\tilde{\mathbf{u}}_k = \mathbf{u}_0(S_N)^{-2j}$. Then calculate $\mathbf{u}_j \tilde{\mathbf{u}}_k = \mathbf{u}_0(S_N)^{2(k-j)}\tilde{\mathbf{u}}_0$. From (3.42) this scalar is $\delta[j-k]$. Use the same argument for the other equations in (3.36).

(10) (a) By (3.10) the detail is zero when $2\mathbf{x}[2n+1] - \mathbf{x}[2n] - \mathbf{x}[2n+2] = 0$ for $n = 0, \ldots, m-1$ and $\mathbf{x}[0] = \mathbf{x}[N]$ by periodicity. This gives the set of homogeneous linear equations

$$\begin{cases} -\mathbf{x}[0] + 2\mathbf{x}[1] - \mathbf{x}[2] & = 0 \\ -\mathbf{x}[2] + 2\mathbf{x}[3] - \mathbf{x}[4] & = 0 \\ \quad\vdots & \vdots\ \vdots \\ -\mathbf{x}[N-2] + 2\mathbf{x}[N-1] - \mathbf{x}[0] = 0 \end{cases}$$

   (b) $\mathbf{x}$ is in the trend space when the detail $\mathbf{d} = V\mathbf{x} = 0$.

(11) (a) $\mathbf{T_a} = \begin{bmatrix} 6 & 6 & -2 & 0 & 0 & -2 \\ 0 & -2 & 6 & 6 & -2 & 0 \\ 0 & 0 & -2 & 6 & 6 & -2 \\ -3 & 3 & -1 & 0 & 0 & 1 \\ 0 & 1 & -3 & 3 & -1 & 0 \\ -1 & 0 & 0 & 1 & -3 & 3 \end{bmatrix}$.   (b) $\mathbf{T_s} = \dfrac{1}{32} \begin{bmatrix} 3 & 0 & 1 & -6 & 0 & 2 \\ 3 & 1 & 0 & 6 & -2 & 0 \\ 1 & 3 & 0 & 2 & -6 & 0 \\ 0 & 3 & 1 & 0 & 6 & -2 \\ 0 & 1 & 3 & 0 & 2 & -6 \\ 1 & 0 & 3 & -2 & 0 & 6 \end{bmatrix}$.

   (c) The coefficients are the entries in $\mathbf{T_a} \begin{bmatrix} 0 \\ 4 \\ 6 \\ 6 \\ 4 \\ 0 \end{bmatrix} = \begin{bmatrix} 12 \\ 56 \\ 12 \\ 6 \\ 0 \\ -6 \end{bmatrix}$.

The projection of **x** onto the trend subspace is

$$\mathbf{x}_s = 12\widetilde{\mathbf{u}}_0 + 56S^2\widetilde{\mathbf{u}}_0 + 12S^4\widetilde{\mathbf{u}} = \frac{3}{8}\begin{bmatrix}3\\3\\1\\0\\0\\1\end{bmatrix} + \frac{14}{8}\begin{bmatrix}0\\1\\3\\3\\1\\0\end{bmatrix} + \frac{3}{8}\begin{bmatrix}1\\0\\0\\1\\3\\3\end{bmatrix} = \frac{1}{8}\begin{bmatrix}12\\23\\45\\45\\23\\3\end{bmatrix}.$$

The projection of **x** onto the detail subspace is

$$\mathbf{x}_d = 6\widetilde{\mathbf{v}}_0 - 6S^4\widetilde{\mathbf{v}}_0 = \frac{3}{16}\begin{bmatrix}-6\\6\\2\\0\\0\\-2\end{bmatrix} - \frac{3}{16}\begin{bmatrix}2\\0\\0\\-2\\-6\\6\end{bmatrix} = \frac{1}{8}\begin{bmatrix}-12\\9\\3\\3\\9\\-12\end{bmatrix} \quad (\mathbf{x} = \mathbf{x}_s + \mathbf{x}_d).$$

The relative compression error is $\|\mathbf{x}_d\|^2/\|\mathbf{x}\|^2 = 7\%$.

(12) (a) Use block multiplication and mutual commutativity of circulant matrices:

$$z^{-1}\begin{bmatrix}A & B\\C & D\end{bmatrix}\begin{bmatrix}D & -B\\-C & A\end{bmatrix} = z^{-1}\begin{bmatrix}AD - BC & 0\\0 & AD - BC\end{bmatrix} = \begin{bmatrix}I & 0\\0 & I\end{bmatrix}.$$

(b) The CDF(3,1) analysis matrix in block form is $\mathbf{T_a} = \begin{bmatrix}A & B\\C & D\end{bmatrix}$ split ,

where $A = (6I - 2S^{-1})/\sqrt{32}$, $B = (6I - 2S)/\sqrt{32}$, $C = -(3I + S^{-1})/\sqrt{32}$, and $D = (3I + S)/\sqrt{32}$. Then $AD - BC = I$ since we know that $\det(\mathbf{T_a}) = 1$. From (a) and $\boxed{\text{split}}^{-1} = \boxed{\text{merge}}$ we get

$$\mathbf{T_s} = \boxed{\text{merge}}\ \frac{1}{\sqrt{32}}\begin{bmatrix}(3I + S) & -(6I - 2S)\\(3I + S^{-1}) & (6I - 2S^{-1})\end{bmatrix}.$$

(13) (a) From the scaling and wavelet vectors construct $\mathbf{T_a} = \begin{bmatrix}a & b & c & d & 0 & 0\\0 & 0 & a & b & c & d\\c & d & 0 & 0 & a & b\\d & -c & b & -a & 0 & 0\\0 & 0 & d & -c & b & -a\\b & -a & 0 & 0 & d & -c\end{bmatrix}.$

The rows of $\mathbf{T_a}$ must have length one, so $a^2 + b^2 + c^2 + d^2 = 1$. The rows will be mutually orthogonal when $ac + bd = 0$.

(b) Yes. The orthogonality conditions involve shifts by two positions (with periodic wraparound), and the equations are the same as in (a).

(14) (a) $\mathbf{Y} = \mathbf{W_a}\mathbf{X}\mathbf{W_a^T} = \frac{1}{4}\begin{bmatrix}1 & 1\\1 & -1\end{bmatrix}\begin{bmatrix}2 & 4\\0 & 8\end{bmatrix}\begin{bmatrix}1 & 1\\1 & -1\end{bmatrix} = \frac{1}{2}\begin{bmatrix}7 & -5\\-1 & 3\end{bmatrix}.$

(b) $\mathbf{X} = \mathbf{W_s}\mathbf{Y}\mathbf{W_s^T} = \mathbf{X_{ss}} + \mathbf{X_{sd}} + \mathbf{X_{ds}} + \mathbf{X_{dd}}$, where $\mathbf{X_{ss}} = \frac{7}{2}\begin{bmatrix}1 & 1\\1 & 1\end{bmatrix}$,

$$\mathbf{X_{sd}} = -\frac{5}{2}\begin{bmatrix}1 & -1\\1 & -1\end{bmatrix},\ \mathbf{X_{ds}} = -\frac{1}{2}\begin{bmatrix}1 & 1\\-1 & -1\end{bmatrix},\ \mathbf{X_{dd}} = \frac{3}{2}\begin{bmatrix}1 & -1\\-1 & 1\end{bmatrix}.$$

(15) (a) $\frac{1}{4}\begin{bmatrix} 1 & 1 \\ 1 & -1 \end{bmatrix}\begin{bmatrix} 0 & 1 \\ 1 & 1 \end{bmatrix}\begin{bmatrix} 1 & 1 \\ 1 & -1 \end{bmatrix} = \begin{bmatrix} 0.75 & -0.25 \\ -0.25 & -0.25 \end{bmatrix}.$

(b) $\mathbf{Y}^{(2)} = \begin{bmatrix} 0.75 & -0.25 & 0 & 0 \\ -0.25 & -0.25 & -1 & 0 \\ 0 & -1 & 0 & 0 \\ 0 & 0 & 0 & -1 \end{bmatrix}.$

(c) Set the three entries -0.25 to zero to get the compressed transform

$\mathbf{Y}_c^{(2)} = \begin{bmatrix} 0.75 & 0 & 0 & 0 \\ 0 & 0 & -1 & 0 \\ 0 & -1 & 0 & 0 \\ 0 & 0 & 0 & -1 \end{bmatrix}.$ The 2D inverse Haar transform of the upper

left-hand block in $\mathbf{Y}_c^{(2)}$ is $\begin{bmatrix} 0.75 & 0.75 \\ 0.75 & 0.75 \end{bmatrix}$. Insert this into $\mathbf{Y}_c^{(2)}$ and take the

2D inverse Haar transform: $\mathbf{X}_c = \begin{bmatrix} 0.75 & 0.75 & -0.25 & -0.25 \\ 0.75 & 0.75 & 1.75 & 1.75 \\ -0.25 & 1.75 & -0.25 & 1.75 \\ -0.25 & 1.75 & 1.75 & -0.25 \end{bmatrix}.$

(d) MSE $= 0.1636$,   PSNR $= 55.99$.

B.4   Solutions to Exercises 4.12

(1) (a) $\mathbf{x}_0 = 2\delta_0 + 4\delta_1$, $\mathbf{x}_1 = 3\delta_{-1} - 5\delta_0$, $\boxed{2\uparrow}\mathbf{x}_0 = 2\delta_0 + 4\delta_2$, $S\boxed{2\uparrow}\mathbf{x}_1 = 3\delta_{-1} - 5\delta_1$. Clearly $\mathbf{x} = \boxed{2\uparrow}\mathbf{x}_0 + S\boxed{2\uparrow}\mathbf{x}_1$.

(b) $\mathbf{X}(z) = 3z + 2 - 5z^{-1} + 4z^{-2}$, $\mathbf{X}_0(z) = 2 + 4z^{-1}$, $\mathbf{X}_1(z) = 3z - 5$, $\mathbf{Y}(z) = 7 + 6z^{-1}$. Since $\mathbf{X}(-z) = -3z + 2 + 5z^{-1} + 4z^{-2}$, we get $1/2\{\mathbf{X}(z) + \mathbf{X}(-z)\} = 1/2\{4 + 8z^{-2}\} = \mathbf{X}_0(z^2)$ and $z/2\{\mathbf{X}(z) - \mathbf{X}(-z)\} = z/2\{6z - 10z^{-2}\} = \mathbf{X}_1(z^2)$. Clearly $\mathbf{X}(z) = \mathbf{X}_0(z^2) + z^{-1}\mathbf{X}_1(z^2)$.

(c) The z-transform is $\mathbf{X}(z)\mathbf{Y}(z) = 21z + 32 - 23z^{-1} - 2z^{-2} + 24z^{-3}$.

(d) The DFT by the Fourier matrix is $\begin{bmatrix} 1 & 1 & 1 & 1 \\ 1 & -i & -1 & i \\ 1 & -1 & 1 & -1 \\ 1 & i & -1 & -i \end{bmatrix}\begin{bmatrix} 7 \\ 6 \\ 0 \\ 0 \end{bmatrix} = \begin{bmatrix} 13 \\ 7 - 6i \\ 1 \\ 7 + 6i \end{bmatrix}.$

(e) Let $\omega = e^{2\pi i/4} = i$. Then $\hat{\mathbf{y}}_{\mathrm{per},4}[k] = \mathbf{Y}(\omega^k)$ (sample the z-transform at the points ω^k on the unit circle). The numerical values in this example are $\hat{\mathbf{y}}_{\mathrm{per},4}[0] = \mathbf{Y}(1) = 13$, $\hat{\mathbf{y}}_{\mathrm{per},4}[1] = \mathbf{Y}(i) = 7 - 6i$, $\hat{\mathbf{y}}_{\mathrm{per},4}[2] = \mathbf{Y}(-1) = 1$, $\hat{\mathbf{y}}_{\mathrm{per},4}[3] = \mathbf{Y}(-i) = 7 + 6i$, which agree with (d).

(f) For any signal $\mathbf{y}$ of length at most 4, $\|\mathbf{y}\|^2 = \|\mathbf{y}_{\mathrm{per},4}\|^2 = (1/4)\|\hat{\mathbf{y}}_{\mathrm{per},4}\|^2$. In this example, $\|\mathbf{y}\|^2 = 7^2 + 6^2 = 85$ and $\|\hat{\mathbf{y}}_{\mathrm{per},4}\|^2 = 13^2 + |7 - 6i|^2 + 1^2 + |7 + 6i|^2 = 340$.

(2) (a) $\boxed{2\downarrow}\mathbf{x} = 2\delta_{-1} + 4\delta_0$ and $\boxed{2\uparrow}\boxed{2\downarrow}\mathbf{x} = 2\delta_{-2} + 4\delta_0$.

(b) $\boxed{2\downarrow}\,\mathbf{y} = \sum_{m\in\mathbb{Z}} \mathbf{y}[2m]\delta_m$. Since $\boxed{2\uparrow}\,\delta_m = \delta_{2m}$ and $\boxed{2\uparrow}$ is a linear transformation, $\boxed{2\uparrow}\,\boxed{2\downarrow}\,\mathbf{y} = \sum_{m\in\mathbb{Z}} \mathbf{y}[2m]\,\boxed{2\uparrow}\,\delta_m = \sum_{m\in\mathbb{Z}} \mathbf{y}[2m]\delta_{2m}$.

(c) By definition of convolution,

$$(\mathbf{h} \star \mathbf{u})[k] = \sum_{n\in\mathbb{Z}} \mathbf{h}[k-n]\mathbf{u}[n] = \sum_{n\in\mathbb{Z}} \mathbf{h}^{\mathrm{T}}[n-k]\mathbf{u}[n]$$
$$= \sum_{n\in\mathbb{Z}} S^k\mathbf{h}^{\mathrm{T}}[n]\mathbf{u}[n] = \langle S^k\mathbf{h}^{\mathrm{T}}, \mathbf{u}\rangle .$$

(d) By definition of convolution and inner product,

$$\langle \mathbf{h} \star \mathbf{v}, \mathbf{w}\rangle = \sum_{k\in\mathbb{Z}}(\mathbf{h} \star \mathbf{v})[k]\mathbf{w}[k] = \sum_{k\in\mathbb{Z}}\left\{\sum_{n\in\mathbb{Z}} \mathbf{h}[k-n]\mathbf{v}[n]\right\}\mathbf{w}[k]$$
$$= \sum_{n\in\mathbb{Z}}\left\{\sum_{k\in\mathbb{Z}} \mathbf{h}^{\mathrm{T}}[n-k]\mathbf{w}[n]\right\}\mathbf{v}[k] = \langle \mathbf{v}, \mathbf{h}^{\mathrm{T}} \star \mathbf{w}\rangle ,$$

where the order of summation was reversed on the last step. This is allowed, since the nonzero terms in the summation only involve a finite set of values of n and k.

(3) (a) In the lazy filter bank, the analysis filter $\mathbf{h}_0$ does nothing to the signal, so $\mathbf{h}_0 = \delta_0$. The analysis filter $\mathbf{h}_1$ shifts the signal backwards one time unit, so $\mathbf{h}_1 = \delta_{-1}$. The synthesis filter $\mathbf{g}_0$ likewise does nothing to the signal, so $\mathbf{g}_0 = \delta_0$. The synthesis filter $\mathbf{g}_1$ shifts the signal forwards one time unit, so $\mathbf{g}_1 = \delta_1$. The z-transforms of these filters are $H_0(z) = 1$, $H_1(z) = z$, $G_0(z) = 1$, and $G_1(z) = z^{-1}$. The analysis modulation matrix is

$$\mathbf{H}_m(z) = \begin{bmatrix} H_0(z) & H_0(-z) \\ H_1(z) & H_1(-z) \end{bmatrix} = \begin{bmatrix} 1 & 1 \\ z & -z \end{bmatrix} .$$

The synthesis modulation matrix is

$$\mathbf{G}_m(z) = \begin{bmatrix} G_0(z) & G_1(z) \\ G_0(-z) & G_1(-z) \end{bmatrix} = \begin{bmatrix} 1 & z^{-1} \\ 1 & -z^{-1} \end{bmatrix} .$$

It is evident that $\mathbf{G}_m(z)\mathbf{H}_m(z) = 2I$. From the flow chart for the lazy filter bank and the definition of the polyphase analysis matrix in Section 4.3, we see that $\mathbf{H}_p(z) = I$ (the 2×2 identity matrix), since the polyphase analysis matrix multiplies $[X_0(z), X_1(z)]^{\mathrm{T}}$ *after* downsampling. Likewise, $\mathbf{G}_p(z) = I$ since the polyphase synthesis matrix multiplies $[Y_0(z), Y_1(z)]^{\mathrm{T}}$ *before* upsampling.

(b) The lowpass condition $H_0(-1) = G_0(-1) = 0$ is not satisfied. The highpass condition $H_1(1) = G_1(1) = 0$ is also not satisfied.

(4) (a) The analysis modulation matrix is

$$\mathbf{H}_m(z) = \begin{bmatrix} H_0(z) & H_0(-z) \\ H_1(z) & H_1(-z) \end{bmatrix} = \begin{bmatrix} (1+z)(1+az) & (1-z)(1-az) \\ (1-z)(1+bz) & (1+z)(1-bz) \end{bmatrix} .$$

Its determinant is $-2z\left(z^2(2ab - a + b) + b - a - 2\right)$ (use a symbolic algebra program for the calculation). This is a nonzero monomial in z in two cases:

i. If $b = a + 2$ with $a \neq -1$, then the determinant is cz^3, where $c = -4(a+1)^2 \neq 0$.

ii. If $b = a/(1+2a)$ with $a \neq -1, -1/2$, then the determinant is cz, where in this case $c = 2(a+1)^2(1+2a)^{-1} \neq 0$.

(b) By Theorem 4.8 the synthesis filters have z-transforms

$$G_0(z) = -zH_1(-z) = -z(1+z)(1-bz) = -z + (b-1)z^2 + bz^3 \, ,$$

$$G_1(z) = z^{-1}H_0(-z) = z^{-1}(1-z)(1-az) = z^{-1} - (a+1) + az \, .$$

Hence the synthesis filters are expressed in terms of unit impulses as

$$\mathbf{g}_0 = -\delta_{-1} + (b-1)\delta_{-2} + b\delta_{-3} \, , \qquad \mathbf{g}_1 = \delta_1 - (a+1)\delta_0 + a\delta_{-1} \, .$$

(5) (a) The analysis modulation matrix $\mathbf{H}_m(z)$ is

$$\begin{bmatrix} H_0(z) & H_0(-z) \\ H_1(z) & H_1(-z) \end{bmatrix} = \begin{bmatrix} (1+z)^3 & (1-z)^3 \\ (1-z)(1+bz+cz^2) & (1+z)(1-bz+cz^2) \end{bmatrix} \, .$$

We calculate with symbolic algebra software that its determinant is $2zf(z)$, where $f(z) = (4-b) + 4(1-6b+c)z^2 + (4c-b)z^4$. Hence the determinant can only be a monomial when $f(z)$ is a monomial. This only occurs in the following cases:

i. $b = 4/23$ and $c = 1/23$. In this case we obtain $f(z) = 88/23$ and $H_1(z) = (1-z)(23 + 4z + z^2)/23$.

ii. $b = 4$, $c = 1$. Then $f(z) = -88z^2$ and $H_1(z) = (1-z)(1+4z+z^2)$.

iii. $b = 4$, $c = 23$. Then $f(z) = 88z^4$ and $H_1(z) = (1-z)(1+4z+23z^2)$.

(b) The highpass synthesis filter has z-transform $G_1(z) = z^{-1}H_0(-z) = z^{-1}(1-z)^3$. The lowpass synthesis filter has z-transform $G_0(z) = -zH_1(-z)$, which by part (a) is one of the following:

i. If $b = 4/23$ and $c = 1/23$, then $G_0(z) = -z(1+z)(23 - 4z + z^2)/23$.

ii. If $b = 4$, $c = 1$, then $G_0(z) = -z(1+z)(1 - 4z + z^2)$.

iii. If $b = 4$, $c = 23$, then $G_0(z) = -z(1+z)(1 - 4z + 23z^2)$.

(6) (a) Get $\mathbf{H}_m(z)$ from $\mathbf{H}_p(z^2)$ as

$$\begin{bmatrix} (1+z^2) & 2 \\ (1-3z^2) & 2 \end{bmatrix} \begin{bmatrix} 1 & 1 \\ z & -z \end{bmatrix} = \begin{bmatrix} (1+2z+z^2) & (1-2z+z^2) \\ (1+2z-3z^2) & (1-2z-3z^2) \end{bmatrix} \, .$$

From the first column of $\mathbf{H}_m(z)$ we get $H_0(z) = 1+2z+z^2 = (1+z)^2$ and $H_1(z) = 1+2z-3z^2$. Clearly $H_0(-1) = 0$ and $H_1(1) = 0$, so the lowpass and highpass conditions are satisfied by these filters.

(b) Perfect reconstruction with FIR filters requires $d(z) = \det \mathbf{H}_m(z)$ to be a nonzero monomial. Use the polyphase analysis matrix to calculate

$$d(z) = \det \mathbf{H}_p(z^2) \det \begin{bmatrix} 1 & 1 \\ z & -z \end{bmatrix} = (2 + 2z^2 - 2 + 6z^2)(-2z) = -16z^3 \, .$$

Now use Theorem 4.7 to get the z-transforms of the synthesis filters:

$$G_0(z) = \tfrac{2}{d(z)}H_1(-z) = \tfrac{1}{8}\left(-z^{-3} + 2z^{-2} + 3z^{-1}\right) \, ,$$

$$G_1(z) = -\tfrac{2}{d(z)}H_0(-z) = \tfrac{1}{8}\left(-z^{-3} + 2z^{-2} - z^{-1}\right) \, .$$

(7) (a) Get analysis modulation matrix from the polyphase analysis matrix by

$$\mathbf{H}_m(z) = \mathbf{H}_p(z^2)\begin{bmatrix} 1 & 1 \\ z & -z \end{bmatrix} = \begin{bmatrix} 1 & (1-z^2) \\ (1+z^2) & (2-z^4) \end{bmatrix}\begin{bmatrix} 1 & 1 \\ z & -z \end{bmatrix}$$

$$= \begin{bmatrix} (1+z-z^3) & (1-z+z^3) \\ (1+2z+z^2-z^5) & (1-2z+z^2+z^5) \end{bmatrix}.$$

From the entries in the first column of $\mathbf{H}_m(z)$ we obtain $H_0(z) = 1+z-z^3$ and $H_1(z) = 1+2z+z^2-z^5$. We have $H_0(-1) = 1$ and $H_1(1) = 3$, so the lowpass/highpass conditions are not satisfied by these filters.

(b) The PR condition is satisfied since $\det\mathbf{H}_m(z)$ is

$$\det\mathbf{H}_p(z^2)\,\det\begin{bmatrix} 1 & 1 \\ z & -z \end{bmatrix} = \left(2-z^4-(1-z^2)(1+z^2)\right)(-2z) = -2z\,.$$

(8) (a) Since k is odd, $H_1(-z) = -z^k G_0(z)$ and $G_1(-z) = -z^{-k}H_0(z)$. Hence the product of the modulation matrices is

$$\mathbf{H}_m(z)\mathbf{G}_m(z) = \begin{bmatrix} H_0(z) & H_0(-z) \\ z^k G_0(-z) & -z^k G_0(z) \end{bmatrix}\begin{bmatrix} G_0(z) & z^{-k}H_0(-z) \\ G_0(-z) & -z^{-k}H_0(z) \end{bmatrix}$$

$$= \begin{bmatrix} G_0(z)H_0(z)+G_0(-z)H_0(-z) & 0 \\ 0 & G_0(-z)H_0(-z)+G_0(z)H_0(z) \end{bmatrix}.$$

Thus the condition for PR is $G_0(z)H_0(z)+G_0(-z)H_0(-z) = 2$ (the half-band condition).

(b) In this case $G_0(z)H_0(z) = (3z+8+6z^{-1}-z^{-3})/8$ has only one term of even degree, which is the constant 1. Hence $H_0(z)G_0(z)+H_0(-z)G_0(-z) = 2$, as needed. The entries in the polyphase analysis matrix are determined by the relations

$$H_{00}(z^2)+zH_{01}(z^2) = H_0(z) = z+2+z^{-1}\,,$$
$$H_{10}(z^2)+zH_{11}(z^2) = H_1(z) = (3z^3-2z^2-z)/8\,.$$

Comparing even and odd terms on each side, we see that

$$H_{00}(z^2) = 2\,, \qquad zH_{01}(z^2) = z+z^{-1} = z(1+z^{-2})\,,$$
$$H_{10}(z^2) = -z^2/4\,, \qquad zH_{11}(z^2) = (3z^3-z)/8 = z(3z^2-1)/8\,.$$

Thus $H_{00}(z) = 2$, $H_{01}(z) = 1+z^{-1}$, $H_{10}(z) = -z/4$, $H_{11}(z) = (3z-1)/8$, and the polyphase analysis matrix is $\mathbf{H}_p(z) = \begin{bmatrix} 2 & (1+z^{-1}) \\ -z/4 & (3z-1)/8 \end{bmatrix}.$

(9) Follow the hints.

(10) (a) From the binomial expansion $G_0(z)$ involves z^k for $0 \le k \le p$. Hence $\mathbf{g}_0$ has length $p+1$.

(b) Since $B_n(y)$ is a polynomial of degree $n-1$ with constant term 1, it follows that $z^{-n}B_n\left(\frac{z+2+z^{-1}}{4}\right)$ involves z^k for $-2n+1 \le k \le -1$. Hence from the binomial expansion we see that $(1+z)^q z^{-n}B_n\left(\frac{z+2+z^{-1}}{4}\right)$ involves z^k for $-2n+1 \le k \le q-1$. Hence $\mathbf{h}_0$ has length $q-1+2n = 2q+p-1$, since $2n = p+q$.

(c) For $(p,q) = (2,2)$ the length of $\mathbf{g}_0$ is $3 = 2 + 1$ and the length of $\mathbf{h}_0$ is $5 = 2 \cdot 2 + 2 - 1$. For $(p,q) = (3,1)$ the length of $\mathbf{g}_0$ is $4 = 3 + 1$ and the length of $\mathbf{h}_0$ is $4 = 3 + 2 \cdot 1 - 1$. For $(p,q) = (6,2)$ the length of $\mathbf{g}_0$ is $7 = 6 + 1$ and the length of $\mathbf{h}_0$ is $9 = 6 + 2 \cdot 2 - 1$.

(11) (a) Since $B_4(y)$ has leading term $20y^3$ and roots r_1, r_2, r_3, it factors as

$$B_4(y) = 20(y - r_1)(y - r_2)(y - r_3) = 20p_1(y)p_2(y).$$

Thus $20p_1(0)p_2(0) = B_4(0) = 1$, so $B_4(y) = p_1(y)p_2(y)/(p_1(0)p_2(0))$.

(b) From (a) we have $G_0(z)H_0(z) = 2^{-7} z^{-4}(1 + z)^8 B_4\left(\frac{-z+2-z^{-1}}{4}\right)$. Now use equation (4.58).

(c) From the definition we have

$$H_0(z^{-1}) = \tfrac{\sqrt{2}}{16} z^2(1 + z^{-1})^4 p_2\left(\tfrac{-z+2-z^{-1}}{4}\right)/p_2(0) = H_0(z),$$

since $z^2(1 + z^{-1})^4 = z^{-2}(1 + z)^4$. Since $p_2(y)$ is a quadratic polynomial, the highest power of z that occurs in $H_0(z)$ is z^4. So by symmetry z^{-4} is the lowest term and $H_0(z)$ has length 9. Likewise,

$$G_0(z^{-1}) = \tfrac{\sqrt{2}}{16} z^2(1 + z^{-1})^4 p_1\left(\tfrac{-z+2-z^{-1}}{4}\right)/p_1(0) = G_0(z)$$

is symmetric. Since $p_1(y)$ is a first-degree polynomial, the highest power of z that occurs in $G_0(z)$ is z^3. So by symmetry z^{-3} is the lowest term and $G_0(z)$ has length 7. Thus

$$G_0(z) = b_0 + b_1\left(z + z^{-1}\right) + b_2\left(z^2 + z^{-2}\right) + b_3\left(z^3 + z^{-3}\right),$$
$$H_0(z) = c_0 + c_1\left(z + z^{-1}\right) + c_2\left(z^2 + z^{-2}\right) + c_3\left(z^3 + z^{-3}\right) + c_4\left(z^4 + z^{-4}\right).$$

When $z = e^{\omega i}$, then from the identity $z^n + z^{-n} = 2\cos(n\omega)$ we can write

$$G_0(e^{\omega i}) = b_0 + 2b_1\cos(\omega) + 2b_2\cos(2\omega) + 2b_3\cos(3\omega),$$
$$H_0(e^{\omega i}) = c_0 + 2c_1\cos(\omega) + 2c_2\cos(2\omega) + 2c_3\cos(3\omega) + 2c_4\cos(4\omega).$$

The coefficients b_k and c_k can be determined (approximately) by MATLAB:

$b_0 = 0.7884856165$	$b_1 = 0.4180922732$
$b_2 = -0.04068941763$	$b_3 = -0.06453888264$
$c_0 = 0.8526986785$	$c_1 = 0.3774028553$
$c_2 = -0.1106244041$	$c_3 = -0.02384946500$
$c_4 = 0.03782845545$	

These numerical values were used to plot the frequency response graphs in Figure 4.9.

(12) (a) From Example 4.16 the modified analysis filters have z-transforms $H_0(z) = \tfrac{\sqrt{2}}{8}\left(-2z^2 + 6z + 6 - 2z^{-1}\right)$ and $H_1(z) = \tfrac{\sqrt{2}}{8}\left(-z^2 + 3z - 3 + z^{-1}\right)$.

(b) By Theorem 4.9 the polyphase analysis matrix $\mathbf{H}_p(z)$ is obtained from the analysis modulation matrix $\mathbf{H}_m(z)$ by the relation

$$\mathbf{H}_p(z^2) = \mathbf{H}_m(z) \begin{bmatrix} 1 & 1 \\ z & -z \end{bmatrix}^{-1} = \frac{1}{2} \begin{bmatrix} H_0(z) & H_0(-z) \\ H_1(z) & H_1(-z) \end{bmatrix} \begin{bmatrix} 1 & z^{-1} \\ 1 & -z^{-1} \end{bmatrix}$$

$$= \frac{1}{2} \begin{bmatrix} (H_0(z) + H_0(-z)) & z^{-1}(H_0(z) - H_0(-z)) \\ (H_1(z) + H_1(-z)) & z^{-1}(H_1(z) - H_1(-z)) \end{bmatrix}.$$

For a Laurent polynomial $f(z)$, the average $(1/2)(f(z) + f(-z))$ gives the *even* terms in $f(z)$, whereas $(1/2)(f(z) - f(-z))$ gives the *odd* terms. So

$$\mathbf{H}_p(z^2) = \frac{\sqrt{2}}{8} \begin{bmatrix} (-2z^2 + 6) & z^{-1}(6z - 2z^{-1}) \\ (-z^2 - 3) & z^{-1}(3z + z^{-1}) \end{bmatrix}$$

$$= \frac{\sqrt{2}}{8} \begin{bmatrix} (-2z^2 + 6) & (6 - 2z^{-2}) \\ (-z^2 - 3) & (3 + z^{-2}) \end{bmatrix}.$$

Hence $\mathbf{H}_p(z) = \dfrac{\sqrt{2}}{8} \begin{bmatrix} (-2z + 6) & (6 - 2z^{-1}) \\ (-z - 3) & (3 + z^{-1}) \end{bmatrix}$. This agrees with the polyphase analysis matrix (3.33), since the shift operator S becomes z^{-1} in the frequency domain.

(13) This is a straightforward matrix calculation.

(14) Multiply the matrices to get the unique solution $f(z) = 2z^{-1}$ and $g(z) = -3z$.

(15) (a) $\det \mathbf{F}(z) = (z + 2)(z + 3 + z^{-1}) - (z + 1)(z + 4 + 2z^{-1}) = 1$.

(b) Take $c = -1$. Then $\mathbf{G}(z) = \begin{bmatrix} 1 & -1 \\ 0 & 1 \end{bmatrix} \mathbf{F}(z) = \begin{bmatrix} 1 & (1 + z^{-1}) \\ (z + 1) & (z + 3 + z^{-1}) \end{bmatrix}$.

(c) Take $f(z) = -(z + 1)$. Then $\begin{bmatrix} 1 & 0 \\ -(z + 1) & 1 \end{bmatrix} \mathbf{G}(z) = \begin{bmatrix} 1 & (1 + z^{-1}) \\ 0 & 1 \end{bmatrix}$.

(d) $\mathbf{F}(z) = \begin{bmatrix} 1 & 1 \\ 0 & 1 \end{bmatrix} \mathbf{G}(z) = \begin{bmatrix} 1 & 1 \\ 0 & 1 \end{bmatrix} \begin{bmatrix} 1 & 0 \\ (z + 1) & 1 \end{bmatrix} \begin{bmatrix} 1 & (1 + z^{-1}) \\ 0 & 1 \end{bmatrix} = U_2 \, P \, U_1$.

(e) The lifting steps are

(First Update) $U_1:$ $\mathbf{s}^{(1)}[n] = \mathbf{x}_{\text{even}}[n] + \mathbf{x}_{\text{odd}}[n] + \mathbf{x}_{\text{odd}}[n - 1]$,

(Prediction) $P:$ $\mathbf{d}[n] = \mathbf{x}_{\text{odd}}[n] + \mathbf{s}^{(1)}[n] + \mathbf{s}^{(1)}[n + 1]$,

(Second Update) $U_2:$ $\mathbf{s}[n] = \mathbf{s}^{(1)}[n] + \mathbf{d}[n]$.

There is no normalization step in this example.

(16) Following Example 3.13, we calculate

$$\langle S^{2m} \mathbf{h}_0^{\mathrm{T}}, \delta_1 \rangle = S^{2m} \mathbf{h}_0^{\mathrm{T}}[1] = \mathbf{h}_0^{\mathrm{T}}[1 - 2m] = \mathbf{h}_0[2m - 1].$$

Thus the trend component of δ_1 is

$$\mathbf{s} = \sum_{m \in \mathbb{Z}} \langle S^{2m} \mathbf{h}_0^{\mathrm{T}}, \delta_1 \rangle S^{2m} \mathbf{g}_0 = \sum_{m \in \mathbb{Z}} \mathbf{h}_0[2m - 1] S^{2m} \mathbf{g}_0.$$

From the formula for $\mathbf{h}_0$ we see that $\mathbf{h}_0[2m - 1] \neq 0$ only for $m = 0, 1$. Hence the trend component of δ_1 is

$$\mathbf{s} = \frac{\sqrt{2}}{8} \{2\mathbf{g}_0 + 2S^2 \mathbf{g}_0\} = \frac{1}{16} \{2(\delta_{-1} + 2\delta_0 + \delta_1) + 2(\delta_1 + 2\delta_2 + \delta_3)\}$$

$$= \frac{1}{8} \{\delta_{-1} + 2\delta_0 + 2\delta_1 + 2\delta_2 + \delta_3\}.$$

Likewise, for the detail component we calculate

$$\langle S^{2m}\mathbf{h}_1^{\mathrm{T}}, \delta_1 \rangle = S^{2m}\mathbf{h}_1^{\mathrm{T}}[1] = \mathbf{h}_1^{\mathrm{T}}[1-2m] = \mathbf{h}_1[2m-1].$$

Thus the detail component of δ_1 is

$$\mathbf{d} = \sum_{m\in\mathbb{Z}} \langle S^{2m}\mathbf{h}_1^{\mathrm{T}}, \delta_1 \rangle S^{2m}\mathbf{g}_1 = \sum_{m\in\mathbb{Z}} \mathbf{h}_1[2m-1] S^{2m}\mathbf{g}_1.$$

From the formula for $\mathbf{h}_1$ we see that $\mathbf{h}_1[2m-1] \neq 0$ only for $m = 0$. Since $\mathbf{h}_1[-1] = \sqrt{2}/2$, the detail component of δ_1 is

$$\mathbf{d} = \tfrac{\sqrt{2}}{2}\mathbf{g}_0 = \tfrac{1}{8}\left\{-\delta_{-1} - 2\delta_0 + 6\delta_1 - 2\delta_2 - \delta_3\right\}.$$

We check that $\mathbf{s} + \mathbf{d} = \delta_1$. But notice that $\mathbf{s}$ and $\mathbf{d}$ are not obtained by shifting the trend and detail components of δ_0 that were calculated in Example 3.13.

(17) Following Example 3.13, we calculate

$$\langle S^{2m}\mathbf{h}_0^{\mathrm{T}}, \delta_0 \rangle = S^{2m}\mathbf{h}_0^{\mathrm{T}}[0] = \mathbf{h}_0^{\mathrm{T}}[-2m] = \mathbf{h}_0[2m].$$

Thus the trend component of δ_0 is

$$\mathbf{s} = \sum_{m\in\mathbb{Z}} \langle S^{2m}\mathbf{h}_0^{\mathrm{T}}, \delta_0 \rangle S^{2m}\mathbf{g}_0 = \sum_{m\in\mathbb{Z}} \mathbf{h}_0[2m] S^{2m}\mathbf{g}_0.$$

From the formula for $\mathbf{h}_0$ in Example 4.16 we see that $\mathbf{h}_0[2m] \neq 0$ only for $m = 0, 1$. Hence the trend component of δ_0 is

$$\begin{aligned}
\mathbf{s} &= \tfrac{\sqrt{2}}{4}\left\{-\mathbf{g}_0 + 3S^2\mathbf{g}_0\right\} \\
&= \tfrac{1}{16}\left\{-(\delta_{-3} + 3\delta_{-2} + 3\delta_{-1} + \delta_0) + 3(\delta_{-1} + 3\delta_0 + 3\delta_1 + \delta_2)\right\} \\
&= \tfrac{1}{16}\left\{-\delta_{-3} - 3\delta_{-2} + 8\delta_0 + 9\delta_1 + 3\delta_2\right\}.
\end{aligned}$$

Likewise, for the detail component we calculate

$$\langle S^{2m}\mathbf{h}_1^{\mathrm{T}}, \delta_0 \rangle = S^{2m}\mathbf{h}_1^{\mathrm{T}}[0] = \mathbf{h}_1^{\mathrm{T}}[-2m] = \mathbf{h}_1[2m].$$

Thus the detail component of δ_0 is

$$\mathbf{d} = \sum_{m\in\mathbb{Z}} \langle S^{2m}\mathbf{h}_1^{\mathrm{T}}, \delta_0 \rangle S^{2m}\mathbf{g}_1 = \sum_{m\in\mathbb{Z}} \mathbf{h}_1[2m] S^{2m}\mathbf{g}_1.$$

From the formula for $\mathbf{h}_1$ in Example 4.16 we see that $\mathbf{h}_1[2m] \neq 0$ only for $m = -1, 0$. Hence the detail component of δ_0 is

$$\begin{aligned}
\mathbf{d} &= \tfrac{\sqrt{2}}{8}\left\{-S^{-2}\mathbf{g}_1 - 3\mathbf{g}_1\right\} \\
&= \tfrac{1}{16}\left\{-(-\delta_{-3} - 3\delta_{-2} + 3\delta_{-1} + \delta_0) - 3(-\delta_{-1} - 3\delta_0 + 3\delta_1 + \delta_2)\right\} \\
&= \tfrac{1}{16}\left\{\delta_{-3} + 3\delta_{-2} + 8\delta_0 - 9\delta_1 - 3\delta_2\right\}.
\end{aligned}$$

We check that $\mathbf{s} + \mathbf{d} = \delta_0$.

(18) (a) From the stem graphs it is evident that if $m \neq 0, 1$ then $S^{2m}\mathbf{h}_0^{\mathrm{T}}[k] = 0$ when $k = 0$ and $k = 1$. Hence $\langle S^{2m}\mathbf{h}_0^{\mathrm{T}}, \mathbf{x} \rangle = 0$ in these cases.

(b) From the stem graphs it is evident that if $m \neq 0, 1$ then $S^{2m}\mathbf{h}_1^{\mathrm{T}}[k] = 0$ when $k = 0$ and $k = 1$. Hence $\langle S^{2m}\mathbf{h}_1^{\mathrm{T}}, \mathbf{x} \rangle = 0$ in these cases.

(c) From the formulas $\mathbf{h}_0^{\mathrm{T}} = \delta_{-1} + 2\delta_0 + \delta_1$ and $S^2\mathbf{h}_0^{\mathrm{T}} = \delta_1 + 2\delta_2 + \delta_3$ we calculate that $\langle \mathbf{h}_0^{\mathrm{T}}, \mathbf{x} \rangle = 2 \cdot 3 + 1 \cdot 4 = 10$ and $\langle S^2\mathbf{h}_0^{\mathrm{T}}, \mathbf{x} \rangle = 1 \cdot 4 = 4$. From the formulas $\mathbf{h}_1^{\mathrm{T}} = \delta_{-1} + 2\delta_0 - 3\delta_1$ and $S^2\mathbf{h}_1^{\mathrm{T}} = \delta_1 + 2\delta_2 - 3\delta_3$ we calculate that $\langle \mathbf{h}_1^{\mathrm{T}}, \mathbf{x} \rangle = 2 \cdot 3 - 3 \cdot 4 = -6$ and $\langle S^2\mathbf{h}_1^{\mathrm{T}}, \mathbf{x} \rangle = 1 \cdot 4 = 4$.

(d) From (a) and (b) we know that $a_m = 0$ if $m < 0$ or $m > 1$, and $a_0 = 10$, $a_1 = 4$. Thus the trend part of $\mathbf{x}$ is

$$\mathbf{x_s} = 10\mathbf{g_0} + 4S^2\mathbf{g_0} = (1/8)\{30\delta_0 + 20\delta_1 + 2\delta_2 + 8\delta_3 - 4\delta_4\}\ .$$

Likewise, $b_m = 0$ if $m < 0$ or $m > 1$, and $b_0 = -6$, $b_1 = 4$. Thus the detail part of $\mathbf{x}$ is

$$\mathbf{x_d} = -6\mathbf{g_1} + 4S^2\mathbf{g_1} = (1/8)\{-6\delta_0 + 12\delta_1 - 2\delta_2 - 8\delta_3 + 4\delta_4\}\ .$$

The sum of trend and detail is $\mathbf{x_s} + \mathbf{x_d} = (1/8)\{24\delta_0 + 32\delta_1\} = \mathbf{x}$.

(19) (a) By formula (4.59) and the relation $(1 + z^{-1})^p = z^{-p}(1 + z)^p$, we have

$$z^{p/2}G_0(z^{-1}) = 2^{-p}\sqrt{2}z^{p/2}(1+z^{-1})^p = 2^{-p}\sqrt{2}z^{-p/2}(1+z)^p = z^{-p/2}G_0(z)\ .$$

Likewise, by formula (4.60) with $n = (p+q)/2$, we have

$$z^{-p/2}H_0(z^{-1}) = 2^{-q}\sqrt{2}z^{n-(p/2)}(1 + z^{-1})^q B_n\left(\frac{-z+2-z^{-1}}{4}\right)$$

$$= 2^{-q}\sqrt{2}z^{-n+(p/2)}(1 + z)^q B_n\left(\frac{-z+2-z^{-1}}{4}\right) = z^{p/2}H_0(z)\ .$$

Here we have used $n - (p/2) - q = -n + (p/2)$ to get the exponent of z.

(b) For the CDF(2, 2) filters

$$z^{-1}G_0(z) = (\sqrt{2}/4)\left(z^{-1} + 2 + z\right)\ ,$$
$$zH_0(z) = (\sqrt{2}/8)\left(-z^2 + 2z + 6 + 2z^{-1} - z^{-2}\right)\ .$$

These are the symmetric filters obtained from the polyphase analysis matrix in Example 4.20. For the CDF(6, 2) filters,

$$z^{-3}G_0(z) = (\sqrt{2}/64)\left(z^3 + 6z^2 + 15z + 20 + 15z^{-1} + 6z^{-2} + z^{-3}\right)\ ,$$
$$z^3 H_0(z) = (\sqrt{2}/64)\left(-5z^4 + 30z^3 - 56z^2 - 14z + 154\right.$$
$$\left. - 14z^{-1} - 56z^{-2} + 30z^{-3} - 5z^{-4}\right)\ .$$

(20) (a) Substitute and expand to get $8Q_3(z) = 3z^4 - 18z^3 + 38z^2 - 18z + 3$.

(b) The coefficients of $Q_3(z)$ are real, so the complex roots occur in complex conjugate pairs. Also by symmetry $Q(1/r) = 0$ whenever $Q(r) = 0$.

(c) By (4.90) we can write $H_0(z) = \kappa(1+z^{-1})^3(1 - r_1 z^{-1})(1 - r_2 z^{-1})$. Since $(1 - r_1 z^{-1})(1 - r_2 z^{-1}) = r_1 r_2 - (r_1 + r_2)z^{-1} + z^{-2}$, the formula for κ now follows from $H_0(1) = \sqrt{2}$. Expanding the formula for $H_0(z)$ and substituting the values of r_1 and r_2, we obtain the filter as[26]

$$\mathbf{h_0} = 0.3327\,\delta_0 + 0.8069\,\delta_1 + 0.4599\,\delta_2 - 0.1350\,\delta_3 - 0.0854\,\delta_4 + 0.0352\,\delta_5\ .$$

---

[26]See [Daubechies (1992), Table 6.1] for filter coefficients to 16 decimal places for the Daub2K filters with $K = 2, \ldots, 10$.

B.5   Solutions to Exercises 5.7

(1) (a) From the graph of f_1, the shifts are by $k = -2, -1, 0, 1$ with corresponding orthonormal basis functions $\phi_{1,k}$ multiplied by $4/\sqrt{2}, 2/\sqrt{2}, 3/\sqrt{2}, 1/\sqrt{2}$. The expansion is $f_1 = (1/\sqrt{2})(4\phi_{1,-2} + 2\phi_{1,-1} + 3\phi_{1,0} + \phi_{1,1})$.

(b) By equations (5.23) and (5.24) we can write

$$f_1 = (1/2)\{4(\phi_{0,-1} + \psi_{0,-1}) + 2(\phi_{0,-1} - \psi_{0,-1})$$
$$+ 3(\phi_{0,0} + \psi_{0,0}) + (\phi_{0,0} - \psi_{0,0})\} = f_0 + g_0,$$

where $f_0 = 3\phi_{0,-1} + 2\phi_{0,0}$ and $g_0 = \psi_{0,-1} + \psi_{0,0}$.

(c) The nonzero values of $f_0(t)$ are 3 on $-1 \le t < 0$ (the average value of $f_1(t)$ on this interval) and 2 on $0 \le t < 0$ (the average value of $f_1(t)$ on this interval). The values of $g_0(t)$ on the subintervals $-1 \le t < -0.5, \ldots, 0.5 \le t < 1$ are the corrections to the values of $f_0(t)$ needed to obtain $f_1(t)$.

(2) (a) From the graph of f_2, the shifts are by $k = 0, 1, 2, 3$ with corresponding orthonormal basis functions $\phi_{2,k}$ multiplied by 2, 1, 1, 1/2. The expansion is $f_2 = 2\phi_{2,0} + \phi_{2,1} + \phi_{2,2} + (1/2)\phi_{2,3}$.

(b) By equations (5.23) and (5.24) we can write

$$f_2 = (1/\sqrt{2})\{2(\phi_{1,0} + \psi_{1,0}) + (\phi_{1,0} - \psi_{1,0})$$
$$+ (\phi_{1,1} + \psi_{1,1}) + (1/2)(\phi_{1,1} - \psi_{1,1})\} = f_1 + g_1,$$

where $f_1 = (3/\sqrt{2})\phi_{1,0} + 3/(2\sqrt{2})\phi_{1,1}$ and $g_1 = (1/\sqrt{2})\psi_{1,0} + (1/2\sqrt{2})\psi_{1,1}$.

(c) By equations (5.23) and (5.24) we can write

$$f_1 = (3/2)(\phi_{0,0} + \psi_{0,0}) + (3/4)(\phi_{0,0} - \psi_{0,0}) = f_0 + g_0,$$

where $f_0 = (9/4)\phi_{0,0}$ and $g_0 = (3/4)\psi_{0,0}$.

(d) On the intervals $0 \le t < 1/4, \ldots, 3/4 \le t < 1$ the values of $g_1(t)$ are $1, -1, 1/2, -1/2$. On the intervals $0 \le t < 1/2, 1/2 \le t < 1$ the values of $g_0(t)$ are $3/4$ and $-3/4$. On the interval $0 \le t < 1$ the value of $f_0(t)$ is $9/4$ (the average value of f_2).

(3) (a) The inner products are integrals over the range $0 < t < 1$. If $\phi_{j,k}(t) = 2^{j/2}\phi_{\text{Haar}}(2^j t - k) \ne 0$ then $0 \le 2^j t - k < 1$, so $0 \le k < 2^j$. The same argument applies to $\psi_{j,k}(t)$.

(b) By evaluating the inner product as an integral we get

$$s_j[k] = 2^j \int_{k2^{-j}}^{(k+1)2^{-j}} t\, dt = 2^{j-1}[2^{-2j}(k+1)^2 - 2^{-2j}k^2] = 2^{-j}[k + (1/2)]$$

when $0 \le k < 2^j$. Since $f(t)$ is a linear function, its average value on the interval is the value at the midpoint $2^{-j}[k + (1/2)]$.

(c) $f_0(t) = s_0[0]\phi_{0,0}(t) = (1/2)\phi_{\text{Haar}}(t)$ and

$$f_1(t) = 2^{-1/2}\{s_1[0]\phi_{1,0}(t) + s_1[1]\phi_{1,1}(t)\}$$
$$= (1/4)\phi_{\text{Haar}}(2t) + (3/4)\phi_{\text{Haar}}(2t - 1).$$

So $f_1(t)$ takes the values $1/4, 3/4$ on the two intervals $0 \le t < 1/2$, $1/2 \le t < 1$ and zero elsewhere.

(d) By evaluating the inner product as an integral we get

$$\mathbf{d}_j[k] = 2^j \int_{k2^{-j}}^{(k+1/2)2^{-j}} t\, dt - 2^j \int_{(k+1/2)2^{-j}}^{(k+1)2^{-j}} t\, dt$$

$$= 2^{-j-1}\left\{\left[(k+1/2)^2 - k^2\right] - \left[(k+1)^2 - (k+1/2)^2\right]\right\} = -2^{-j-2}.$$

Note that the detail coefficients depend on the level j but not the shift k.

(e) $g_0(t) = \mathbf{d}_0[0]\psi_{0,0}(t) = -(1/4)\psi_{\text{Haar}}(t)$ has values $-1/4$, $1/4$ on the two intervals $0 \le t < 1/2$, $1/2 \le t < 1$ and zero elsewhere.

$$g_1(t) = 2^{-1/2}\left\{\mathbf{d}_1[0]\psi_{1,0}(t) + \mathbf{d}_1[1]\psi_{1,1}(t)\right\}$$

$$= -(1/8)\left\{\psi_{\text{Haar}}(2t) + \psi_{\text{Haar}}(2t-1)\right\}$$

has values $-1/8$, $1/8$, $-1/8$, $1/8$ on the four intervals $0 \le t < 1/4$, $1/4 \le t < 2/4$, $2/4 \le t < 3/4$, $3/4 \le t < 1$ and zero elsewhere.

(4) It suffices to check the case $f(t) = t^n$ for $a2^{-j} < t < b2^{-j}$. The calculation is similar to Exercise 3(d): If $a \le k < b$ then

$$\mathbf{d}_j[k] = 2^j \int_{k2^{-j}}^{(k+1/2)2^{-j}} t^n\, dt - 2^j \int_{(k+1/2)2^{-j}}^{(k+1)2^{-j}} t^n\, dt$$

$$= c\left\{2(k+1/2)^{n+1} - (k+1)^{n+1} - k^{n+1}\right\}$$

$$= -c\left\{2k^{n+1} + (n+1)k^n - 2k^{n+1} - (n+1)k^n + p(k)\right\} = -cp(k)$$

by the binomial expansion, where $p(k)$ is a polynomial in k of degree $n-1$ and $c = 2^{-j}/(n+1)$.

(5) (a) Write $\langle f,g\rangle - \langle f,g_n\rangle = \langle f, g-g_n\rangle$ and apply the Cauchy–Schwarz inequality to get $\lim_{n\to\infty} |\langle f,g\rangle - \langle f,g_n\rangle| = 0$.

(b) Write $\|g - g_n\|^2 = \|g\|^2 - 2\langle g,g_n\rangle + \|g_n\|^2$. When $n \to \infty$ then (a) gives $0 = -\|g\|^2 + \lim_{n\to\infty}\|g_n\|^2$.

(6) (a) $\phi(2t) = 1/3$ for $0 \le t < 3/2$ and is otherwise zero, while $\phi(2t-3) = 1/3$ for $3/2 \le t < 3$ and is otherwise zero. Hence $\phi(2t) + \phi(2t-3) = \phi(t)$ for all t, which is the two-scale equation associated with the filter $\mathbf{g}_0$.

(b) $\mathbf{g}_1 = (1/\sqrt{2})(\delta_0 - \delta_3)$, so $\psi(t) = \phi(2t) - \phi(2t-3) = (1/3)\psi_{\text{Haar}}(t/3)$ by (a).

(c) Since S^k shifts graphs by k units and $\phi(t)$, $\psi(t)$ are zero except when $0 \le t < 3$, all these inner products are obviously zero when $|k| \ge 3$. It's clear from the graphs that $\langle S^k\phi, \phi\rangle \ne 0$ for $|k| \le 2$. We calculate $\langle S^{\pm 1}\psi, \psi\rangle = 0$, $\langle S^{\pm 2}\psi, \psi\rangle = -1/9$, $\langle \psi,\phi\rangle = 0$, and $\langle S^k\psi, \phi\rangle = -1/9$ for $k = \pm 1, \pm 2$. Note that $\langle S^k f, f\rangle = \langle f, S^{-k}f\rangle$ for any $f \in L^2$.

(7) By the two-scale equation

$$\sum_{n\in\mathbb{Z}} \phi^{(j+1)}(t+n) = \sum_{k\in\mathbb{Z}}\sum_{n\in\mathbb{Z}} \sqrt{2}\,\mathbf{g}_0[k]\,\phi^{(j)}(2t+2n-k)\,.$$

Write $m = 2n - k$. Then m has the same parity as k and we get all integers m when we sum over $n \in \mathbb{Z}$. Thus we can split the double sum on the right into two parts:

$$\sum_{m \text{ even}}\left\{\sum_{k \text{ even}} \sqrt{2}\,\mathbf{g}_0[k]\phi^{(j)}(2t+m)\right\}$$

$$+ \sum_{m \text{ odd}}\left\{\sum_{k \text{ odd}} \sqrt{2}\,\mathbf{g}_0[k]\phi^{(j)}(2t+m)\right\}.$$

Each of the two summations over k give 1 by the sum rule. So the whole sum over m even and odd is $\sum_{n \in \mathbb{Z}} \phi^{(j)}(2t + n)$. By induction on j, this sum is 1 for all t.

(8) (a) If $\mathbf{v} = \mathbf{0}$, then $\phi(n) = 0$ for all integers n since $\phi(0) = \phi(3) = 0$. Hence $\phi(k/2^j) = 0$ for all j, k by the two-scale equation. Since $\phi(t)$ is assumed continuous, this makes $\phi(t) = 0$ for all t, a contradiction. The two-scale equation gives $\phi(1) = a\phi(2) + b\phi(1)$ and $\phi(2) = c\phi(2) + d\phi(1)$. Hence $A\mathbf{v} = \mathbf{v}$.

(b) By the partition of unity property with $t = 0$ we know that $\sum_{n \in \mathbb{Z}} \phi(n) = 1$. Hence $\phi(2) = 1 - \phi(1)$ since the other terms in the sum are zero. The equation $A\mathbf{v} = \mathbf{v}$ gives $\phi(1) = a/(1 + a - b)$ and $\phi(2) = (1 - b)/(1 + a - b)$.

(c) The two-scale equation and the vanishing of $\phi(1)$ outside $0 < t < 3$ gives

$$\phi(1/2) = a\phi(1), \quad \phi(3/2) = b\phi(2) + c\phi(1), \quad \phi(5/2) = d\phi(2). \text{ So } B = \begin{bmatrix} a & 0 \\ c & b \\ 0 & d \end{bmatrix}.$$

(d) From (b) we obtain $\phi(1) = (1 + \sqrt{3})/2 \approx 1.366$ and $\phi(2) = (1 - \sqrt{3})/2 \approx -0.366$. From (c) we then get $\phi(1/2) = (2 + \sqrt{3})/4 \approx 0.933$, $\phi(3/2) = 0$, and $\phi(5/2) = (2 - \sqrt{3})/4 \approx 0.067$. These values are consistent with approximate graph of ϕ in Figure 5.6.

(9) (a) Use the two-scale equation for ϕ to write the moment integral as

$$\mu_n = \int_{-\infty}^{\infty} \sum_{k=0}^{L} \sqrt{2}\, \mathbf{g}_0[k]\, t^n\, \phi(2t - k)\, dt$$

$$= \frac{1}{2^n} \int_{-\infty}^{\infty} \sum_{j=0}^{n} \binom{n}{j} \left\{ \frac{1}{\sqrt{2}} \sum_{k=0}^{L} k^{n-j}\, \mathbf{g}_0[k] \right\} s^j\, \phi(s)\, ds$$

$$= \frac{1}{2^n} \sum_{j=0}^{n} \binom{n}{j} \mu_j\, \sigma_{n-j}\,.$$

Here we made the change of variable $s = 2t - k$, $ds = 2dt$ in the integral and used the binomial expansion of $(s + k)^n$. Now use $\sigma_0 = 1$ to get the formula for μ_n.

(b) Use the two-scale equation for ψ and same method as in (a), replacing $\mathbf{g}_0$ by $\mathbf{g}_1$ and using $\tau_0 = 0$.

(10) (a) Let $p(t)$ be the Taylor polynomial of order $N - 1$ for $f(t)$ centered at $t = k2^{-j}$. Then $\int_0^L p\left(2^{-j}(k + x)\right) \phi(x)\, dx = p(k2^{-j}) = f(k2^{-j})$ since the moments of ϕ of order $1, \ldots, N - 1$ are all zero. Hence, as in Lemma 5.48,

$$s_j[k] - f(k2^{-j}) = \int_0^L \left\{ f\left(2^{-j}(k + x)\right) - p\left(2^{-j}(k + x)\right) \right\} \phi(x)\, dx. \quad (\star)$$

By Taylor's theorem $\left| f\left(2^{-j}(k + x)\right) - p\left(2^{-j}(k + x)\right) \right| \le M2^{-jN} |x|^N / N!$ for $0 \le x \le L$. Now use this estimate and the Cauchy–Schwarz inequality in $(\star)$, as in Lemma 5.48. The constant $C = (1/N!)\left\{ \int_0^L x^{2N}\, dx \right\}^{1/2}$.

(b) Let $p(t)$ be the Taylor polynomial of $f(t)$ as in (a). Since all moments of ψ of order less than N are zero, $\int_0^L p\left(2^{-j}(k + x)\right) \psi(x)\, dx = 0$. Hence

$$\mathbf{d}_j[k] = \int_0^L \left\{ f\left(2^{-j}(k + x)\right) - p\left(2^{-j}(k + x)\right) \right\} \psi(x)\, dx.$$

Now follow the same method as in (a).

Bibliography

Boggess, A. and Narcowich, F. J. (2001). *A First Course in Wavelets with Fourier Analysis* (Prentice Hall).

Broughton, S. A. and Bryan, K. (2009). *Discrete Fourier Analysis and Wavelets: Applications to Signal and Image Processing* (John Wiley & Sons).

Burrus, C. S., Gopinath, R. A., and Guo, H. (1998). *Introduction to Wavelets and Wavelet Transforms* (Prentice-Hall).

Cohen, A., Daubechies, I., and Feauveau, J. C. (1992). *Biorthogonal bases of compactly supported wavelets*, Comm. Pure and Appl. Math. **44**, pp. 485–560.

Cohen, A. and Ryan, R. D. (1995). *Wavelets and Multiscale Signal Processing* (Chapman & Hall).

Daubechies, I. (1988). *Orthonormal bases of compactly supported wavelets*, Comm. Pure and Appl. Math. **41**, pp. 909–996.

Daubechies, I. (1992). *Ten Lectures on Wavelets* (SIAM, Philadelphia).

Daubechies, I. (1993). *Orthonormal bases of compactly supported wavelets, II: Variations on a theme*, SIAM J. Math. Anal. **24**, pp. 499–519.

Daubechies, I. and Sweldens, W. (1998). *Factoring Wavelet Transforms into Lifting Steps*, J. Fourier Anal. Appl. **4**, 3, pp. 247–269.

Frazier, M. W. (1999). *An Introduction to Wavelets Through Linear Algebra* (Springer).

Gundy, R. F. (2007). Probability, ergodic theory, and low-pass filters, in J. M. Rosenblatt, A. M. Stokolos, and A. I. Zayed (eds.) *Topics in harmonic analysis and ergodic theory*, Contemp. Math. **444** (Amer. Math. Soc., Providence, RI), pp. 53–87.

Hubbard, B. Burke (1998). *The World According to Wavelets*, 2nd ed. (A K Peters).

Jaffard, S., Meyer, Y., and Ryan, R. D. (2001). *Wavelets: Tools for Science & Technology* (SIAM, Philadelphia).

Mallat, S. (1989) *Multiresolution approximation and wavelets*, Trans. Amer. Math. Soc. **315**, pp. 69–88.

Mallat, S. (1999). *A Wavelet Tour of Signal Processing*, 2nd ed. (Academic Press).

Mulcahy, C. (1996). *Plotting and scheming with wavelets*, Mathematics Magazine **69**, 5, pp. 323–343.

Moler, C. B. (2004). *Numerical Computing with* MATLAB (SIAM, Philadelphia).

Jensen, A. and la Cour-Harbo, A. (2001). *Ripples in Mathematics: The Discrete Wavelet Transform* (Springer).

Strang, G. (2006). *Linear Algebra and its Applications*, 4th ed. (Thomson Brooks/Cole).

Strang, G. and Nguyen, T. Q. (1997). *Wavelets and Filter Banks*, rev. ed. (Wellesley-Cambridge Press, Wellesley, MA).

Strichartz, R. S. (1993). *How to Make Wavelets*, Amer. Math. Monthly **100**, pp. 539-556.

Van Fleet, P. J. (2008). *Discrete Wavelet Transformations: An Elementary Approach with Applications* (Wiley-Interscience).

Walker, J. S. (2008). *A Primer on Wavelets and their Scientific Applications*, 2nd ed. (Chapman & Hall/CRC).

Walnut, D. F. (2002). *An Introduction to Wavelet Analysis* (Birkhäuser).

Index